MICROCLIMATE AND LOCAL CLIMATE

A microclimate is a canopy-scale atmospheric zone where the climate differs from the surrounding area. The term may refer to areas as small as a few square meters or as large as several square kilometers. Microclimates exist in the atmospheric boundary layer and the upper soil. Local climates are related to landscape elements (hill slopes, valley bottoms and ridge tops). They occur near water bodies, which may affect local temperatures and precipitation, in heavily urban areas, as a result of rapidly changing topography, or in the shadow of a mountain. Although general, larger-scale climate patterns are clearly important, local micrometeorological factors may have impacts on many elements, including local weather, ecological systems, soil climate, and agriculture. The topic of small-scale climate is therefore one that is of interest not only to climatologists but also to geographers, biologists, crop scientists, environmental scientists, hydrologists, and urban planners.

This book provides an up-to-date, comprehensive treatment of the variables and processes of microclimate and local climate, including radiation balance and energy balance. It describes and explains the climate within the lower atmosphere and upper soil, the region critical to life on Earth. Topics that are covered include not only the physical processes that affect microclimate, but also biological processes that affect vegetation and animals, including people. A geographic tour of the microclimates of the major ecosystems around the world is included. All major biomes and surface types are examined, including urban areas, and the effects of climate change on microclimate are described. This book is invaluable for advanced students and researchers in climatology in departments of environmental science, geography, meteorology, agricultural science and forestry.

Roger G. Barry was Director of the National Snow and Ice Data Center (NSIDC) at the University of Colorado, Boulder from 1977 to 2008, and Professor of Geography from 1968 to 2010. He was appointed a University of Colorado Distinguished Professor in 2004. From 2012 to 2014 he was Director of the International Climate Variability (CLIVAR) Project Office, National Oceanography Centre, Southampton, UK. He is a Fellow of the American Geophysical Union. His awards include the Founder's Medal of the Royal Geographical Society, the Nobel Peace Prize (as part of the Intergovernmental Panel on Climate Change team), a Guggenheim Fellow, a Fulbright Fellow, a Humboldt Prize Fellow, and a Foreign Member of the Russian Academy of Environmental Sciences (RAEN). He has directed 31 Masters and 36 PhD degrees. He is the authored or co-author of more than 250 peer-reviewed papers and many textbooks, including: *Atmosphere, Weather and Climate* (with R.J. Chorley, ninth edition, 2010, Routledge); *Mountain Weather and Climate* (third edition, 2008, Cambridge University Press), *Synoptic and Dynamic Climatology* (with A.M. Carleton, 2011, Routledge); *The Arctic Climate System* (with M. C. Serreze, second edition, 2014, Cambridge University Press); *The Global Cryosphere: Past, Present and Future* (with T.Y. Gan, 2011, Cambridge University Press); and *Essentials of the Earth's Climate System* (with E. A. Hall-McKim, 2014, Cambridge University Press).

Peter D. Blanken is a Professor and former Chair of the Department of Geography at the University of Colorado, Boulder. He has published more than 100 peer-reviewed papers, several book chapters, and is a co-author of *Straits of Mackinac Weather* (with Sandy Planisek, 2015, Sandy Planisek Publishing), and has served on the editorial board of the *Bulletin of the American Meteorological Society* for over a decade.

MICROCLIMATE AND LOCAL CLIMATE

ROGER G. BARRY

University of Colorado, Boulder

PETER D. BLANKEN

University of Colorado, Boulder

CAMBRIDGE
UNIVERSITY PRESS

CAMBRIDGE
UNIVERSITY PRESS

32 Avenue of the Americas, New York NY 10013

Cambridge University Press is part of the University of Cambridge.

It furthers the University's mission by disseminating knowledge in the pursuit of education, learning and research at the highest international levels of excellence.

www.cambridge.org
Information on this title: www.cambridge.org/9781107145627

First published 2016

Printed in the United States of America by Sheridan Books, Inc.

A catalog record for this publication is available from the British Library.

Library of Congress Cataloging in Publication Data
Names: Barry, Roger G. (Roger Graham), 1935– | Blanken, Peter D. (Peter David) 1965–
Title: Microclimate and local climate / Roger G. Barry, University of Colorado, Boulder, Peter D. Blanken, University of Colorado, Boulder.
Description: New York, NY : Cambridge University Press, 2016. | Includes bibliographical references and index.
Identifiers: LCCN 2015050661 | ISBN 9781107145627 (hardback)
Subjects: LCSH: Microclimatology. | Climatology. | Micrometeorology. | Ecology
Classification: LCC QC981.7.M5 B37 2016 | DDC 551.6/6–dc23
LC record available at http://lccn.loc.gov/2015050661

ISBN 978-1-107-14562-7 Hardback

Contents

Preface

This book aims to provide a concise overview of microclimates and local climates – climatic conditions on scales of centimeters to meters and 10 m to 1 km, respectively. There has been no recent treatment of this topic. It has been a decade or more since these topics have been addressed in a textbook, and there have been numerous advances in methodology and understanding over that interval. New instrumentation and increased resolution of satellite and airborne remote sensing have greatly expanded the possibilities for observations and have been accompanied by improved models and data assimilation techniques.

The text is aimed at upper level undergraduates and beginning graduate students in environmental sciences, physical geography, climatology, and biological sciences. The topic of small-scale climate is one that is of interest not only to climatologists but also to biologists, crop scientists, and specialists in landscape design. There are texts that approach the subject from an agricultural perspective using a theoretical framework, and there are texts that either are outdated or have regional coverage. Our book fulfills the need for an up-to-date text that bridges both theory and practice, and encompasses a global geographic perspective. Its scope covers the physical principles that determine micro- and topoclimate.

The text begins with definitions of microclimate and local climate (or topoclimate), a brief history of the two areas, and a case study. Chapter 2 describes each of the microclimatic elements in turn (temperature, moisture, and wind) including carbon dioxide, photosynthesis and respiration, the nitrogen cycle, and pollutants. Chapter 3 summarizes methods of observation and instrumentation. Chapter 4 treats solar, infrared, and net radiation. Chapter 5 describes the energy balance components – soil heat flux, momentum and mass exchange, sensible and latent heat fluxes, and advective effects. Chapter 6 addresses the monitoring and modeling of radiation and energy balance via remote sensing and land surface models. Chapter 7 deals with the microclimates of different vegetated environments – arctic and alpine tundra, grassland, farmland, wetlands, and coniferous, deciduous, and tropical forests. Chapter 8 similarly treats the microclimates of physical systems – lakes, rivers, snow cover, mountains, and cities.

Chapter 9 examines human and animal bioclimatology. Part II opens with a discussion of urban climates in Chapter 10. Chapter 11 then analyzes topoclimatic effects on microclimates. Part III examines environmental change; Chapter 12 looks at the effects of climate change on microclimates. The book is completed by Problems, References, the glossary, Symbols, SI Units and Conversions, and the Index.

The scope of the book is ambitious in addressing the full range of natural and built environments. While the underlying physical principles remain the same, their application to diverse environments necessitates differences in approach.

Acknowledgments

We express our thanks to the following:

Our parents, families, and teachers for providing the opportunities to make this possible.

Professor Jeff Dozier, University of California, Santa Barbara, for assistance with the B-Directional Reflectance Functions (BDRFs).

Dr. John Kimball, Montana State University, for AMSR-E data on lake surface area.

Professor Ann Nolin, Oregon State University, Corvallis, for assistance with BRDF of snow.

Gloria Hicks, Librarian in the Archives and Resource Center at NSIDC, for assistance with references.

Undergraduate assistants in the NSIDC Message Center for help scanning figures.

American Geophysical Union:

Eos 96(2015), p. 13, Fig. 2b.
Journal of Geophysical Research 114(2009): p. 4, Fig. 2.
Reviews of Geophysics 37(1999), p. 193, plate 4.
*Water Resources Research,*20(1984), p. 70, fig. 3; 50 (2014), p. 8436, fig. 13.

American Meteorological Society:

Bulletin of the American Meteorological Society 93(2012), p. 812, fig. 1 and p. 815, fig. 3; 96 (2015), p. 706, fig. 3.
*Journal of Applie*d Meteorology 13(1974), p. 860, fig. 6; 28(1980), p. 608, fig. 2; 33(1993), 38, p. 928, fig. 2; p. 410, fig. 5; 41(2002), p. 793, fig. 2.
Journal of Applied Meteorology and Climatology 17(1978), p. 1164, fig. 2; 46(2007), p. 2048, fig. 7; 54(2015), p. 150, fig. 8.
Journal of Climate 8 (1995), p. 1270, fig. 5; p. 1277, fig. 10; p. 2718, fig. 1; 16(2004), p. 139, fig. 3; 27(2014), p. 5114, fig. 2.
Journal of the Atmospheric Sciences 37(1980), p. 2739, fig. 4.
Journal of Hydrometeorology 4 (2003), p. 687, fig. 10; 9 (2008), p. 168, fig. 3.

Cambridge University Press:

The Global Cryosphere: Past, Present and Future, R. G. Barry and T. Y. Gan, 2011, p. 194, fig. 6.3.
Essentials of the Earth's Climate System, R. G. Barry and E. Hall-McKim, 2014, p. 87, fig 4.5.

Copernicus:

Geoscience and Model Development. 4(2011), p. 680, fig. 1.

Elsevier:

Agricultural and Forest Meteorology 144(2007), p. 19, fig. 1.
P. D. Blanken, 2015. *Canopy Processes.* In G. R. North, J. Pyle, and F. Zhang (eds.). *Encyclopedia of Atmospheric Sciences,* 2nd ed., Vol. 3: p. 246, fig. 2; p. 248, fig. 4; p. 250, fig. 6.
Vegetation and Atmosphere, J. L. Monteith(ed.), 1975, A. S. Thom, p. 80, fig. 1.
Journal of Great Lakes Research 37 (2011), p. 710, fig. 2.
National Research Council of Canada.
Institute for Research in Construction, Ottawa. *Canadian Building Digest* 1976 No. 180, fig. 2.

Royal Meteorological Society:

Quarterly Journal of the Royal Meteorological Society, 97 (1971), p. 549, fig. 1; 103 (1977), p. 400, fig. 1.

Springer:

Biogeochemistry 122(2015), fig. 2.
International Journal of Biometeorology 58(2014), p. 1991, fig. 5

Taylor and Francis:

R. G. Barry and R. J. Chorley 1987, *Atmosphere, Weather and Climate,* 5th ed., Routledge, p. 22, fig. 1.16.
R. G. Barry and R. J. Chorley 2010 *Atmosphere, Weather and Climate*, 9th ed., p. 422, fig. 12.27.
M. Kirkham, 2011. *Elevated Carbon Dioxide Impacts on Soil and Plant Water Relations*, p. 256, fig. 12.8 (CRC Press).

World Meteorological Organization:

1998, WMO/TD 1872 (Report No. 67), p. 3, fig. 1.2.1.

1

Introduction

This chapter starts by defining micro- and local climates. It then outlines the scope of the book, and this section is followed by a survey of the historical development of these fields of study. The central problem of scales of local and microclimates is then addressed, and the chapter concludes with the results of a case study of microclimatic variations.

A. Definitions

Climate is typically described from standard measurements of temperature and humidity recorded by instruments exposed in a meteorological shelter (Stevenson screen) at a height of about 1.5–2 m. Wind velocity is recorded on a mast at a height of 10 m. However, conditions change rapidly in the underlying air layer immediately above ground where plants, animals, and insects live. In the case of forest canopies, the effective surface on which atmospheric processes operate is elevated well above the ground. The climate of these complex environments cannot be determined from the standard meteorological measurements, and hence special observations and a whole theoretical framework are required to describe them. This is the domain of micrometeorology and microclimatology.

Microclimatology is the study of climates near the ground and in the soil, the factors that affect them, and the relationships and interactions between plants, insects, and other animals and their local environment. Microclimates are usually defined vertically in terms of the plant canopy height and so range from a few centimeters in tundra areas to about 50 m in tropical rain forests. Horizontally, they may extend between a meter and several kilometers. Micrometeorological processes operate on a time scale of seconds to minutes and spatial scales of centimeters to a few hundred meters (see Figure 1.1).

Microclimatic phenomena involve climatic averages (hours to years) on these same spatial scales. Scale concepts in climatology (Barry, 1970) illustrate the spatial and temporal characteristics of microclimates.

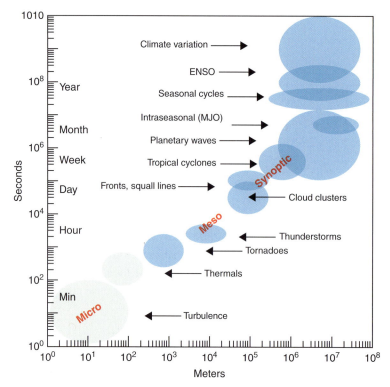

Figure 1.1. The space and time scales of meteorological phenomena illustrating the micro, meso, and synoptic scales.
Source: The COMET program, UCAR

Baum and Court (1949) defined microclimatology as the study of the geographical distributions, both horizontally (up to a few kilometers) and vertically, of air layers near the ground and the understanding of their physics. However, this appears to merge micro- and topoclimatology. Contemporary definitions of microclimate are focused on the critical layer of the interface between the surface and atmosphere, where the boundary conditions have a large impact not only on the troposphere, but also on the abiotic conditions for life itself. This layer where the majority of energy and mass exchange occurs extends from a few meters above to a few meters below the air-surface interface. Because of this relatively small scale, and the large variation that can occur spatially and temporally within this critical zone, the term "microclimate" has often been adopted in reference to the study of the long-term atmosphere-surface properties within this zone.

Microclimates have typically been studied empirically in specific situations. Most studies examine differences within a vegetation or crop canopy, or around a boulder, as well as conditions in the upper soil layers. Microclimatology therefore involves meteorology, soil physics, and the ecology and physiology of plants and animals. We are concerned primarily with the exchanges of energy and water between the atmosphere and different natural and built surfaces and their orientation and slope.

Local climates, also called topoclimates, are identified on a horizontal scale of hundreds of meters to ~10 km (areas < 100 km^2). They are primarily determined by elements of topography – valley bottoms, slopes, ridge tops – as well as the built environments of cities. Their vertical dimension is the planetary boundary layer that varies diurnally between about 500 and 1,500 m. In 1949 Knoch introduced the concept of landscape climate (*Geländeklima*) on a scale of 1:25,000. He defined it as the local climate under the influence of the local relief. Troll (1950) referred to it as the "topographical climate," but it has subsequently been translated as topoclimate. That term was introduced by Thornthwaite (1953), who defined it as the climate of a very small space.

B. The Scope of the Book

This text covers microclimates in Part I and local climates in Part II. Chapter 2 examines the microclimatic elements – temperature, moisture, wind, carbon dioxide, the carbon and nitrogen cycles, and pollutants. Chapter 3 describes the theory and methods for observing microclimates. Chapter 4 presents the basics of radiation. Chapter 5 describes the terms of the energy balance at the surface. Chapter 6 treats the monitoring and modeling of radiation and the energy balance through remote sensing and land surface models. Chapter 7 describes microclimatic conditions in the major types of land cover (tundra, grassland, farmland, wetland, and forests) while Chapter 8 extends this to lakes, rivers, snow cover, mountains, and cities. We illustrate the characteristics of microclimates in different types of canopy and different macroclimates and their variations with season and time of day. Chapter 9 on bioclimatology focuses on how microclimate affects living organisms, especially humans. Part II presents the characteristics of the major categories of local climate. Chapter 10 treats urban climates. Chapter 11 examines topoclimatic effects on microclimates. Part III (Chapter 12) examines the impacts of climate change on microclimates.

C. Historical Development

Microclimatology had its origins in the later nineteenth century with temperature observations relating to agriculture. One of the earliest microclimatic studies was carried out on soil temperatures by Kerner (1891) in the Inntal and Gschnitztal, Austria. In Finland, Homén (1893, 1897) published extensively on soil and air temperatures, giving special attention to night frosts, and a detailed paper on soil temperatures and radiative exchanges. Kraus (1911) wrote a book on soils and small-scale climate based on observations near Karlstadt-am-Main, Germany. Baum and Court (1949)

Table 1.1. Number of Citations of Microclimatological
Temperature Studies by Region between the Years 1920–9
and 1930–9

Region	1920–9	1930–9
Austria and Germany	29	49
North America	13	10
England and India	12	10
Russia	6	10
All others	6	9

Source: From Baum and Court (1949).

list citations of microclimatological temperature studies by region for 1920–9 and 1930–9, showing the dominance of the German-speaking world (Table 1.1):

Rudolf Geiger wrote the first book (in German) on *The climate near the ground* in 1927. He published a fourth English edition in 1961, and this was updated and extended by Geiger et al. (2003). O. G. Sutton (1953) published *Micrometeorology*, treating the physical processes. Slatyer and McIlroy (1961) at CSIRO in Australia published *Practical microclimatology*. Berényi (1967) published a text in German on microclimatology. Gol'tsberg (1969) published *Microclimates of the USSR*. N. Rosenburg et al. (1983), at the University of Nebraska, published *Microclimate: The biological environment*. Bailey et al. (1997) published *Surface climates of Canada*. Ecological works have appeared in recent years. Jones (1992) published *Plants and microclimate*. Campbell and Norman (1998) published *An introduction to environmental biophysics*, which has chapters on plants, animals, and humans and their environments. Bonan (2008) wrote an *Ecological climatology* and Monson and Baldochhi (2014) published *Terrestrial biosphere-atmosphere fluxes*, both from a plant ecological background. Monteith and Unsworth (2014) published a fourth edition of *Principles of environmental physics*: *Plants, animals and the atmosphere*.

Books on local or topoclimates include M. M. Yoshino's (1975) *Climate in a small area*, *The atmospheric boundary layer* by J. R. Garratt (1994), and *Boundary layer climatology* by T. R. Oke (2002).

D. A Central Problem of Microclimatology

The need to relate microclimates and macroclimates quantitatively has been pointed out many times, as processes acting on the small scale should integrate to the larger scale. Holmes and Dingle (1965) addressed this question and noted that there are three broad approaches to the problem. The first is the analog method. Small-scale conditions can be extrapolated if radiation, shade, slope, moisture, and so on, are comparable between locations. The second method is by linear correlation, with which the degree of dependence of microclimatic variables on macroclimatic factors is determined. Multiple and partial correlation techniques can be employed to determine the effect of a single variable independently of the effect of other variables. Non-linear data have to be linearized before being analyzed. The third approach is by building physical models.

Scale issues are of great importance in microclimatology. Franklin et al. (2013) produced climate data sets by separately downscaling 4 km climate models to three finer resolutions based on 800, 270, and 90 m digital elevation models and deriving bioclimatic predictors from them. As climate-data resolution became coarser, statistical downscaling models (SDMs) predicted larger habitat area with diminishing spatial congruence between fine- and coarse-scale predictions. These trends were most pronounced at the coarsest resolutions and depended on climate scenario and species' range size. On average, SDMs projected onto 4 km climate data predicted 42 percent more stable habitat (the amount of spatial overlap between predicted current and future climatically suitable habitat) compared with 800 m data. They found only modest agreement between areas predicted to be stable by 90 m models generalized to 4 km grids compared with areas classified as stable on the basis of 4 km models, suggesting that some climate refugia captured at finer scales may be missed using coarser scale data.

Sears et al. (2011) point out that models used to predict shifts in the ranges of species during climate change rarely incorporate data resolved to <1 km^2, although most organisms integrate climatic drivers at much smaller scales. Furthermore, variation in abiotic factors ignores thermoregulatory behaviors that many animals use to balance heat loads.

A recent paper sets out the problem of relating microclimatic conditions experienced by fauna and flora to standard climatic data (Potter et al., 2013). The researchers illustrate the problem by infrared mapping of surface temperature in the Loire River valley, at scales ranging from 10 to 150 cm, over a single patch of grass and forbs. Higher resolution data show a shift in the frequency distribution toward higher temperatures. They also show that there is a scale difference of the order of three orders of magnitude between the size of plants, and four orders of magnitude between organisms, on the one hand, and conventional climatic data (1 or 10 km), on the other. This raises the question of how best to assess the microclimatic conditions affecting the organisms without specifically tailored measurements. We can illustrate this by a case study carried out in the 1940s.

E. A Case Study

For a small valley in central Ohio, Wolfe et al. (1943) found large local differences in temperature conditions in woodrat (*Neotoma*) habitats (see Tables 1.2, 1.3 and 1.4).

The range in absolute minima was from − 25 °C in a frost pocket to +32.5 °C in leaf litter in a cove in January with corresponding values in September of 29 °C and 52 °C.

Absolute monthly maxima had a 19 °C range between sites in January, but a 41 °C range in August. The frost-free period ranged from 124 days in the frost pocket to 235

Table 1.2. Absolute Monthly Minimum Temperatures (°C) at *Neotoma* Habitats

	JAN	SEPT
Frost pocket	−25.0	29.0
Lower NE slope	−19.5	38.0
Upper NE slope	−15.5	42.5
Ridge top	−19.0	44.5
Covehead	−16.0	40.0
Crevice	+10.0	54.0
Leaf-litter (cove)	+32.5	52.0

Table 1.3. Absolute Monthly Maximum Temperatures (°C) at *Neotoma* Habitats and at Lancaster Weather Service Office 25 km from the *Neotoma* Sites

	J	F	M	A	M	J	J	A	S	O	N	D
Lancaster	13	16	17	31	33	35	39	29	33	22	22	19
Cliff top	20	22	29	39	43	43	45	47	46	42	29	23
Cove	9	6	9	27	20	23	24	24	25	23	14	13
Red maple	13	12	16	32	30	31	31	31	28	27	21	17
Chestnut oak	12	16	18	32	32	31	32	34	30	30	24	19
Hemlock	15	12	14	28	31	29	32	32	28	29	24	18

Table 1.4. Length of the 1941 Frost-free Period at *Neotoma* Sites and at Lancaster

	Spring Last Frost	Fall First Frost	Frost Free Days
Frost pocket	25-May	26-Sep	124
Lower NE slope	14-May	11-Oct	150
Upper NE slope	22-Apr	29-Oct	190
Cove	3-Apr	11-Nov	209
Crevice	3-Apr	26-Nov	235
Lancaster	5-May	26-Sep	144

in a crevice. The number for the Lancaster weather station was only 144 days, close to the 150-day value for the lower northeast slope site (Table 1.4).

A further issue in describing ecological niches arises in the temporal domain. Kearney et al. (2012) argue that data collection at a daily time interval is required to prevent biases in environmental predictions. In some cases, hourly data may be important.

Suggitt et al. (2011) note that species sometimes survive where the average background climate appears unsuitable, and equally may be eliminated from sites within apparently suitable grid cells where microclimatic extremes are intolerable. Local vegetation structure and topography can be important determinants of fine-resolution microclimate. They show that habitat type (grassland, heathland, deciduous woodland) is a major modifier of the temperature extremes experienced by organisms. They recorded differences among these habitats in north Yorkshire, United Kingdom, of more than 5 °C in monthly temperature maxima and minima, and of 10 °C in thermal range in September and January, on a par with the level of warming expected for extreme future climate change scenarios. Temperature minima were around 5 ° C lower in grassland and heathland than under woodland canopy in both September and January, and maxima were around 5 ° C higher in these habitats in September, when the woodland was in leaf. Comparable differences were found in relation to variation in local topography (15° slope and south/north aspect) in Lake Vyrnwy, Wales, and the Peak District of Derbyshire. Maximum temperatures in September were 6–7 °C higher on south- compared with north-facing slopes. However, mean temperatures were not affected.

References

Bailey, W. G., Oke, T. R., and Rouse, W. R. 1997. *Surface climates of Canada*. Montreal: McGill-Queen's Press.

Barry, R. G. 1970. A framework for climatological research with particular reference to scale concepts. *Inst. Brit. Geogr. Trans.* 49, 61–70.

Baum, W. A., and Court, A. 1949. Research status and needs in microclimatology. *Trans. Amer. Geophys. Union*, 30, 488–93.

Berényi, D. 1967. *Mikroklimatologie: Mikroklima der bodennahen Atmosphäre*. Stuttgart: Gustav Fischer Verlag.

Bonan, G. 2008. *Ecological climatology: Concepts and applications*. Cambridge: Cambridge University Press.

Campbell, G. S., and Norman, J. M. 1998. *An introduction to environmental biophysics*, 2nd ed. New York: Springer-Verlag.

Franklin, J. et al. 2013. Modeling plant species distributions under future climates: How fine scale do climate projections need to be? *Global Change Biol.* 19, 473–83.

Garratt, J. R. 1994. *The atmospheric boundary layer*. New York: Cambridge University Press.

Geiger, R., Aron R. H., and Todhunter, P. 2003. *The climate near the ground*, 6th ed. Lanham, MD: Rowman & Littlefield.

Gol'tsberg, I, A. (ed.) 1969. *Microclimate of the USSR*. Jerusalem: Israel Program for Scientific Translations.

Holmes, R. M., and Dingle, A. N. 1965. The relationship between the macro- and microclimate. *Agric. Met.* 2, 127–33.

Homén, T. 1893. Bodenphysikalische und meteorologische Beobachtungen mit besonderer Berücksichtigung des Nachtfrostphänomens. *Finlands Natur och Folk* 53–4, 187–415.

Homén, T. 1897. Der tägliche Wärmeumsatz im Boden und die Wärmestrahlung zwischen Himmel und Erde. *Acta Soc. Sci. Fenn.* 23(2):1–147.

Jones, H. G. 1992. *Plants and microclimate: A quantitative approach to environmental plant physiology*. Cambridge: Cambridge University Press.

Kearney, M. R., Matzelle, A., and Helmuth, B. 2012. Biomechanics meets the ecological niche: The importance of temporal data resolution. *J. Exper. Biol.* 215, 922–33.

Kerner, A. 1891. Die Änderumg der Bodentemperatur mit der Exposition. *Sitzungsbericht d. Akademie d. Wissensschaft, Wien* 100, 704–29.

Knoch, K. 1949. Die Geländeklimatologie, ein wichtiger Zweig der angewandten Klimatologie. *Berichte deutsch. Landeskunde* 7:115–23.

Kraus, G. 1911. *Boden und Klima auf kleinem Raum*. Jena: Fischer.

Monson, R., and Baldochhi, D. 2014. *Terrestrial biosphere-atmosphere fluxes*. New York: Cambridge University Press.

Monteith, J. L. (ed.) 1975. *Vegetation and the atmosphere*. Vol. 1. *Principles*. London: Academic Press.

Oke, T. R. 2002. *Boundary layer climatology*. London: Routledge.

Potter, K. A., Woods, H. A., and Pincebourde, S. 2013. Microclimatic challenges in global change biology. *Global Change Biol*. 19, 2932–9)

Rosenburg, R. J., Blad, B. L., and Verma, S. B. 1983. *Microclimate: The biological environment*. New York: J. Wiley & Sons.

Sears, M. W., Raskin, E., and Angilletta, M. J., Jr. 2011. The world is not flat: Defining relevant thermal landscapes in the context of climate change. *Integrat. Compar. Biol.* 51, 666–75.

Slatyer, R. O., and McIlroy, I. C. 1961. *Practical microclimatology*. Melbourne, Australia: CSIRO.

Suggitt, A. J. et al. 2011. Habitat microclimates drive fine-scale variation in extreme temperatures. *Oikos* 120, 1–8.

Sutton, O. G. 1953. *Micrometeorology*. New York: McGraw-Hill.

Thornthwaite, C. W. 1953. Topoclimatology. *Proceedings Toronto Meteorological Conference.* London: Royal Meteorological Society, pp. 227–32.

Troll, C. 1950. Die geographische Landschaft und ihre Erforschung. *Studium generale* 3, 163–81.

Wolfe, J. N. et al. 1943. The microclimates of a small valley in central Ohio. *Trans. Amer. Geophys. Union* 24, 154–66.

Yoshino, M. M. 1975. *Climate in a small area: An introduction to local meteorology*. Tokyo: University of Tokyo Press.

Part I

Controls of Microclimate

Chapter 2 introduces the microclimatic elements and the methods and instruments used to observe them are described in Chapter 3. The following two chapters treat radiation and the energy balance. Chapter 6 examines monitoring and modeling of radiation and energy balance via remote sensing and land surface models. Chapters 7 and 8 discusses microclimates of different vegetated environments and of physical systems. Chapter 9 describes the field of bioclimatology.

2

Microclimatic Elements

In this chapter we discuss the basic characteristics of the major climatic elements – temperature, moisture in the air and soil, wind, and carbon dioxide – in both a global and a microclimatological context. We also consider the processes of photosynthesis, respiration and carbon exchange, and the nitrogen cycle. Finally, pollutants and aerosols are discussed. This provides a basic introduction to the elements that are used in characterizing micro- and local climates.

A. Temperature

Air Temperature

Temperature is a measure of the kinetic energy of an object. Air temperatures have traditionally been measured by mercury or alcohol in glass thermometers at a height of 1.5–2.0 m above ground in a louvered box known as a Stevenson screen or weather shelter (to prevent solar heating). Air temperatures at the Earth's surface generally range between about –50 °C and +50 °C. The coldest locations, outside the Greenland and Antarctic ice sheets, are in northeastern Siberia in winter, and the hottest ones are in the subtropical deserts of North Africa and the Arabian Peninsula in summer.

At meteorological stations, readings are typically made every six hours (00, 06, 12, and 18 UTC, Coordinated Universal Time), but at climatological stations there may be only readings of the daily maximum and minimum temperatures. Temperature measurements are widely made around the world (more than 20,000 stations) and in Europe there are data going back more than 300 years. Most instrumental records begin around 1850.

As well as the dry-bulb temperature reading, a wet-bulb is operated: the thermometer bulb is wrapped in a muslin bag that is kept wet. The depression of the wet-bulb reading with respect to the dry-bulb is used to determine the dew point temperature and relative humidity. This instrument, known as a psychrometer, is described in Chapter 3, Section H.

The true daily mean temperature is obtained as the arithmetic mean of 24 hourly readings, but often the average of the daily maximum and minimum temperatures using a "max-min thermometer" is used since this requires a reading only once per 24 hours. This gives a biased value, however, often negative in winter and positive in summer, with a higher standard deviation. Variables of interest include the daily and monthly mean values, the absolute monthly maxima and minima, and derived indices such as frequency distributions and freezing and thawing indices. Figure 2.1 shows the seasonal mean surface air temperatures over the globe and Figure 2.2 shows mean maximum values, which reach 40 °C in the subtropics. This distribution is a reflection of the pattern of average solar incoming radiation.

A frequency distribution is a tabulation of the frequency of each value or range of values. The normal (or Gaussian) distribution has a symmetrical bell-shaped curve with the mean, median, and mode that are equal and at the center of the distribution; 68, 95.4, and 99.7 percent of the values lie within ±1, 2, and 3 standard deviations of the mean, respectively. The population standard deviation (σ) is determined from

$$\sigma = \left[1/n \sum_{i=1}^{n} \left(x_i - x \right) \right]^{0.5} \qquad\qquad 2.1$$

where x is the arithmetic mean value of the individual observation i, and n is the total number of observations. Mean air temperature data are more or less normally distributed, but maximum (minimum) values tend to be skewed with a tail toward high (low) values. Other measures of the central tendency are the median, which is the central value of a frequency distribution, with half the observations above it and half below it, and the mode, or most frequent value or category if the data are grouped. The calculation of the median and mode has the advantage of not being susceptible to extreme values that lie far outside the distribution. Examples of basic statistics that describe the diurnal air temperature hourly observations are shown in Tables 2.1 and 2.2.

In addition to the median, a frequency distribution can be used to determine the 25 and 75 percent values (sometimes referred to as percentiles), with the range between them known as the interquartile deviation, and the upper and lower deciles, or 90 percent and 10 percent frequency values.

In dry air, daytime temperature decreases on average at the rate of 0.98 °C per 100 m increase in height, the dry adiabatic lapse rate, or DALR. On calm, clear nights there is often an increase of temperature in the lowest 10 to 100 m above the ground surface. This is termed an "inversion," and over snow surfaces in winter it can amount to 10–25 °C. To permit comparisons of air temperature at different heights, temperature is often expressed as potential temperature, θ (K);

$$\theta = T_a \left(\frac{P_o}{P} \right)^{R/c_p} \qquad\qquad 2.2$$

where T_a is the air temperature in Kelvin, P_o is the standard reference air pressure (usually 100 kPa), P is the ambient air pressure (kPa), R is the specific gas constant

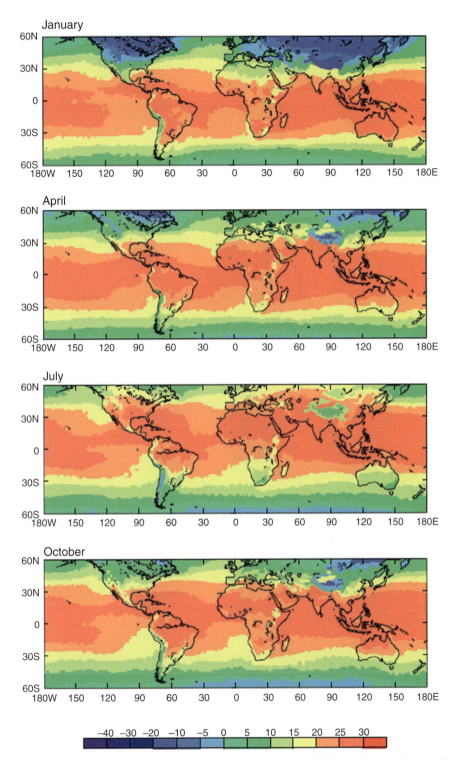

January

April

July

October

Figure 2.1. Seasonal mean temperatures over the globe (°C) (a) January (b) April (c) July (d) October 1961–90 (Jones et al., 1999).

Source: *Reviews of Geophysics,* 37: p. 193, plate 4 (Courtesy of American Geophysical Union).

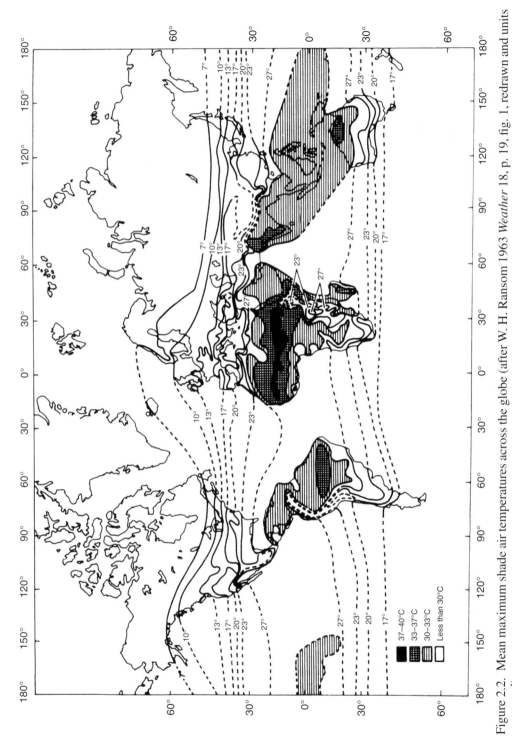

Figure 2.2. Mean maximum shade air temperatures across the globe (after W. H. Ransom 1963 *Weather* 18, p. 19, fig. 1, redrawn and units converted).

Source: Barry and Chorley, 1987, *Atmosphere, weather and climate*, 5th ed., p. 22, fig. 1.16, London: Taylor and Francis.

14

Table 2.1. Example of Hourly (Hr:Min) Observations of Air Temperature (T_a; °C)

Time	1:00	2:00	3:00	4:00	5:00	6:00
T_a	3	1	0	−2	−4	−3
Time	7:00	8:00	9:00	10:00	11:00	12:00
T_a	0	2	6	10	14	18
Time	13:00	14:00	15:00	16:00	17:00	18:00
T_a	22	23	24	23	21	20
Time	19:00	20:00	21:00	22:00	23:00	24:00
T_a	15	12	10	8	6	5

Table 2.2. Statistics Calculated Using T_a Data in Table 2.1

Measures of central tendency	
Arithmetic mean	9.75
(Tmax + Tmin)/2	10
Median	9
Mode	0
Measures of dispersion	
Population standard deviation	9.26
Maximum	24
Minimum	−4
Range	28

for dry air (287 J kg⁻¹ K⁻¹), and c_p is the specific heat capacity for air at constant pressure (1003.5 J kg⁻¹ K⁻¹ for dry air at sea level at 0 °C); it is the temperature an air parcel would acquire if altered adiabatically to the reference pressure. The rising or sinking of the air parcel does not affect its value. For example, air at 25 °C (25 °C + 273.15 = 298.15 K) at a pressure of 85 kPa if taken adiabatically to sea level would have an equivalent temperature of

$$\theta = 298.15 \text{ K} \left(\frac{100 \text{ kPa}}{85 \text{ kPa}} \right)^{(287 \text{J kg}^{-1}\text{K}^{-1} / 1003.5 \text{ J kg}^{-1}\text{K}^{-1})} = 312.34 \text{ K or } 39 \text{ °C}.$$

In cases when heat transfer involves buoyancy due to humidity, it is convenient to use virtual temperature, T_v, the temperature at which a sample of dry air would have the same density and pressure as one of moist air at a particular temperature, T:

$$T_v = \frac{T\left(1 + r_v / \varepsilon\right)}{\left(1 + r_v\right)} \qquad 2.3$$

where r_v is the water vapor mixing ratio (g kg⁻¹) and ε is the ratio of the gas constants for air and water vapor (~0.622). For typical conditions, T_v can be approximated by

$$T_v \approx T\left(1 + 0.61 r_v\right). \qquad 2.4$$

Temperature Extremes

An extreme high air temperature of 56.7 °C was recorded in Death Valley, California, on July 10, 1913. In the Northern Hemisphere, minima of –68 °C have been recorded at both Omaikon (February 6, 1933) and Verkhoyansk (February 7, 1892), in Yakutia, northeastern Russia. In Antarctica, an extreme of – 94.7 °C was observed in August, 2010 by satellite in the vicinity of Dome A at an elevation of about 4050 m. A lower limit to the decrease in surface air temperature on a clear night is provided by the sky radiant temperature, which in midlatitudes is typically about 20 °C below the surface temperature, in the absence of an inversion (Ramsey et al., 1982).

Temperature extremes can be described in terms of the frequency of departures ≥ 2 and 3 standard deviations from the mean, representing ±2.3 percent and ±0.6 percent of the distributions, respectively. To screen data for errors, a 4 standard deviation threshold is sometimes applied to identify outliers for closer scrutiny.

A common index of air temperature is the growing degree-day (GDD) obtained by accumulating the values of mean daily air temperature above a threshold for plant growth, typically 6 °C or 10 °C. For example, a day with a mean air temperature of 25 °C would have a GDD of 15 using a 10 °C threshold; three days with mean air temperatures of 8, 16, and 20 °C would have a GDD of 16. A mean air temperature less than the threshold are set equal to the threshold. Analogous freezing and thawing degree-days (FDDs and TDDs) for negative and positive temperatures, respectively, with respect to 0 °C are also commonly used in studies of seasonal cold and warmth. Degree-day indices provide a measure of the duration and magnitude of temperature anomalies and provide a useful climate-based index for various stages of plant phenology such as flowering, germination, and maturity. For example, with reference to a base threshold of 10 °C, wheat (*Triticum aestivum*) requires 145– 175 GDD to emergence and 1550–1680 to maturity. GDDs are also used to determine developmental events in insects (Damos and Savopoulou-Soultani, 2012). FDDs and TDDs have been widely applied in studies of snow and ice melt and ice accumulation. It should be noted that degree-day indices assume a constant relationship between increments of temperature and the response of the variable that is of interest. While this linear relationship may hold over much of the time range involved, it is unlikely to be true over the whole range.

Near-Surface Profiles of Temperature

The largest changes in temperature, humidity, and wind speed typically occur close to the Earth's surface. For example, profiles of temperature, vapor content, and wind speed above a grass surface in summer at Davis, California, are shown in Figure 2.3. There is a nocturnal inversion of about 5 °C that disappears by 7 a.m. Afternoon temperatures are a few degrees higher at the surface, but humidity values are considerably higher than in the air above.

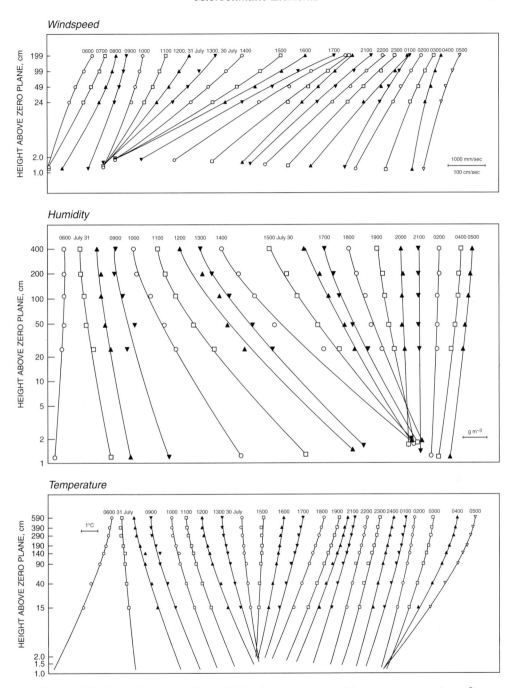

Figure 2.3. Logarithmic profiles of (a) air temperature (b) vapor content (g m^{-3}) and (c) wind speed (cm s^{-1}) above 0.12 m high grass at Davis, California, on 30–31 July 1962 (from Brooks et al., 1963).

The diurnal variation of air temperature typically lags 1–2 hours behind the maximum of incoming solar radiation at solar noon because of the time required for heat absorbed at the surface to be transferred to the overlying air. The daily minimum air temperature typically occurs shortly after sunrise as a result of the continuous emission of infrared radiation by the surface, until it is compensated by absorbed solar radiation. The annual cycle of air temperature over land similarly exhibits a lag of 1–2 months behind the June and December solstices. In the oceans the sea surface temperature lag may be 2–3 months as a result of the high heat capacity of the water and the deeper mixing of the warmed/cooled water. Variability about the mean diurnal and annual temperature cycles is a result of many factors. Hourly variations are affected by large-scale and mesoscale weather systems and especially by variations in cloud cover. The annual cycle is perturbed over weeks and months by the occurrence of persistent modes of atmospheric circulation (see Barry and Chorley, 2010).

An infrared thermometer can measure soil surface temperature if the soil emissivity is known or estimated. Since an infrared thermometer measures the "skin" or radiative temperature, this temperature is typically much higher by day and lower by night than the air temperature, or even the soil temperature measured just below the surface (e.g., Figure 2.4). This was first reported by Schübler in Tübingen, Germany, in the 1790s (Parkes, 1845, p. 138); he recorded an average surface temperature around noon on fine summer days of 55.2 °C compared with a shade average of 21.3 °C. Monteith and Szeicz (1962) cite a maximum surface temperature in England of 48 °C with an air temperature of 30 °C according to H. Penman. At an elevation of 3480 m on Mt. Wilhelm, Papua New Guinea, Barry (1978) recorded a surface temperature of 60 °C with an air temperature of 15 °C in September 1975. At Tucson, Arizona, Sinclair (1922) reported a temperature of 71.5 °C at 4 cm depth with an air temperature of 42.5 °C on June 21, 1915.

The diurnal range of soil temperatures is modulated by the heat capacity of surface vegetation cover, and by depth into the soil. Measurements by Kristensen (1959) near Copenhagen are illustrated in Table 2.3. There is almost no diurnal variation at 50 cm depth. During September 1954–September 1957 the soil temperature at 2.5 cm ranged between 0 °C in February and 22.6 °C in July. This compared with air temperature values of -3.3 °C in February and 17.1 °C in July. At 5 m depth the minimum of 6.5 °C occurred in May and the maximum of 10.6 °C in October.

Much earlier Homén (1893) compared soil temperatures beneath six different surface types in southern Finland at Härjänvatsa, 60° 17'N, 23° 40'E in the summer of 1892. Data for August 12–13, 1892, showed the amplitude of soil temperature variations beneath each type of surface (Table 2.4).

The open heath has the largest range and the wooded heath the least. Measurements were made at the surface, 1, 2, 5, 10, 20, 30, 40, 50, and 60 cm beneath granite rock, sandy heath, and moor meadow in 1893 (Homén, 1897). For August 10–16, 1893, the ranges of maximum and minimum temperatures in the air at 2 m, at the surface, and at 60 cm were as indicated in Table 2.5.

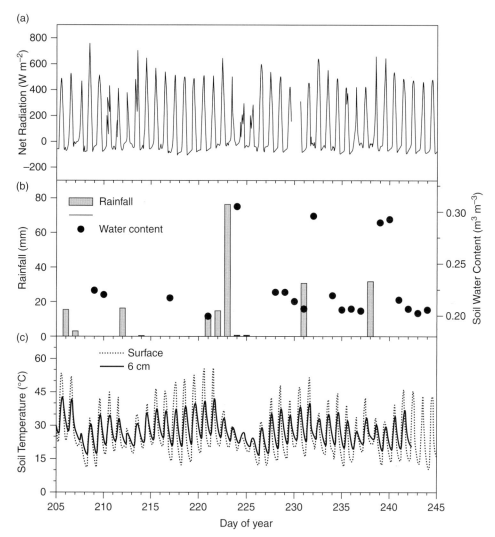

Figure 2.4. Diurnal variations in net radiation, soil moisture, precipitation, and soil temperature measured in a bare field in central Iowa. Soil surface temperature measured with an infrared thermometer varied from 10 to 55 °C compared to a range of 16 to 41 °C only 6 cm below the surface (from J. L. Heitman et al., 2008).
Source: *Journal of Hydrometeorology*, 9: p. 168, Fig. 3. Courtesy American Meteorological Society.

The sandy heath recorded the highest surface temperature and the largest range. It also had the lowest minima. The granite rock had the highest minima.

Temperature Profiles in Plant Canopies

The air temperature profile in plant canopies depends on the trapping of air between the vegetation stalks (or trunks) and the poor conductivity of the soil layers beneath

Table 2.3. Soil Temperatures (°C) Near Copenhagen for 20–22 July 1955

Depth (cm)	Bare Soil		Grass Cover	
	Tmax	Tmin	Tmax	Tmin
2.5	36.6	16.3	31.9	18.6
50	19.5	18.6	19.7	19.1
Air (screen at 2 m)			25.0	15.0

Source: From Kristensen, 1959, table 1, p. 107.

Table 2.4. Soil Temperatures (°C) under Different Vegetation Covers during the Summer in Finland

Depth	Open heath	Wooded heath	Loam field	Moor meadow	Moor field (full grown)	Moor woodland
Surface	18.4	7.1	8.5	18.1	12.0	8.2
10 cm	6.4	1.5	2.4	2.1	1.1	0.8
40 cm	0.6	0.2	0.3	0.0	0.0	0.0

Source: From Homén, 1893.

Table 2.5. Range of Maximum and Minimum Soil Temperatures (°C) under Different Surface Covers during the Summer in Finland

	Maximum	Minimum
Air	14.0–23.6	3.4–11.1
	Granite rock	
Surface	19.6–35.4	8.1–15.9
60 cm	16.3–20.9	16.1–19.4
	Sandy heath	
Surface	21.0–45.7	2.0–10.1
60 cm	Missing–14.2	Missing–14.0
	Moor meadow	
Surface	15.6–30.0	1.3–7.8
60 cm	Missing–15.8	Missing–11.6

Source: From Homén, 1893.

(Brunt, 1953). The first factor is dominant in tall grasses and plants, while the second is important where roots form a horizontal mat. The absorption of both short- and longwave radiation and wind (momentum) by canopy elements (leaves, branches, stems, trunks) also influence the in-canopy air temperature profile. With a short grass cover, the highest afternoon temperatures and the lowest night temperatures are at about 10–20 cm above the ground, where the grasses are dense enough to absorb most of the incoming radiation by day and serve as the most effective surface for outgoing infrared radiation at night.

In a forest with a dense canopy of leaves, the uppermost canopy layer is the effective absorbing and radiating layer. In clear weather the air at the ground is at all times cooler than the air at the top of the canopy, creating a temperature inversion within the forest. This often results in downward sensible heat flux within the canopy and an increase in wind speed within the trunk space. On a clear night the air that is cooled by contact with the upper canopy sinks to the ground. In a forest with gaps, some sunlight penetrates to the ground and is absorbed by the tree trunks. Daytime temperatures are then intermediate between those in a dense forest and those in the clearing.

As shown for the grass cover, soil temperatures are also greatly influenced by canopy cover such as a forest. Properties of the forest canopy such as height, stand density, leaf area index, and leaf type all influence the amount of solar radiation reaching the forest floor, thus affecting soil temperature. Other factors at the forest floor that influence soil temperature include soil type (organic matter), moisture content, and wind, all of which are affected by the canopy properties. For example, in a montane forest near Bailey, Colorado, tremendous differences in soil temperature were recorded when the forest canopy was removed by fire (Figure 2.5).

Soil temperature is also affected by the insulating properties of snow cover. Since snow is an effective insulator, soil temperatures are influenced by snow depth. As snow depth increases, the diurnal fluctuations in soil temperature decrease and the depth where freezing occurs increases (Hardy et al., 2001), conditions much more tolerable for subnivean life. If the soil does freeze, however, the soils remain at 0 °C until all the ice has melted since the recipient energy must first be used in the solid-liquid phase latent heat phase change (334 kJ kg^{-1}) before warming can occur (Figure 2.6).

B. Moisture

Moisture in the ground, at the surface, and in the atmosphere are all components of the hydrological cycle. This is summarized in Box 2.1 for convenience.

Atmospheric Moisture

Moisture in the atmosphere can be expressed in four ways. The vapor pressure (*e* in unit of pressure, typically pascals, or kPa) is the partial pressure exerted by water vapor as a gas. The vapor content is given as either the specific humidity (*q*), the weight of water in the air per weight of moist air (g kg^{-1}), or the humidity mixing ratio (g kg^{-1}), the weight of water in dry air. The two values are closely similar except in hot, humid conditions. A further measure is the relative humidity (percentage) – the ratio of the vapor pressure in the air to its saturated value at the same temperature. The saturation vapor pressure (e_s; kPa) with respect to pure water or ice is represented

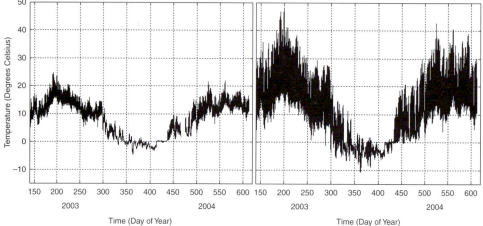

Figure 2.5. Simultaneous soil temperature measured at a depth of 2 cm in an un
burned (left) and burned (right) montane forest near Bailey, Colorado.
Source: P.D. Blanken. Redrawn after T. Oakley, (2004). Figure 5.23a, p. 116, *The spatial and temporal variability of soil moisture and temperature in response to fire in a
montane Ponderosa pine forest, Bailey, Colorado*. M.A. Thesis. Boulder: University
of Colorado.

by the Magnus formula (Magnus, 1844), also referred to as the Clausius-Clapeyron
equation

$$e_{s} = a\exp\left(\frac{bT}{T+c}\right) \qquad\qquad 2.5$$

where $a = 0.61121$, $b = 17.502$, and $c = 240.97$ with respect to water, and $a = 0.61115$,
$b = 22.452$, and $c = 272.55$ with respect to ice, with temperature (T) in degrees Celsius
(°C) in both cases (Buck, 1981).

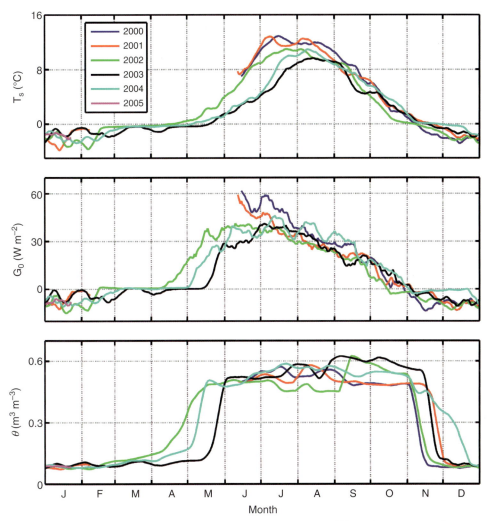

Figure 2.6. Effect of snow cover on soil temperature (T_s), soil heat flux at the surface (G_o), and volumetric moisture (θ) at a high elevation wetland in central Colorado. Note both the soil temperature and soil heat flux did not exceed 0 °C and 0 W m^{-2} (respectively) until after snowmelt occurred as indicated by the sharp increase in soil moisture (from Blanken 2014).
Source: *Journal of Geophysical Research: Biogeosciences*, 119(7): p. 1360, fig. 2. Courtesy American Geophysical Union.

This relationship is important because it indicates the decrease in air temperature and/or the increase in absolute humidity required for the air parcel to reach saturation (100 percent relative humidity) (Figure 2.8). This basic property is the reason that cold air is usually dry, and warm air usually moist. The relative humidity is less useful as it is inversely related to the air temperature and therefore undergoes diurnal variations unrelated to the actual moisture content, whereas the dew point temperature is an indicator of absolute humidity that does not vary with temperature and pressure.

Box 2.1. The Global Hydrologic Cycle and Rivers in the Atmosphere

The hydrologic cycle refers to the continual cycling of water among the oceans, land, and atmosphere. Water is evaporated from the oceans and land into the atmosphere, where it is transported often large distances before it condenses into clouds and is precipitated out. Over land it either runs off in streams and rivers back to the ocean or soaks into the ground, adding to the groundwater in aquifers. The timescales involved range over 10^5 orders of magnitude. The residence time of water in the atmosphere is about 10 days, in the soil it is ~1–3 years, in deep aquifers it may be 1000–10,000 years, and in the oceans it is about 3000 years. Some of the precipitation falls as snow, where it may reside from 1–9 months, except where it lands on ice caps and ice sheets and eventually turns into ice. In Antarctica, this ice may persist for 10^5–10^6 years.

Figure 2.7 shows the exchanges (10^3 km^3 yr^{-1}) and storage (10^3 km^3) components of global water budget. The Earth holds about 1.4×10^9 km^3 of water, about 97 percent of which is in the oceans. The total *freshwater,* about 35 million km^3, is mainly locked up in the Greenland and Antarctic ice sheets, or in deep groundwater. Lakes, swamps, rivers, and soil moisture account for about 120,000 km^3 and ground ice for 300,000 km^3. The atmosphere holds about 13,000 km^3, equivalent to water with a depth of 2.23 mm if spread over the entire Earth.

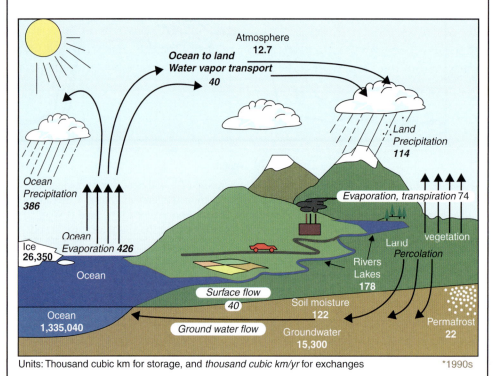

Figure 2.7. The hydrologic cycle based on re-analysis data for 2002–2008 (Trenberth and Fasullo, 2012). Exchanges (10^3 km^3 yr^{-1}) and storage components (10^3 km^3).

The exchanges of water involve evaporation from the oceans (426,000 km³ yr⁻¹) and land (74,000 km³ yr⁻¹), atmospheric moisture transport (40,000 km³ yr⁻¹), precipitation on the oceans (386,000 km³ yr⁻¹) and land (114,000 km³ yr⁻¹), and river runoff (40,000 km³ yr⁻¹).

Recent work shows that narrow streams, or filaments, called "atmospheric rivers," affect much of the horizontal transport of atmospheric water vapor. First reported by Newell et al. (1992), these have been widely studied subsequently (Ralph et al., 2004; Neiman et al., 2008: Gimeno et al., 2014). They are about 2000 km in length, but only about 400 km wide and with a core only 150–175 km wide, and 4 km deep. At any one time there are three to five such plumes within each hemisphere flowing southwest to northeast in the North Atlantic and North Pacific and northwest to southeast in the South Atlantic, South Indian Ocean, and South Pacific. They are linked to warm conveyor belts in midlatitude frontal cyclones. Walliser et al. (2012) determined that there were about 130 systems annually. They give rise to major precipitation and flooding events on midlatitude west coasts.

Vapor pressure typically ranges between about 1 and 10 hPa. Vapor content is in the range 1–10 g kg⁻¹. The total moisture in an atmospheric column is expressed as the precipitable water, which ranges from about 1 to 6 mm.

The dew point temperature is the temperature at which the air becomes saturated when cooled at constant pressure (Figure 2.8). Readings of air temperature and the dew point are used to calculate the relative humidity. Especially for plants (transpiration) the "vapor pressure deficit," the difference between the saturation

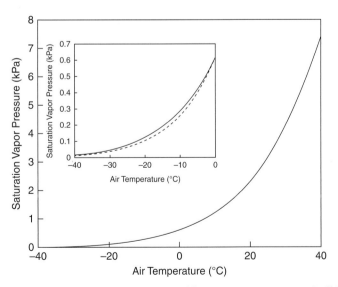

Figure 2.8. Saturation vapor pressure curve with respect to pure water (solid line) and ice (insert: dashed line).

and ambient vapor pressures, can be a useful way to express the dryness of the atmosphere.

Soil Moisture

Soil moisture is the water contained in the unsaturated (or vadose) soil zone. It is a volumetric measure of cubic meters of water per cubic meter of soil, or can be expressed by weight, or gravimetrically, as the mass of water per mass of dry soil. Volumetric and gravimetric measures can be converted knowing the dry bulk density of the soil. Surface soil moisture is the water contained in the upper 5–10 cm of soil, whereas root zone soil moisture is that which is available to plants, generally considered to be in the upper 200 cm of soil. The underlying saturated zone is defined by the water table. The maximum soil water content of a given soil volume defines the porosity, which ranges between 0 and 1, where organic soils tend to have a much higher porosity than mineral soils. The maximum volume of water actually available to plants is the field capacity minus the permanent wilting point. Field capacity is the amount of water that remains in the soil a few days after free drainage following wetting.

The volumetric soil moisture content remaining at field capacity is about 15 to 25 percent for sandy soils, 35 to 45 percent for loams, and 45 to 55 percent for clay soils. Above the field capacity, water cannot be held against gravitational drainage, and below the wilting point, the water is tightly held by the soil matrix and is unavailable to plants. The permanent wilting point is the water content of a soil when most plants growing in it wilt and fail to recover their turgor upon rewetting. The volumetric soil moisture content at the wilting point will have fallen to around 5 to 10 percent for sandy soils, 10 to 15 percent in loams, and 15 to 20 percent in clay soils.

Soil moisture can also be given as a potential (or soil tension) in hectopascals (hPa) (see Box 2.2). This can be thought of as the suction (negative pressure) required to

---

Box 2.2. Water Potential

Especially when dealing with vegetation, knowing the amount of energy required to extract water from soils is often more important than knowing the quantity of water in the soil itself. Whereas the quantity of water can be expressed as either the volume or mass of water per volume or mass of dry soil (volumetric or gravimetric, respectively), the energy required for a plant (or any other pump) to extract water from the soil matrix is expressed as joules per kilogram ($J\ kg^{-1}$). The forces that must be overcome to transport water from the soil to a leaf include the force of gravity, the binding force of the water within the soil (matrix potential), chemical action due to ion concentrations (solute potential), and the liquid-air interface within the stomata.

The total water potential, therefore, refers to the potential energy required to extract this water and has positive values if the water is flowing under pressure (energy is released, not required) and has negative values to indicate the energy deficit that must be matched before water will flow. This is almost always the case in soils and the soil-plant-atmosphere continuum, an exception being water flowing from an artesian well.

The relationship between the volumetric water content and water potential in a soil is known as the soil water retention curve and is very non-linear. These plots are the basis for the permanent wilting point and field capacity determinations, and they show the importance of soil texture through the matrix potential. For example, at equivalent moisture contents, more energy (suction) must be exerted to extract water from the loam soil than the sandy-clay soil (Figure 2.9).

extract water from the soil, where the water is "held" in the soil. The plot of soil water content against soil water potential is known as the soil water retention curve (Figure 2.9). As the plot shows, it is much harder to extract water from a clay soil than a sandy soil even at the same volumetric water content. Thus, soil texture, the mass fraction of sand, silt, and clay, plays an important role in plant-water relationships. Soil moisture potentials of −1 hPa, −100 to 300 hPa, and −1.5 MPa are typical of the saturation point, field capacity, and permanent wilting point, respectively (Senaviratne et al., 2010).

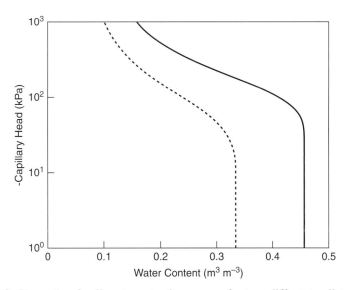

Figure 2.9. Example of soil water retention curves for two different soil textures, a Pachapa loam (solid line) and a Pachapa fine sandy clay (dashed line) (created using Eq. 21 with absolute values for ψ and ψ_L and data presented in table 3 in Assouline and Tessier, 2010).

Darcy's law characterizes the movement of water in soils. This is described in Box 2.3.

Box 2.3. Water Flow in Soils: Darcy's Law

The flow of water in a porous medium such as soils depends on the difference in water pressure and the resistance to water flows between two points of interest. This framework of the flux, which is proportional to difference in concentration (pressure, in this case) divided by a resistance, is analogous to the Ohm's law analog for sensible heat or evaporation, or Fick's law for the diffusion of a gas. Henry Darcy (1803–58) is credited with framing the equation that governs water flow through soils. His construction of a gravity-feed, sand-filtered water system for Dijon, France, led him to develop the well-known equation of Darcy's law:

$$q = \frac{\Delta P}{H_R} \qquad\qquad 2.6$$

where q is the water flux (meters per second, m s^{-1}), ΔP is the change in water pressure between two locations (the pressure gradient; Pa m^{-1}), and H_R is the hydraulic resistance, equal to $\mu \kappa^{-1}$ where μ is the fluid's dynamic viscosity (Pa s) and κ is the medium's permeability (m^2). To quantify q in terms of the more typical units of discharge (Q; m^3 s^{-1}), Eq. (2.1) is simply multiplied by the applicable cross sectional flow area (A; m^2) and a "negative" is often inserted to convey direction of the flow from areas of high to low pressure (e.g., over the distance L, the water pressure at location 1 is greater than the water pressure at location 2; thus Q is positive and water flows from location 1 to 2):

$$Q = -A \frac{\kappa}{\mu} \frac{P_2 - P_1}{L} \qquad\qquad 2.7$$

Note that Darcy's law is applicable only when the flow is laminar (not turbulent), as is almost always the case for water flow in both the unsaturated (vadose) and saturated (groundwater) zones. Most often the soil's hydraulic conductivity (K; m s^{-1}) is measured or estimated, then related to κ knowing μ and the fluid's density (ρ; kg m^{-3}):

$$\kappa = K \frac{\mu}{\rho g} \qquad\qquad 2.8$$

where g is the acceleration due to gravity (9.81 m s^{-2}). Using a hydraulic conductivity of 1×10^{-5} m s^{-1} (a sandy soil), and a value for water's μ at 20 °C of ~1 mPa s (1×10^{-3} Pa s or kg m^{-1} s^{-1}), κ would be

$$\kappa = 1 \times 10^{-5} \, \text{m s}^{-1} \, \frac{1 \times 10^{-3} \, \text{kg m}^{-1} \text{s}^{-1}}{1000 \, \text{kg m}^{-3} \times 9.81 \, \text{m s}^{-2}} = 1.02 \times 10^{-12} \, \text{m}^2.$$

If the pressure at one end of a 5-m-long cylinder of this soil with a radius of 4 m were 0.10 Pa and 0.05 Pa at the other end, the discharge would be

$$Q = -A \frac{\kappa}{\mu} \frac{P_2 - P_1}{L}$$

$$= -\pi \times (4 \, \text{m})^2 \, \frac{1.02 \times 10^{-12} \, \text{m}^2}{1 \times 10^{-3} \, \text{Pa s}} \, \frac{0.05 \, \text{Pa} - 0.10 \, \text{Pa}}{5 \, \text{m}}$$

$$= 5.13 \times 10^{-10} \, \text{m}^3 \text{s}^{-1}$$

This is a small number, but when converted to cubic centimeters per day, it is equal to 44 (same as 44 ml of water flow per day).

Although the use of Darcy's law to predict water flow in soils and other porous mediums has largely been in engineering and geological groundwater applications, it does have micrometeorological applications, especially when modeling water flow through vegetation. For example, Darcy's law has been used successfully to help understand and model water flow through trees in the Colorado Rocky Mountains (McDowell et al., 2008) and old-growth coniferous forests in the Pacific Northwest in Oregon and Washington (Warren et al., 2007).

C. Evaporation

Evaporation strictly refers to the process of a phase change in water from a liquid to a gas. Therefore, the term "evaporation" is used to include evaporation of transpired water from within the stomata in plant leaves. The process in which vegetation internally transports water within vascular tissues (e.g., the xylem) from the roots to the stomata, where the water then evaporates, is termed transpiration. It has been a long-standing challenge in micrometeorology to separate transpiration from evaporation, but when this partitioning is not possible, the term "evapotranspiration" (ET) is often used. On the basis of isotopic ratios (see Box 2.4), we know that transpiration is the larger component of ET over land surfaces. Recent estimates claim that up to 80–90 percent of the total evaporation from terrestrial surface is transpiration, according to Jasechko et al. (2013). However, others consider these data to be biased; as a fraction of ET, transpiration could account for a far lower amount, 38–77 percent, based on global scale field measurements (Wang et al., 2014). Some of this large range in estimates of the fraction of ET that is transpiration stems from the variation in transpiration among vegetation species, but also the non-linear relationship between the area of leaves (the leaf area index, LAI, square meters of leaves per square meter of ground). For example, recent derived relationships between LAI (x) and T/ET (y) are $y = 0.91\ x^{0.07}$ for agricultural systems, $y = 0.77\ x^{0.10}$ for natural systems, and $y = 0.91\ x^{0.08}$ for the overall global data (Wang et al., 2014). Even under low LAI conditions (e.g., LAI = 0.5), T/ET value could be up to 0.72 and 0.90 for natural and agricultural systems, respectively.

Ambrose and Sterling (2014) apply statistical analysis and models to a new global ET database (Global ET Assembly 2.0, GETA 2.0) to determine how the annual ET varies with land cover (LC) type using 5-minute resolution data. They derive global fields for each LC using linear mixed effect models (LMMs) that incorporate geographical and meteorological variables as possible independent regression variables. The majority of the ET lie between 0.3 and 1.5 m yr^{-1} with the observed mean equal to 0.84 m yr^{-1} and the median equal to 0.69 m yr^{-1}. Table 2.6 provides the mean values from the models and the observed data at 2363 points across the globe.

Table 2.6. Land Cover (LC) Type and Estimated Annual Actual ET (m yr^{-1})

LC Type	ET_LMM	ET_OBS
Evergreen broadleaf forest	1.21	1.20
Deciduous broadleaf forest	0.75	0.71
Evergreen needle leaf forest	0.39	0.56
Deciduous needle leaf forest	0.28	0.47
Mixed forest	0.34	0.66
Savannah	0.78	0.88
Grassland	0.42	0.58
Shrubland	0.31	0.39
Barren land	0.07	0.32
Wetlands	0.83	1.06
Lakes and reservoirs	0.56	1.61
Irrigated cropland	0.93	1.14
Non-irrigated cropland	0.65	0.62
Tree plantations	0.67	0.83
Grazing	0.66	0.77
Urban and built-up	0.48	0.52

Notes: ET_LMM refers to ET statistics derived from the ET_LMM method. ET_OBS refers to ET statistics derived from the ET_OBS.
Source: From Ambrose and Sterling, 2014, table 1.

Transpiration is the evaporation that takes place inside a leaf within pores known as stomata (Figure 2.10). These are typically more numerous on the underside of foliage, and the aperture of the stoma is regulated by two guard cells through changes in the turgor of these cells. Stomata are up to 5 µm wide and 10–20 µm long. There may be 100 to 500 per square millimeter of leaf surface, but the total pore area is only 1 percent of the leaf area. Plants require liquid water to assimilate carbon in the form of carbon dioxide; thus the stoma "open" to sequester carbon, but at the same time suffer a loss of water through transpiration. This is sometimes referred to as the "transpiration dilemma," and the optimum situation is to maintain open stomata to assimilate carbon without a large water loss (the ratio of the mass of the carbon gained to water lost is the plant water use efficiency). Therefore, stomata respond to the same environmental variables that drive photosynthesis: light (photosynthetically active radiation [PAR] that provides the energy for photosynthesis), temperature (that regulates the rate of chemical reactions), and available water (to dissolve carbon and transport nutrients).

There are three main photosynthetic pathways; hence vegetation and its response to microclimate variables are represented by these three groups. The most common are C3 species (85 percent of the world's vegetation, including small grains, legumes, and all trees); these plants use only the Calvin (or C3) cycle requiring light to sequester atmospheric CO_2. Vegetation found in highly stressed hot and dry

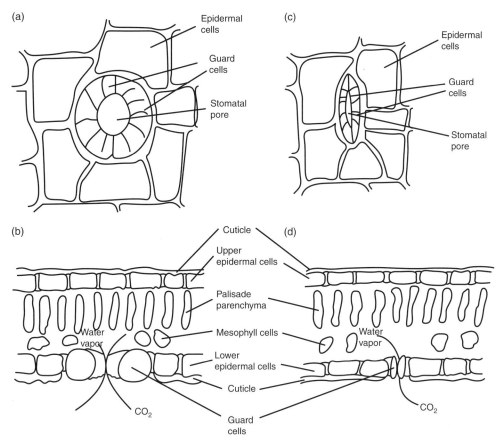

Figure 2.10. Schematic top (a and c) and cross section (b and d) of an open (a and b) and closed (c and d) stoma and guard cells (from Blanken, 2015).
Source: Canopy Processes. In G. R. North, J. Pyle, and F. Zhang (eds.). *Encyclopedia of Atmospheric Sciences, 2nd Edition*, Vol. 3: p. 246, Fig. 2. Courtesy Elsevier.

conditions (e.g., tropical grasses, maize, sorghum, sugarcane, dry-prairie grasses) recycle some of the respired CO_2 in specialized cells located in the bundle sheath and hence are referred to as C4 species. Last, crassulacean acid metabolism (CAM) species, found in arid conditions, can reverse the typical diurnal stomata cycle by forming an acid using the daytime solar energy, while keeping the stomata closed, and then metabolizing this acid at night to open stomata to admit CO_2 when transpiration water losses are minimal. Stomata typically open in response to light and close in response to dark. They also close when temperatures are above or below some optimum range and in response to a high potential for desiccation. The typical stomatal response to the ambient microclimatic conditions for a C3 species is shown in Figure 2.11.

Transpiration serves to cool the plant by evaporative cooling. Transpiration rates vary from < 100 mm a^{-1} to 1300 mm a^{-1}. Crop plants transpire between 200 and 1000 kg of water for every kilogram of dry matter they produce.

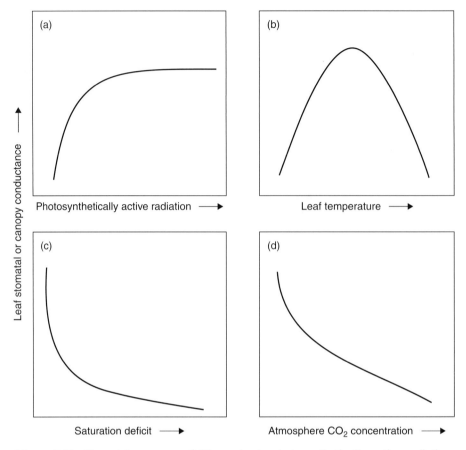

Figure 2.11. Stomatal response of C3 species to photosynthetically active radiation (a), leaf temperature (b), saturation deficit (c), and atmospheric CO_2 concentration (d) (from Blanken, 2015).
Source: Canopy Processes. In G. R. North, J. Pyle, and F. Zhang (eds.). *Encyclopedia of Atmospheric Sciences, 2nd Edition*, Vol. 3: p. 250, Fig. 6. Courtesy Elsevier.

Box 2.4. Isotopic Ratios

Oxygen has two primary stable isotopic forms: oxygen-16 (^{16}O) and oxygen-18 (^{18}O). Their respective abundances (percentages) are 99.76 and 0.1995.

The lighter ^{16}O is preferentially evaporated into the atmosphere and the heavier ^{18}O is preferentially precipitated after condensation. Thus, water vapor as it moves poleward attains an isotopically lighter signature. This fact is utilized in the analysis of ice cores from polar ice sheets and high elevation ice caps. Ice in Greenland and Antarctica has large deficits (-30 to -50) of $\delta^{18}O$ as it has been progressively rained out as the air moved poleward. Conversely, sea water is enriched in $\delta^{18}O$ by two to three parts during glacial times.

When water is transpired, fractionation does not occur, thus there is a disctinct isotopic signature between evaporated and transpired water. Therefore, if the stable isotopes of water for both oxygen $^{18}O/^{16}O$ and hydrogen $^{2}H/^{1}H$ are measured in a plant canopy, then transpiration component of the total evapotranspiration would be known.

At night, condensation may occur on grass and leaves, leading to the deposition of water droplets known as dew, or frost if below 0 °C. The role of radiation in cooling the surface and air above it was first pointed out by W. C. Wells (1814) in an essay on dew (Parkes, 1845, p. 130). When the dew point temperature is < 0 °C, ice, termed hoar frost, is deposited on the surface. Bare soil surfaces more rarely have dew as heat is transferred from the soil layers to the surface. Dewfall amounts rarely exceed 0.5 mm per night. In saturated air under clear skies, the rate of dew formation is about 0.06–0.07 mm hr^{-1}.

Fog (< 1 km visibility) forms when the surface air is cooled to its dew-point temperature and there is a light breeze to keep the condensed droplets in suspension. The cooling may be radiative under clear skies at night, forming radiation fog that in winter can persist during the day. Advection fog forms when moist air is advected over a cooler surface, especially cold ocean currents. This is common over the Grand Banks of Newfoundland, over the California Current, and over the Benguela Current off the coast of Namibia. Fog may give rise to light precipitation (drizzle).

Fog deposition on plants and trees is a major contributor to the water balance in tropical cloud forests and coastal desert environments. Slinn (1982) states that fog water deposition flux, F (kg m^{-2} s^{-1}), is a product of the deposition velocity, V_d (m s^{-1}), and the liquid water content, L_{WC} (kg m^{-3}):

$$F = V_d L_{WC} \qquad 2.9$$

Katata (2014) shows nearly linear dependence of V_d on wind speed above the canopy (u)

$$V_d = A\, u \qquad 2.10$$

where A (nondimensional) is the removal efficiency of fog droplets by the canopy depending on vegetation characteristics. For coniferous trees, Katata (2014) proposed

$$A = 0.0164\, (\text{LAI} / h)^{-0.5} \qquad 2.11$$

where h is the canopy height (m). The relationship has been modified for sparse canopies. If the ratio of the removal efficiency of fog droplets by a 20 m coniferous canopy is 1.0, the value for 20-m-high broadleaf forest is 0.826. For 0.5-m-high cropland and grassland the efficiency is 0.217. Literature values of V_d range from 2.1 to 8.0 cm s^{-1} for short vegetation, whereas the value is 8–92 cm s^{-1} and 0 to 20 cm s^{-1} for forests measured by throughfall-based methods and the eddy covariance method, respectively.

When air temperatures are below freezing, supercooled fog droplets encountering obstacles (trees, fences) are deposited on their windward side as ice crystals known as rime. The ice crystals build out into the wind, and accumulations of many centimeters' thickness can form. Such ice fogs are a major hazard when the ice accumulates on power lines or aircraft wings. Since the saturation vapor pressure over ice is always less than that over liquid water (Figure 2.8), the humidity immediately around rimed areas decreases, leading to a clearing of the air under calm conditions. This effect was first noted by Tor Bergeron (1922), leading to his theory of the Bergeron ice crystal effect (1933), important in the precipitation formation process described next.

D. Precipitation

Daily rainfall is measured, usually manually, in a rain gauge. There at least 50 different designs in use around the world, which make for major difficulties in comparing amounts recorded. Different heights above the surface, placement, gauge diameter, and rim design all create problems, and wind flow over the gauge may require the incorporation of a wind shield to improve the catch efficiency. Rainfall rates are measured less commonly by the use of a recording rain gauge. There are more than 200,000 gauges operating around the world, mostly located in urban areas and airports.

Rainfall data are bounded by zero and are best described by a gamma distribution with appropriate scale (β) and shape (α) parameters. For $\alpha < 1$, the maximum of variable x occurs near zero, while for values of $\alpha > 20$ the distribution approaches a normal curve. The scale parameter, β, stretches or shrinks the distribution on the x-axis.

The gamma distribution is expressed as

$$f(x) = \frac{(x/b)^{\alpha-1} e^{-x/b}}{\beta \Gamma(\alpha)}$$
                                                                             2.12

where $\Gamma(\alpha) = \int e^{-1} t^{\alpha-1} \, dt$.

Figure 2.12 illustrates the effects of changing the shape and scale parameters.

Studies have been performed using the gamma distribution to describe monthly rainfall in Africa (Husak et al., 2007) and daily rainfall in the United States (Becker, 2009).

The wettest places in the world on an annual basis are Mawsynram and Cherrapunji in Meghalya State, India; Tutendo, Colombia; and Cropp River on the western slope of the Southern Alps, New Zealand, where more than 11.5 m of precipitation is recorded annually. The heaviest fall in 24 hours was 1825 mm at Réunion in the Southwest Indian Ocean during tropical cyclone Denise on January 7–8, 1966.

The envelope of maximum amounts in a given time interval is described by the equation

$$R \text{ (mm)} = 421.6 \, D^{0.475}$$
                                                                             2.13

where D is the duration in hours (Paulhus, 1965).

The mean rainfall per rain day (> 1 mm) averages 5 mm in London, England, and 22 mm in Mumbai, India.

The principal spatial characteristic of rainfall is its high variability. Convective precipitation falls from cumulus clouds in brief, intense showers over small areas.

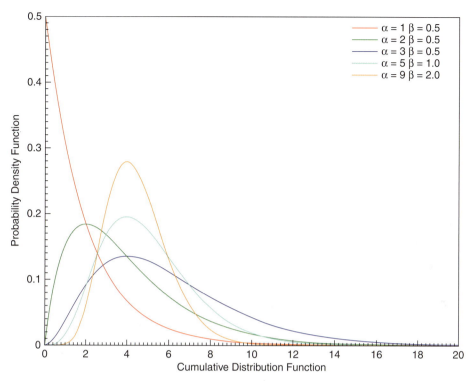

Figure 2.12. The gamma distribution for varying shape and scale parameters. *Source*: Wikipedia, http://upload.wikimedia.org/wikipedia/commons/e/e6/Gamma_distribution_pdf.svg.

Cyclonic precipitation is mostly from stratiform cloud decks that extend over vast areas and last for many hours.

A deficiency of rainfall over an extended time interval, usually of several months, is termed a drought. Drought can be defined in several ways depending on whether the focus is meteorological, hydrological, or agricultural. The numerical thresholds that are used to characterize meteorological drought vary with the climate of the region. In tropical and subtropical climates, drought conditions, without rainfall, may last for a major part of the year, or even for several years. In midlatitudes, droughts may develop within a few months. Hydrological drought occurs when stream flow volume drops and the water stored in aquifers, lakes, and reservoirs falls substantially below the long-term mean amount. These typically begin after several months of meteorological drought. Agricultural drought affects crop productivity and plant growth. It occurs when the soil moisture content reaches the wilting point for plants that are no longer able to recover their turgidity.

A useful measure of drought intensity is the Palmer Drought Severity Index (PDSI) developed by Palmer (1965). It adopts zero as normal, and drought is shown by negative numbers; for example, –2 is moderate drought, –3 is severe drought, and –4 is extreme drought. Positive numbers indicate wetness. Palmer computed potential

evapotranspiration using the Thornthwaite heat index. Soil moisture depletion is estimated for two soil layers. Recharge of soil moisture is determined from precipitation
and runoff is also calculated. The detailed equations are provided by Karl (1986).

The PDSI is calculated routinely for the United States and made available by
the National Oceanic and Atmospheric Administration (NOAA). Karl et al. (1983)
have calculated the PDSI back to 1895. However, Karl (1986) shows that the PDSI
is sensitive to the base period used. The standard period, 1931–60, which was used
traditionally, was anomalously hot and dry. The calibration period should be at
least 50 years. Monthly gridded PDSI data for 1850–2010 are available at www
.esrl.noaa.gov/psd/data/gridded/data.pdsi.html

Snowfall

Ice crystals grow by deposition of water vapor on a freezing nucleus – typically a
clay mineral particle. Typically the arms of the hexagonal crystals entwine because
of turbulent motion in the cloud to build a snowflake. Snow will reach the ground
whenever the freezing level is lower than about 200 m, implying a surface temperature ≤1 °C. A record snowfall of 28.5 m was measured over a 12-month period in
1971–2 at 1650 m on Mt. Rainier, Washington. This can be compared with an annual
average of 16.3 m.

Accurate measurement of snowfall is a major problem because of wind effects
that can lead to 50 percent under catch by unshielded gauges. Catch accuracy can
be greatly improved by the installation of a double snow fence structure around a
shielded gauge. The WMO Double Fence Intercomparison Reference (DFIR) gauge
has been shown to match snowfall closely at sheltered sites (see Chapter 3, Section
G, Measuring Precipitation).

Snow cover is described by the spatial extent of the snow covered area (SCA) and
depth on the ground, or preferably by the snow water equivalent (SWE) – the liquid
water obtained by melting a column of the snowpack. In the Northern Hemisphere
in January, snow cover averages 47 million km^2. In July in the Southern Hemisphere
there is about 1 million km^2. Snow has a density in the range 100–400 kg m^{-3}, with an
average value of about 300 kg m^{-3}. Site measurements of SWE are readily obtained,
but spatial data are difficult to obtain. Passive microwave data have an upper limit of
about 0.8 m of snow depth and are biased by frozen crusts, depth-hoar layers, and
especially the screening effect of evergreen coniferous trees.

The liquid water content of the snowpack is important in many regions that rely
upon the spring melt for water resources. One such region is the Intermountain
Region of the western United States, where a network of ~700 measurement sites
have been established to measure SWE. These SNOwpack TELemetry (SNOTEL)
sites calculate SWE by measuring the hydrostatic pressure (P_h) created by the
product of the snow density (ρ_s) and depth of the overlying snow (h_s; $P_h = \rho_s h_s g$
where g is the acceleration due to gravity) on a 3-m by 3-m snow pillow (a cushion

Figure 2.13. Snow drifts created in the lee of topography (background) and krumm-holz vegetation (foreground) just above the alpine treeline in Colorado.
Source: Photograph provided by P.D. Blanken.

filled with antifreeze). This hydrostatic pressure, when divided by g, is equivalent to SWE, knowing the density of liquid water, ρ_w ($SWE = h_s\rho_s/\rho_w = (P_h/g)(1/\rho_w)$). SNOTEL data are transmitted via meteor burst technology (bouncing radio signals from meteor trails in the ionosphere), to receiving stations in Boise, Idaho, and Portland, Oregon.

Snow depth is highly variable over short distances as a result of drifting around small-scale topography and clumps of vegetation. Sampling studies using airborne lidar in the Colorado Rocky Mountains indicate that above 15–40 m resolution there is a nearly random distribution of snow depths whereas below that range there is a scale break to a structured system (Fassnacht and Deems, 2006). Sampling needs to be at half the scale break distance in order to capture small-scale variability adequately. Earlier, Shook and Gray (1996) found a cutoff length of 30 m between the two regimes for shallow snow covers on wheat stubble and fallow surfaces in the Canadian prairies.

Plant canopies also interact with wind to alter the snow cover and snow depth distribution. In alpine regions, for example, high wind speeds often scour the snow-pack, completely removing it until a reduction in wind speed occurs as a result of topography or vegetation (e.g., krummholz; Figure 2.13). This reduction in wind speed deposits the snow, providing an insulation cover in winter and a source of water in spring. So-called living snow fences have been designed using vegetation in drift-prone areas. Studies have shown that a 79 percent trapping efficiency can be achieved by a properly designed living snow fence (Blanken, 2009).

Hail is composed of ice particles that have accreted around a graupel particle (soft hail comprising a ball of rime ice) as alternate transparent and opaque ice layers as the hail is swept up in the rising air column of a cumulonimbus cloud and successively melted and refrozen. Hailstones may be up to 20 cm in diameter with a record weight of 1 kg. Hail of golf ball size (4-cm diameter) and larger can be very destructive to buildings, cars, and crops. Hail that is 8 cm in diameter has a fall speed of almost 50 m s^{-1}, for example.

In the United States, hailstorms are common between May and September over the high plains. Between 2003 and 2012 there were six to eight hailstorms annually per 260 km^2 in this area, according to insurance reports. They are also common in northern India, China, central Europe, and southern Australia. In July 2010 a hailstorm in the mountains near Nederland, Colorado, produced an accumulation of almost a meter of hail. Using high-resolution Next-generation Weather Radar (NEXRAD) data, Cintineo et al. (2012) analyzed the years 2007–10. They found a triangular corridor of severe hail (≥29 mm diameter) from southwest Texas, extending east to Missouri, and north to South Dakota. The monthly hail maps show enhanced hail frequency in the Great Plains during March–September with June the most active month. In March–May, the southern plains and parts of the southeastern United States exhibit higher hail frequency, whereas July–September shows higher hail frequency in the central and northern plains.

E. Wind

Wind velocity is a vector that has speed and direction. It is typically measured at a standard height of 10 m above the ground by a cup anemometer.[1] The direction is that from which the wind blows – a north wind is one from 360°, an east wind from 90°. Commonly in meteorology, wind is separated into u (west-east) and v (south-north) components. The vertical (upward) component is w. In a three-dimensional reference framework, u, v, and w refer to the horizontal, cross-wind (horizontally perpendicular to u), and vertical directions (vertically perpendicular to u), respectively.

Wind speed increases with height above the surface, and there is a clockwise (counterclockwise) rotation in the Northern (Southern) Hemisphere as the frictional effect of the surface diminishes and the wind blows parallel to the isobars (lines of equal of air pressure) – a geostrophic wind, where there is a balance between the pressure gradient and the Coriolis force due to the Earth's rotation. The wind rotation is known as the Ekman spiral, and it defines the depth of the planetary boundary layer (PBL).

Upward air motion (convection) arises as a result of the spatial differences in heating of the Earth's surface. Hence there is a near-constant horizontal transfer of heat, generally referred to as advection. At the surface the wind blows at an angle toward low pressure. The angle ranges from about 10° over open water to 30° over hilly terrain. The ground also slows the air movement by the form drag of obstacles and skin friction, so that over rough terrain or a city the speed may be only 40–50 percent of the geostrophic value at 1500 m height. The PBL becomes deeper by day as a result of surface heating and convection, but at night radiative cooling decouples the surface winds from those above the boundary layer. The height at which the surface has a negligible effect on the wind, known as the "gradient height," ranges from about 275 m over open terrain to 450 m over cities.

The world's highest recorded wind gust is 113.2 m s^{-1} measured on Barrow Island, Australia. The gust occurred on April 10, 1996, during passage of the eyewall of tropical cyclone Olivia.

The wind profile of the atmospheric boundary layer (surface to around 1500 meters) is generally logarithmic and the appropriate equation takes account of surface roughness and atmospheric stability. It is often approximated by the wind profile power law relationship; this is

$$u/u_r = (z/z_r)^\alpha \qquad\qquad 2.14$$

where u is the wind speed (m s^{-1}) at height z (m), and u_r is the known wind speed at a reference height z_r. The exponent (α) is an empirically derived coefficient that varies depending upon the stability of the atmosphere. For neutral stability conditions, α is approximately 1/7.

Monteith (1973, p. 88) shows that the wind speed at height z over a crop increases linearly with $\ln(z-d)$ where d is a reference level termed the zero plane displacement; it is typically 0.6–0.8 of the height of the canopy, but it depends on the spacing of the roughness elements and on the ratio of the accumulated area of each element to unit area of the underlying surface (see Table 3.2).

Where the stream lines are more or less parallel to the ground surface the flow is said to be laminar. The layer where this occurs is of limited depth, because the flow becomes unstable and breaks down into chaotic eddies known as turbulence. Here the flow is highly irregular with rapid variations in velocity and pressure in space and time. Accordingly, it is generally treated statistically.

F. Carbon Dioxide

The atmospheric concentration of carbon dioxide (CO_2) is a primary climatological variable as a result of its absorption of longwave radiation and its consequent effect on the radiation balance and global temperatures (see Chapter 12). The carbon cycle involves the biogeochemical transfers of carbon among the biosphere, geosphere, hydrosphere, and atmosphere (see Figure 2.14). The atmosphere contains 720 Gt of carbon that exists in gaseous form as CO_2 and methane (CH_4). Carbon dioxide leaves the atmosphere primarily through photosynthesis and net ocean uptake. The terrestrial biosphere includes organic carbon in living and dead organisms and organic

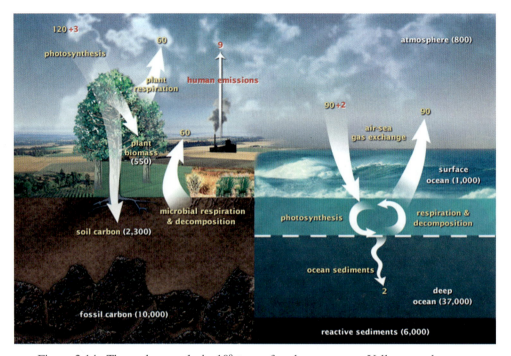

Figure 2.14. The carbon cycle in 10^9 tons of carbon per year. Yellow numbers are natural fluxes; red are human contributions. White numbers indicate stored carbon. *Source*: U.S. Department of Energy, Biological and Environmental Research Information System.

and inorganic carbon stored in soils. There are some 600–1000 Gt in living organisms and 1200 Gt in dead biomass. The carbon reservoir in frozen yedoma (loess) in Siberia is estimated to be ~500 Gt (50×10^3 Tg), nearly equal to that in vegetation (650 Gt), but another ~400 Gt of carbon is stored in non-yedoma permafrost (Zimov et al., 2006).

Forests contain about 86 percent of terrestrial aboveground carbon and forest soils have 73 percent of soil carbon. Annual net primary production over land is about 57 Gt. The oceans contain about 38,000 Gt of mostly inorganic carbon, primarily in the deep ocean, while sediments in the lithosphere have at least 75,000 Gt.

The CO_2 flux at maximum solar irradiance of 80 mol m^{-2} d^{-1} or 1800 μmol m^{-2} s^{-1} (the photosynthetic capacity) averages 28.0 for crops, 23.8 for grasslands, and 20.2 μmol m^{-2} s^{-1} for forests (Ruimy et al., 1995).

The global mean atmospheric concentration of CO_2 exceeded 400 ppm in June, 2015, a 40 percent increase from the preindustrial level. During the Pleistocene ice ages the concentration varied between about 180 ppm in glacial times and 280 ppm in interglacials. The concentration has not been as high as at present since the Miocene period 4 million years ago. Methane concentration in the atmosphere is currently 1800 ppbv. It has increased 2.5 times from the preindustrial value of 722 ppbv. Methane is 29 times more potent as a greenhouse gas than carbon dioxide.

G. Photosynthesis, Respiration, and Carbon Exchange

This section briefly examines the processes of photosynthesis and respiration and the factors that influence their rates, as well as the characteristics of carbon exchange.

Photosynthesis

Green plants extract solar energy through the process of photosynthesis, which drives the synthesis of organic compounds (carbohydrates), enabling plant growth and the development of leaves, stems, roots, and fruits. Photosynthesis involves the conversion of light energy into chemical energy. Light reactions produce primary sugars in plants that contain chlorophyll – a green pigment. It absorbs blue and red wavelengths, but not green wavelengths, which are reflected, hence, the color of leaves. The photosynthetic process was first described in 1845 by J. R. von Mayer. The basic equation is

$$CO_2 + 2H_2O + \text{light energy} \rightarrow CH_2O + H_2O + O_2 \qquad 2.15$$

where the right-hand side represents carbohydrate plus water and oxygen. A detailed account of the biochemical processes involved may be found in Whitmarsh (2009): www.life.illinois.edu/govindjee/paper/gov.html.

Photosynthetic enzymes (RuBisCO) process CO_2 optimally at about 30 °C. Carbon dioxide fixation produces simple sugars from carbon dioxide and water. The reverse process involving oxidation of the carbon compounds, known as respiration, releases energy in the form of heat, and oxygen. Carbon dioxide is supplied by the atmosphere, soil, and living roots. While photosynthesis begins at sunrise (only a quarter of full photosynthetically active radiation, PAR, is needed for full photosynthesis), it is some time before the CO_2 fixation rate exceeds that of the respiration. The point at which the two are equal is termed the light compensation point. Above this point, photosynthesis increases hyperbolically with an increase in incoming solar radiation. At some level of light saturation, photosynthesis becomes independent of the radiation level. Moss (1965) shows that the leaves of shade and woody species are light saturated at about 350 W m^{-2}, and orchard grass and red clover at about 700 W m^{-2}, whereas corn (maize) leaves and sugarcane are still not saturated at 1400 W m^{-2}.

Water can be a limiting factor in photosynthesis. Low soil moisture or very low atmospheric humidity can create water stress that affects the efficiency of plant photosynthesis. Stomata in the leaves begin to close in response to water stress, increasing the resistance to the diffusion of carbon dioxide into the leaves. This is to prevent excessive (negative) xylem pressure potentials, which, if exceeded, can result in cavitation of the water column within the xylem. Typically, plants have their highest stomatal conductance just after sunrise when PAR is sufficient and the VPD is minimal. A midday stomatal reduction often occurs (even in some wetland species; Blanken and Rouse, 1996) in response to excessive xylem pressure potentials. Air temperature generally has a limited effect on photosynthesis, but wind and turbulence are important in the resupply of carbon dioxide that has been depleted by the action of plant photosynthesis.

Respiration

Respiration involves the emission of CO_2 from the soil and flora and fauna. The oxidation of the carbon compounds that is involved in respiration releases energy to perform chemical and biological work by the plant. The equation that describes this is the opposite of that given for photosynthesis:

$$C_6H_{12}O_6 + 6O_2 \rightarrow 6CO_2 + 6H_2O + \text{heat} \qquad 2.16$$

Living roots and organisms in the soil together with decaying organic matter (humus) consume oxygen and respire CO_2. Soil respiration is temperature dependent and can be described by the equation

$$R_T = R_0 \, Q_{10}^{\,(T-T_0)/10} \qquad 2.17$$

where R_T and R_0 are the respiration rates at temperature T, and at a reference temperature T_0. Q_{10} is a factor accounting for the change in respiration rate for a 10 °C change of temperature. Rosenberg et al. (1983, p. 112) cite values of Q_{10} of about 3 for a barley crop, 2.8 for an alfalfa field, and 1.6–2.4 for soybeans. Lloyd and Taylor (1994) review the various relationships between respiration and temperature that have been proposed in the literature. They show that the exponential (Q_{10}) relationship underestimates (overestimates) respiration at low (high) temperatures and that the same applies to a relationship proposed by S. Arrhenius in 1889. They suggest that the respiration rate R relative to that at 10 °C, R_{10}, can be determined from

$$R = R_{10} \, e^{\,306.56\,(1/56.02 - 1/(T - 227.13))} \qquad 2.18$$

without bias across a wide range of ecosystem types and soil temperatures.

Abiotic factors in addition to soil temperature can affect soil respiration, such as the soil moisture. For example, in alpine tundra in Colorado, soil respiration was found to reach a maximum at volumetric soil moisture content of roughly 38 percent (Figure 2.15). When dry, the lack of soil moisture limited soil respiration; when wet, excess soil moisture limited the availability of oxygen and decreased the diffusion of any CO_2 that was respired (Knowles et al., 2015). This finding, and others, have shown that there is often a great deal of spatial variability in soil respiration in response to both biotic and abiotic factors, so the application of generic empirical relationships should be used with caution.

Both C3 and C4 plants respire via an almost identical biochemical pathway by day and night. Since it is not light dependent, it is termed "dark respiration." C3 plants have an additional mechanism that is controlled by light and oxygen availability. This is

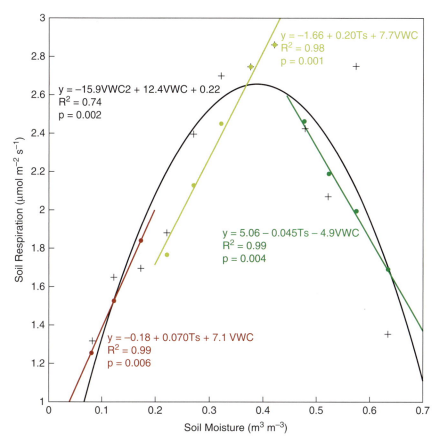

Figure 2.15. Summer-time soil respiration from alpine tundra in Colorado under a range of soil moisture conditions, from dry (brown), mesic (yellow), and wet (green) sites. Average relationship is shown by the black line (from Knowles et al., 2015). *Source*: *Biogeochemistry*, 122, Fig. 2. DOI: 10.1007/s10533-015-0122-3. Courtesy Springer.

termed "photorespiration" and occurs only by day. The preceding temperature-based respiration equations apply to dark respiration as well as the respiration due to humus oxidation in the soil, to living roots, and to the metabolic activity of flora and fauna. It has been shown in many studies that respiration increases at an accelerating rate with temperature.

Soil moisture availability affects respiration from plants and from the soil. Water stress in leaves leads to decreased photosynthesis that in turn reduces dark respiration. Rainfall can reverse this and lead to increased photosynthesis and, in turn, greater respiration. Chen et al. (2014) summarize global data on soil respiration (R_s). Global R_s values range from 75 to 98 Pg C yr^{-1}, more than 10 times the amount of fossil fuel combustion. Rates of R_s are most closely related to annual precipitation, mean annual temperature, and soil organic carbon (or net primary productivity), which together account for 50 percent of the variance of annual R_s. Annual R_s rates range from 0.05 to 2.75 kg C m^{-2} yr^{-1}, with an average of 0.87 kg C m^{-2} yr^{-1}. The highest annual rate was from an evergreen needle leaf forest in Brazil. The lowest R_s was from grassland at Moab, Utah.

Respiration rates were measured in a Ponderosa pine forest in Oregon by Law et al. (1999) using soil chambers. The stands were 34 m high and old (250 years), 10 m high and young (45 years), and mixed age. The independent contributions of soil surface CO_2 efflux (F_s), wood respiration (F_w), and foliage respiration (F_f) were measured and normalized to 10 °C. Mean foliage respiration was 0.20 μmol m^{-2} (hemi-leaf surface area, HSA) s^{-1} and reached a summer maximum of 0.24 μmol m^{-2} HSA s^{-1}. Mean wood respiration was 5.9 μmol m^{-3} (sapwood) s^{-1} and did not differ significantly between old and young stands. Soil surface respiration ranged from 0.7 to 3.0 μmol m^{-2} (ground) s^{-1}, with the lowest rates in winter and highest rates in late spring. Annual CO_2 flux from soil surface, foliage, and wood was 683, 157, and 54 g C m^{-2} y^{-1}, respectively, with soil fluxes responsible for 76 percent of ecosystem respiration.

Carbon Exchange

Photosynthetic organisms remove about 100×10^{15} g (100 Pg) of carbon annually from the atmosphere. The carbon released by biota into the atmosphere is $1–2 \times 10^{15}$ g, compared with 5×10^{15} g from fossil fuel burning. About 30 percent of this amount is taken up by the oceans and eventually stored in sediments on the seabed. Carbon organic compounds are also stored in living biomass (450–650 Pg or Gt), litter and soils (1500–2400 Pg) (Batjes et al., 1996), 300–700 Pg in wetlands (Bridgham et al., 2006), and 1670 Pg to 3-m depth in permafrost (Tarnocai et al., 2009).

The atmospheric carbon content increases by about 3×10^{15} g annually. This has resulted in atmospheric CO_2 concentration increasing from about 285 ppm in 1850 to 400 ppm in 2015. Note that 1 ppm of CO_2 corresponds to 1.97 mg m^{-3}. The annual rate of increase of atmospheric CO_2 rose from about 0.75 ppm yr^{-1} in 1960 to about 2.0 ppm yr^{-1} in 2010. The 40 percent increase in concentration since 1850 was the cause of the global warming of more than 1 °C since the Industrial Revolution in the mid-nineteenth century. The radiative forcing of atmospheric CO_2 is now 1.7 W m^{-2}. Methane (CH_4), which has a concentration of almost 1800 ppb, has a radiative forcing of 0.48 W m^{-2} as it is 25 times more potent as a greenhouse gas.

The flux density of CO_2 (F_c; g m^{-2} s^{-1}) can be expressed using Ohm's law, or electrical circuit, analogy (see Box 2.5):

$$F_c = \frac{c_i - c_a}{r_s + r_a}$$
2.19

where c_i is the CO_2 concentration (g m^{-3}) within the leaf and c_a is the ambient CO_2 concentration (g m^{-3}), r_s is the stomatal resistance (s m^{-1}), and r_a the aerodynamic resistance (s m^{-1}) (see Box 2.5).

Box 2.5. Ohm's Law Analogy

Since water and carbon cycle through the climate system in a closed circuit, it is often convenient to express this cycling using an electrical circuit analogy. There are many similarities. The concentration gradient of mass (water or carbon) between two locations is analogous to the voltage (V) potential of electrons between two locations in a closed cir-

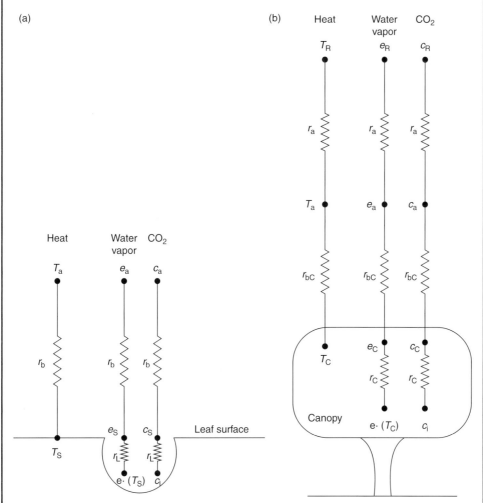

Figure 2.16. An Ohm's law analogy illustrating the pathway of heat, water vapor, and CO_2 exchange between the leaf (a) or canopy (b) and the atmosphere (from Blanken, 2015).
Source: Canopy Processes. In G. R. North, J. Pyle, and F. Zhang (eds.). *Encyclopedia of Atmospheric Sciences, 2nd Edition*, Vol. 3: p. 248, Fig. 4. Courtesy Elsevier.

cuit. The stomatal resistance (the inverse of the conductance) is analogous to the electrical resistance (R). The flux of water or carbon is the analogous to the current (I) that flows through the circuit (see Figure 2.16).

Box 2.5 (*cont.*)

Ohm's law, $V = I/R$, conveniently relates all of these variables to allow complex flows of energy and mass in the microclimate system to be conceptualized as an electrical circuit that can be easily added together to form surprisingly realistic and accurate representations of physical systems. For example, restrictions on water flow through a leaf, such as the stomatal resistance, can be represented as a variable resistor. The leaf's laminar boundary layer can be added as a resistor in-series to the stomatal resistance. The stomatal resistance and boundary-layer resistors from individual leaves can be placed in-parallel to model an entire canopy. Such an Ohm's law analogy is currently used in most of the climate models to represent biosphere-atmosphere interactions.

H. The Nitrogen Cycle

Nitrogen (N_2) is the most abundant gas in the atmosphere (78 percent). The nitrogen cycle is an essential element of ecosystems that involves the atmosphere, plants, organisms, and the soil. Key processes in the nitrogen cycle include the chemical processes of fixation, ammonification, nitrification, and denitrification. Natural fixation (through bacterial conversion) is necessary to convert gaseous nitrogen into compounds such as nitrate or ammonia, which can be used by plants. Nitrogen, in fact, is required by plants to assimilate carbon as N is used in the formation of chlorophyll. The abundance or scarcity of this reactive nitrogen often limits plant growth in both natural and managed environments. Bacteria have the nitrogenase enzyme that combines gaseous nitrogen with hydrogen to produce ammonia, which is converted by the bacteria into other organic compounds. Plants take up nitrogen from the soil by absorption through their roots in the form of either nitrate (NH_3) or ammonium (NH_4) ions. When a plant dies, the initial form of nitrogen is organic. Bacteria or fungi convert the organic nitrogen back into ammonium, a process called ammonification. The oxidation of ammonium is performed by bacteria that convert ammonia to nitrites (NO_2). Other bacteria species are responsible for the oxidation of the nitrites into nitrates (NO_3). The nitrogen cycle is completed when denitrification – the reduction of nitrates back into nitrogen gas (N_2) – takes place. In waterlogged soils, the anaerobic conditions enable bacteria to use nitrates for respiration and the release of nitrogen gas to the atmosphere.

Galloway et al. (2004) document the past, present, and future global nitrogen budget. For the early 1990s, they tabulate the creation of reactive nitrogen (N_r), atmospheric emission and deposition, riverine fluxes, and dentrification. The amounts in teragrams (Tg) of nitrogen per year are shown in Table 2.7.

The anthropogenic creation of N_r globally is nowadays nearly equal to the natural contributions, primarily as a result of industrial ammonia production. The largest sources of terrestrial emission to the atmosphere are NH_3 and fossil fuel combustion releasing NO_x. Riverine fluxes play a significant role in taking up N_r and in dentrification.

Table 2.7. The Annual Exchanges of Reactive Nitrogen in Teragrams in the Earth's Climate System

N creation			
Natural		Anthropogenic	
Lightning	5	Haber-Bosch	100
BNF terrestrial	107	BNF cultivation	31.5
BNF marine	121	Fossil fuel combustion	24.5

Atmospheric emission	
NO_x	
Fossil fuel combustion	24.5
Lightning	5
Other emissions	16
NH_3	
Terrestrial	53
Marine	6
N_2O	
Terrestrial	11
Marine	4

Atmospheric deposition	
NO_y	
Terrestrial	25
Marine	21
NH_x	
Terrestrial	39
Marine	18

Riverine fluxes	
Nr input into rivers	118
Nr export to inland systems	11
Nr export to coastal areas	47

Denitrification	
Continental:	
Terrestrial	67
Riverine	48
Estuary and shelf:	
Riverine nitrate	48
Open ocean nitrate	145
Open ocean	129

BNF = Biological N_2 fixation.
Haber-Bosch is the main industrial process to produce ammonia (used in fertilizers) by the reaction of nitrogen and hydrogen.
Source: After Galloway, *Biogeochemistry*. 2004, table 1, p. 159. Courtesy of Springer.

I. Pollutants

Pollutants are substances in the air that can adversely affect humans and the ecosystem. They can be solid particles, liquid droplets, or gases, of natural or anthropogenic origin. Primary pollutants are produced directly by a natural or anthropogenic process, whereas secondary pollutants form when primary pollutants react or interact with one another. Major primary pollutants are sulfur and nitrogen oxides, volatile organic compounds (VOCs), carbon monoxide, particulates, metals, chlorofluorocarbons, and ammonia. Secondary pollutants are particulates, ground-level ozone, and peroxyacetyl nitrate (PAN).

Sulfur dioxide (SO_2) is a chemical compound produced in volcanic eruptions and industrial processes. Oxidation of SO_2 produces sulfuric acid, H_2SO_4, or acid rain. Other components of acid rain are carbonic acid, H_2CO_3, and nitric acid, HNO_3. The pH of acid rain may be below 2.4, where a pH of 7 is neutral. Acid deposition occurs mainly through wet deposition during rainfall and snowfall. Dry deposition occurs when particles and gases stick to plants and the ground surface. It can account for 20–60 percent of total acid deposition. Acid deposition effects can be especially important in high-altitude forests as clouds and hill fog have higher levels of acidity than precipitation.

Nitrogen oxides (NO_x) form a brown haze above and downwind of cities. Los Angeles, California, and Denver, Colorado, are well-known examples. Atmospheric particulate matter, or fine particles, are tiny particles of solid or liquid suspended in the atmosphere. In contrast, "aerosol" refers to combined particles and gas. Aerosols include mineral dust and organic carbon. The oceans are a major source of sea salt aerosols. Mineral dust from desert surfaces in North Africa, Arabia, and central Asia can cross the North Atlantic westward to the Caribbean and the North Pacific eastward to North America.

Particle concentrations in polluted continental air reach 50,000 cm^{-3}, of which about two-thirds is soot, compared with less than 3000 cm^{-3} in clean continental air. An interactive map of 17,000 sources of air pollution in the USA is available at www.npr.org/news/graphics/2011/10/toxic-air/#4.00/39.00/-84.00.

In the 1990s, amendments to the Clean Air Act in the United States began to reduce significantly the emissions of sulfur dioxide and nitric oxides. Regulations in Europe and the ending of burning of lignite in Central and Eastern Europe had similar beneficial effects.

Aerosols have important effects on global and regional climate. They give rise to direct radiative forcing through the scattering and reflection of solar radiation and the absorption of it by black carbon. Longwave terrestrial and atmospheric radiation is also absorbed and emitted by aerosols. Indirect radiative forcing involves clouds. The mean cloud droplet radius decreases proportionately with the aerosol concentration, and in turn the number of cloud droplet increases. This can increase the cloud albedo, leading to enhanced reflection and cooling at the surface, as well as affecting the cloud lifetime. The total radiative forcing by aerosols is approximately -1.7 W m^{-2}.

Note

1 Conversion of wind speed measured at 10 m to equivalent at 2-m height: $u_2 = u_{10}$ [4.87 / ln(67.8 (10) – 5.42)] = 0.75 u_{10}.

References

Ambrose, S.M. and Sterling, S.M. 2014. Global patterns of annual actual evapotranspiration with land-cover type: knowledge gained from a new observation-based database. *Hydrol. Earth System Sci., Discuss.,* 11: 12, 103–135.

Assouline, S., and Tessier, D. 2010. A conceptual model of the soil water retention curve. *Water Resources Res.* 34(2), 223–1.

Barry, R. G. 1978. Diurnal effects on topoclimate on an equatorial mountain. Arbeiten Zentralanstalt f. Meteorologie Geodynamik (Vienna), Publ. 32(72): 1–8.

Barry, R. G., and Chorley, R. J. 2010. *Atmosphere, weather and climate,* 9th ed. London: Routledge.

Batjes, N. H. 1996. Total carbon and nitrogen in the soils of the world. *Eur. Soil Sci.* 47: 151–63,

Becker, E. J. 2009. The frequency distribution of daily precipitation over the United States. Ph.D. dissertation. College Park: University of Maryland.

Blanken, P. D. 2009. Design of a living snow fence for snow drift control. *Arctic Antarct., Alp. Res.* 41(4): 418–25.

Blanken, P. D. 2014. The effect of winter drought on evaporation from a high-elevation wetland. *J. Geophys. Res. Biogeosci.* 119–1369.

Blanken, P. D. 2015. Canopy Processes. In Gerald R. North (editor in chief), John Pyle and Fuqing Zhang (editors), *Encyclopedia of atmospheric sciences,* 2nd ed., Vol. 3, pp. 244–255.

Blanken, P. D., and Rouse, W. R. 1996. Evidence of water conservation mechanisms in several subarctic wetland species. *J. Appl. Ecol.* 33(4): 842–50.

Bridgham, S. D. et al. 2006. The carbon balance of the North American wetlands. *Wetlands* 26L, 889–916.

Brooks, F. A. 1959. *An introduction to physical microclimatology.* Syllabus No. 397. Davis: University of California.

Brooks, F. A. et al. 1963. Investigation of energy and mass transfers near the ground including the influences of the soil-plant-atmosphere system. Davis: Department of Agricultural Engineering and Department of Irrigation, University of California. Contract DA-36-039-SC-80334 U.S. Army Electronic Proving Ground Fort Huachuca, AZ.

Brunt, D. 1953. Bases of microclimatology. *Nature* 171: 322–3.

Buck, A. L. 1981. New equations for computing vapor pressure and enhancement factor. *J. Appl. Meteorl.* 20, 1527–32.

Chen, S-T. et al. 2014. Global annual soil respiration in relation to climate, soil properties and vegetation characteristics: Summary of available data. *Agric. For. Met.* 198–9: 335–48.

Cintineo, J. L. et al. 2012. An objective high-resolution hail climatology of the contiguous United States. *Weath. Forecasting* 27: 1235–48.

Damos, P., and Savopoulou-Soultani, M. 2012. Temperature-driven models for insect development and vital thermal requirements. *Psyche* 2012, Article ID 123405.

Fassnacht, S. R., and Deems, J. S. 2006. Measurement sampling and scaling for deep montane snow depth data. *Hydrol. Proc.* 20: 829–38.

Galloway, J. N. et al. 2004. Nitrogen cycles: Past, present, and future *Biogeochem.* 7-0:153–226.

Gimeno, L. et al. 2014. Atmospheric rivers: A mini review. *Frontiers Earth Science* 2: 2. doi: 10.3389/feart.2014.00002

Hardy, J. P. et al. 2001. Snow depth manipulation and its influence on soil frost and water dynamics in a northern hardwood forest. *Biogeochem.* 56(2): 151–74.

Heitman, J. L. et al. 2008. Sensible heat observations reveal soil-water evaporation dynamics. *J. Hydrometeor* 9: 165–71.

Homén, T. 1893. Bodenphysikalische und meteorologische Beobachtungen mit besonderer Berücksichtigung des Nachtfrostphänomens. *Finlands Natur och Folk* 53–4: 187–415.

Homén, T. 1897. Der tägliche Wärmeumsatz im Boden und die Wärmestrahlung zwischen Himmel und Erde. *Acta Soc. Sci. Fenn.* 23(2): 1–147.

Husak, G. J., Michaelsen, G., and Funk, C. 2007. Use of the gamma distribution to represent monthly rainfall in Africa, for drought monitoring applications. *Int. J. Climatol.* 27: 935–44.

Jasechko, S. et al. 2013. Terrestrial water fluxes dominated by transpiration. *Nature* 496: 347–5

Jones, P. D. et al. 1999. Surface air temperzture and its changes over the past 150 years. *Rev. Geophys.* 37: 173–99.

Karl, T. R. 1986. The sensitivity of the Palmer Drought Severity Index and Palmer's Z-Index to their calibration coefficients including potential evapotranspiration. *J. Climate Appl. Met.* 25: 77–86.

Karl, T. R. et al. 1983. Statewide average climatic history. *Hist. Clim. Series* 6.1. Asheville, NC: National Climate Data Center.

Katata, G. 2014. Fogwater deposition modeling for terrestrial ecosystems: A review of developments and measurements. *J. Geophys. Res. Atmos.* 119. doi:10.1002/2014JD02166.

Knowles, J. F., Blanken, P. D., and Williams, M. W. 2015. Soil respiration variability across a soil moisture and vegetation community gradient within a snow-scoured alpine meadow. *Biogeochemistry*. doi: 10.1007/s10533-015-0122-3.

Kristensen, K. J. 1959. Temperature and heat balance of soil. *Oikos* 10:103–20.

Law, B. E., Ryan, M. G., and Anthoni, P. M. 1999. Seasonal and annual respiration of a ponderosa pine ecosystem. *Global Change Biol.* 5: 169–82.

Lloyd, J., and Taylor, J. A. 1994. On the temperature dependence of soil respiration. *Functional Ecol.* 8: 315–23.

Magnus, G. 1844. Versuche über die Spannkrafte des Wasserdampfes. *Ann. Phys. Chem. (Poggendorff)* 61, 225.

McDowell, N. G., White, S., and Pockman, W. T. 2008. Transpiration and stomatal conductance across a steep climate gradient in the southern Rocky Mountains. *Ecohydrol.* 1: 193–204. doi: 10.1002/eco.20.

Wells W. C. 1814. *An essay on dew*. London: Taylor & Hessay.

Monteith, J. L. 1973. *Principles of environmental physics*. Sevenoaks, UK. E. Arnold.

Monteith, J. L., and Szeicz, G. 1962. Radiative temperature in the heat balance of natural surfaces. *Quart. J. Roy. Met. Soc.* 88: 496–507.

Moss, D. N. 1965. Capture of radiant energy in plants. In P. E. Waggoner(ed.), *Agricultural meteorology*. Met. Monogr. 6(28). Boston: American Meteorological Society, pp. 90–108.

Neiman, P. J. et al. 2008. Meteorological characteristics and overland precipitation impacts of atmospheric rivers affecting the West Coast of North America based on eight years of SSM/I satellite observations. *J. Hydromet.* 9: 22–47.

Newell, R. E. et al. 1992, Tropospheric rivers? A pilot study. *Geophys. Res. Lett.* 19: 2401–4.

Palmer, W.C. 1965. *Meteorological drought*. Weather Bureau Res. Paper, no. 45. Washington, DC. Dept. of Commerce.

Parkes, J. 1845. On the influence of water on the temperature of soils. *Roy. Agr. Soc. Eng. J.* 5: 119–46.

Paulhus, J. L. H. 1965. Indian Ocean and Taiwan rainfall set new records. *Mon. Wea. Rev.,* 93(5): 331–35.

Ralph, F. M., Neiman, P. J., and Wick, G. A. 2004: Satellite and CALJET aircraft observations of atmospheric rivers over the eastern north Pacific Ocean during the winter of 1997/98. *MWR* 132: 1721–45.

Ransom, W. H. 1963. Solar radiation and temperature. *Weather* 18: 18–23.

Senaviratne, S. I. et al. 2010. Investigating soil moisture – climate interactions in a changing climate: A review. *Earth-Sci. Rev.* 99: 125–61.

Ramsey, J. W., Chiang, H. D., and Goldstein, R. J. 1982. A study of the incoming long wave atmospheric radiation from a clear sky, *J. Appl. Met.* 21: 566–78.

Ruimy, A. et al. 1995. CO_2 fluxes over plant canopies and solar radiation: A review. *Adv. Ecol. Res.* 26: 1–68.

Slinn, W. G. N. 1982. Predictions for particle deposition to vegetative canopies. *Atmos. Environ.* 16, 1785–94.

Shook, K., and Gray, D. M. 1996. Small-scale spatial structure of shallow snowcovers. *Hydrol. Proc.* 10: 283–92.

Sinclair, J. J. 1922. Temperature of soil and air in a desert. *MWR* 50: 142–4.

Tarnocai, C. et al. 2009. Carbon pools in the northern circumpolar permafrost region. *Global Biogeochem. Cycles:* 23: GB003327.

Trenberth, K. E. and Fasullo, J. T. 2012. Atmospheric moisture transport from ocean to land in reanalyses. *GEWEX News, World Climate Research Programme.* 22(1): 8–10.

Waliser, D. E. et al. 2012. The year of tropical convection (May 2008–April 2010): climate variability and weather highlights. *Bull. Amer. Met. Soc.* 93: 1189–1218.

Wang, L-X., Good, S. P., and Caylor, K. K. 2014. Global synthesis of vegetation control on evapotranspiration partitioning. *Geophys. Res. Lett.* 41: 6753–7.

Warren, J. M., Meinzer, F. C., Brooks, R., Domec, J-C., and Coulombe, R. 2007. Hydraulic redistribution of soil water in two old-growth coniferous forests: Quantifying patterns and controls. *New Phytologist* 173: 753–65.

Wells W. C. 1814. *An essay on dew.* London: Taylor & Hessay.

Whitmarsh, J. 2009. The photosynthetic process. In G. S. Singhal et al. (eds.), *Concepts in photobiology: Photosynthesis and photomorphogenesis.* New Delhi: Narosa and Dordrecht, Netherlands: Kluwer Academic. pp. 11–51.

Zimov, S. A. et al. 2006. Permafrost and the global carbon budget. *Science* 313: 1612–13.

3

Methods of Observation and Instrumentation

A. Introduction

The task of observing the microclimate of a region or object in a quantitative manner is important to the discipline for many reasons. As a consequence, many of the innovations and developments in meteorological instrumentation stem from the microclimate community. Initially (and currently), the agricultural community had a great interest in measuring the key variables that were (are) important for crop growth and yields: solar radiation (photosynthetic active radiation, or PAR), air and soil temperature, precipitation, and soil properties (texture for proper drainage and water retention, water content). As such, instruments and techniques for measuring the spatial and temporal variability of these and other crop variables were developed within agriculture and soil physics departments at many universities. Given the need to understand the spatial aspects of microclimate, such as the spatial interpolation between measurement locations, geography departments also often fostered and refined microclimate observations and instrumentation.

This chapter provides a review of some of the main techniques, instrumentation, and theory behind micrometeorological observations. The fundamental principles governing these observations are first discussed. This is followed by discussions of the measurement of temperature, soil properties (temperature, moisture, and heat flux), radiation, wind, precipitation, total and partial atmospheric pressures, and turbulent fluxes.

B. Fundamentals

Any measurement strives to quantify reality. The measurement's accuracy refers to the difference between reality and what the instrument actually detects. For example, if the actual air temperature is 25 °C, a thermometer reading 24 °C is more accurate than one reading 28 °C. In addition to accuracy, a well-designed instrument should have a high precision, meaning that the instrument should read the same value if the conditions are not changing. For example, an object is at a steady temperature of 10 °C. A thermometer that reads 10 °C for five consecutive measurements would be

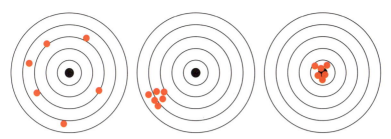

Figure 3.1. Assuming the bull's-eye represents reality, each figure represents (a) low accuracy and low precision, (b) low accuracy and high precision, (c) high accuracy and high precision.
Source: Figure created by P.D. Blanken.

said to have a high precision and high accuracy. If the thermometer "drifted" so that the five consecutive readings were, 10, 9, 8, 7, 6 °C, respectively, the thermometer would have a low precision and low accuracy. Note that an instrument could be precise, but not accurate (see Figure 3.1). In the previous example, consecutive readings of 4 °C while the actual temperature is 10 °C would indicate a high-precision, but low accuracy, thermometer. Ideally an instrument should be accurate and precise.

How well the instrument can resolve the actual reality, the resolution, is also important. A thermometer that can detect temperatures of 0.01 °C has a higher resolution than one that can detect only 0.5 °C. The resolution of the instrument (as do accuracy and precision) depends on the research questions. Sometimes instrument drift (lack of precision) can be tolerated in exchange for increased resolution. For example, when using the eddy covariance method (described later) to measure fluxes, it is the *variance* in the scalar quantities that is important, not the *absolute* quantities.

Finally, for micrometeorologial instruments, the time constant (τ), how fast the sensor responds to ambient changes in the quantity it is measuring, is important to know. When designing an instrument, it is important to match the response of the sensor to the variable one is trying to measure. For example, measuring soil temperature would not require a fast-response, fine-wire thermocouple. Sampling soil temperature at 10 or 20 Hz would be oversampling, whereas 10 or 20 Hz is required to capture the high-frequency fluctuations observed near the surface. An instrument's time constant is determined by measuring the time a sensor takes to decrease to 1/e (36.8 percent) of the initial value in response to a step-change decrease in what it is measuring. The response time to an increase in the stimulus should also be determined since the step-down and step-up responses are seldom the same. The time constant for the step-up response (increase in value) is the time it takes to reach 1 − 1/e (63.2 percent) of the initial value.

For example, consider a pyranometer that is exposed to an energy source of 1200 W m^{-2}, and that energy source is turned off. It takes 5 seconds for the pyranometer to read 0 W m^{-2}. When the energy source is turned back on, it takes it takes 3 seconds for the pyranometer to again read 1200 W m^{-2}. Assuming a linear response, we can calculate τ as

Pyranometer's step-down time constant:

$$\frac{1200\,\text{W m}^{-2}}{5\,\text{s}} = 240\,\frac{\text{W m}^{-2}}{\text{s}};\ (0.368)\left(1200\,\text{W m}^{-2}\right) = 441.6\,\text{W m}^{-2};\ \frac{441.6\,\text{W m}^{-2}}{240\,\dfrac{\text{W m}^{-2}}{\text{s}}} = 1.84\,\text{s}$$

Pyranometer's step-up time constant:

$$\frac{1200\,\text{W m}^{-2}}{4\,\text{s}} = 300\,\frac{\text{W m}^{-2}}{\text{s}};\ (0.632)\left(1200\,\text{W m}^{-2}\right) = 758.4\,\text{W m}^{-2};\ \frac{758.4\,\text{W m}^{-2}}{300\,\dfrac{\text{W m}^{-2}}{\text{s}}} = 2.53\,\text{s}$$

If, for example, sun flecks (patches of full sun followed by full shade due to leaf movement) in the understory of an aspen forest were observed to occur every 1 second, then a pyranometer with a faster response (i.e., a smaller τ) would be required.

An instrument's accuracy, resolution, and time constant are determined by the instrument's design. The precision is also determined by the instrument's design and can be checked and usually adjusted by routine and proper calibration.

The final two considerations that will be discussed here concerning the fundamentals of observations and instrumentation are the sampling interval and the calculation of basic statistics from the observation's time series. As shown by the time constant calculation example, it is critical that the instrument's sampling interval at least match that of what is being measured. Sometimes this is difficult to determine a priori, so the rule of thumb is to sample at the Nyquist rate, which is twice the maximum frequency of the measurement's time series. Sampling faster or slower than this rate can introduce errors.

Once the time series is collected, for example, solar radiation sampled every 1 second, 30-min statistics are typically calculated. These include measures of central tendency (mean, median, mode) and measures of variance (range, standard deviation, or standard error). If a long-term, continuous measurement record is available (typically a minimum of 30 years to qualify as a "climatology"), then various time-series analysis tools can be employed to detect any significant changes over time (time-series analysis). The reader is referred to texts such as von Storch and Zwiers (2002) or Mudelsee (2014) for details on the statistical analysis of climatological data.

Several techniques have been developed to interpolate microclimate measurements spatially, from basic Thiessen polygons originally developed by the American meteorologist Alfred H. Thiessen or the popular technique known as Kriging. The reader is referred to Storch and Navarra (1999) or Navarra and Simoncini (2010) for details on spatial interpolation techniques. More advanced techniques integrate meteorological observations with weather forecast models to produce a hybrid, historical reanalysis of climate data; these are known as "reanalysis" products. Reanalysis products, however, usually do not have sufficient spatial resolution to resolve microclimate processes of interest.

C. Measuring Temperature

Temperature, as a measure of kinetic energy and the solution to the energy balance equation, is the fundamental measurement in climatology and meteorology. Therefore, there have been many developments in making this measurement accurately. The temperature of an object is expressed numerically with the reference points for the three main scales based on kinetic energy (Kelvin; K), the freezing and boiling points of water (Celsius; °C), or the average temperature of a healthy human (Fahrenheit, °F). The Kelvin scale is absolute, with 0 K representing no molecular motion (absolute zero; no kinetic energy so negative values are not possible); thus the "degrees" symbol should not be used. The Celsius scale is relative to 0 °C and 100 °C, representing the freezing and boiling points of pure water at 1 atmosphere pressure (i.e., at sea level). The conversion from K to °C is simple offset, so 1 °C = 1 K:

$$K = °C + 273.15 \qquad\qquad 3.1$$

where 273.15 is the triple point of water (temperature at which all three phases of water coexist). The Fahrenheit scale is seldom used except by the general public in the United States.

To measure temperature, it is first important to make sure that the temperature recorded is in fact the temperature of the desired object. For example, to measure air temperature properly, the thermometer must be shaded and ventilated so that the reading represents the temperature of the air, not the thermometer itself or the stagnant air around it. For soils or water, the depth of the reading must be specified, as should the height aboveground for air temperature. Infrared thermometers, for example, detect the "skin" of the object whose temperature can be quite different from that just below the surface. Changes in temperature with height and depth are usually large near the surface, and spatial variation is often large even on small spatial scales (Figure 3.2).

The simplest, cheapest, and perhaps most accurate way to measure the temperature with direct contact with the object is through the use of a thermocouple (Figure 3.3). As the name implies, a thermocouple consists of two dissimilar metal wires connected at both ends. A small voltage is created in this circuit if the two junctions are at different temperatures; this is known as the Seebeck effect. Alternatively, a temperature difference between the two junctions can be created if a voltage is applied to the circuit. In microclimatology, copper-constantan (type T; 43 μV °C^{-1}) or chromel-constantan (type E; 68 μV °C^{-1}) is commonly used. The advantages of using thermocouples are their low cost and fast response. The response time decreases as the thermocouple thickness deceases, so very thin thermocouples can be used to resolve the air temperature of small turbulent eddies in the atmosphere (Figure 3.3), and unattended recording of data is made possible by recording their voltage output. The main disadvantage is that the temperature at one of the two junctions must be known independently, or else just a relative temperature difference between the two junctions is provided. The thermocouple must also be isolated from any unwanted, external electrical currents

(a)   (b)

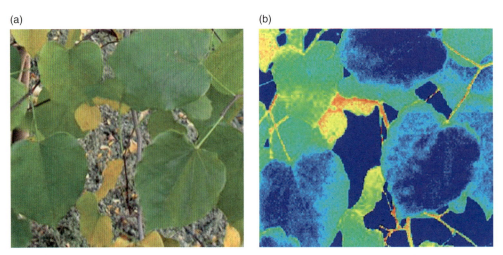

Figure 3.2. Comparison between a regular photograph (A) visible wavelengths and a thermal image (B) infrared wavelengths of poplar shows that the spatial variation in temperature can be large.
Source: Looking at the fall colors in a different way, The GLOBE Program Scientists' Blog, Margaret LeMone, posted October 18, 2005.

that could mistakenly be interpreted as temperature difference. A device known as a thermistor often provides determination of the so-called reference temperature.

Instead of producing a voltage that is proportional to a temperature difference, a thermistor imposes an electrical resistance that is proportional to temperature. For microclimate studies, the electrical resistance of platinum is typically used, since platinum is a stable, unreactive, soft metal. The platinum is formed into a wire to provide a 100 ohm resistance at 0 °C, with a change of 0.4 ohm per 1 °C. This is commonly referred to as a PT100 sensor. Given that a reference temperature is not required for measuring temperature with a thermistor, thermistors are commonly used despite being more expensive than simple thermocouples.

The voltage generated by a thermocouple is small: less than 0.0001 volt for a 1 °C difference in temperature between the two junctions. To amplify this response, often thermocouples are connected in series to increase the voltage output, known as a thermopile (see Box 3.1).

Box 3.1. Rules for Series versus Parallel Circuits

Simple electrical circuits are used in micrometeorological instrumentation and also can be used as a simple model to represent flows of energy and mass. Most of us experience the difference in two common circuit types, series and parallel, with common batteries. When batteries are connected with the positive end of one battery joined to the negative end of the other (and so on, if more than two batteries are used), a series circuit is formed. In a series circuit, the total voltage of the circuit is equal to the sum of the voltage of each component, without any change in the current (amperage). For example, two 1.5 V batteries connected in series would produce 3.0 V. The same rule applies for resistors placed in series: the total resistance is equal to the sum of the individual resistors.

In a parallel circuit, there are two (or more) parallel paths for the current to flow. Two 1.5 V batteries with the negative ends connected, then the positive ends connected, would result in 1.5 V (no change), but the total current would equal the sum of the current provided by each battery. Parallel resistors would create a total resistance equal to the sum of the reciprocals of each individual resistor.

This concept can be applied to increase the total voltage output of individual thermocouples without increasing the total current. For example, 50 type E thermocouples each producing 68 μV $°C^{-1}$ would produce 3400 μV $°C^{-1}$ when connected in series.

Thermopiles are commonly used to measure the soil heat flux in a soil heat flux plate, and in radiometers to measure shortwave (pyranometer), longwave (pyrgeometer), or net radiation (net radiometer). Although thermocouples and thermistors are commonly used in micrometeorology, there are other instruments used to measure temperature, such as the traditional glass-bulb thermometer. Glass-bulb thermometers are largely obsolete in research since it is not possible to automate readings, and they have a slow response time (a long time-constant). As shown in Figure 3.3, a sonic anemometer can also be used to measure air temperature by measuring changes in the transit time of sound (sonic) waves between opposing pairs of transducers. The speed of sound varies with both air temperature and humidity, so a sonic anemometer actually measures what is referred to as the virtual temperature, T_v, which must be corrected to account for the effects of humidity on temperature $(T_a\,(°C) \approx T_v\,/\,(1+0.61r_v)$ where r_v is the water vapor mixing ratio).

Commonly used in remote sensing, an infrared thermometer can measure the skin or surface temperature of an object without the need for direct contact. The ability to measure temperature without direct contact is advantageous when measuring objects that are difficult or impossible to attach directly to a thermocouple: snow, water, ice, animals, leaves, hot or remote objects. An infrared thermometer measures the surface upwelling infrared radiation ($L\uparrow$) emitted by the object (typically in the 8–14 μm wavelengths) and converts this to an effective radiative temperature using the Stefan-Boltzmann law when solved for temperature (T_s; K):

$$L\uparrow = \varepsilon\sigma T_s^4; T_s = \left(\frac{L\uparrow}{\varepsilon\sigma}\right)^{1/4} \qquad 3.2$$

where ε is the surface emissivity (estimated if not known) and σ is the Stefan-Boltzmann constant (5.67×10^{-8} W m^{-2} K^{-4}). For example, if a melting snowpack with an emissivity of 0.98 is emitting 309 W m^{-2} of longwave radiation, the infrared thermometer should measure a surface temperature of 0 °C:

$$T_s = \left(\frac{L\uparrow}{\varepsilon\sigma}\right)^{1/4} = \left(\frac{309\,\mathrm{W\,m^{-2}}}{(0.98)(5.67\times10^{-8}\,\mathrm{W\,m^{-2}K^{-4}})}\right)^{1/4} = 273\,K = 0\,°C \qquad 3.3$$

(a)

(b)

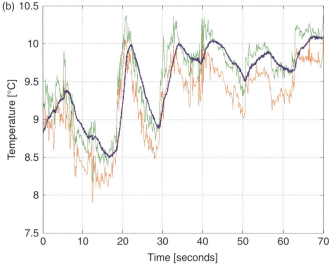

Figure 3.3. (A) Two chromel-constantan (type E) thermocouples (center of photograph) colocated with a sonic anemometer (larger, white instrument). (B) The smaller diameter thermocouple (0.0254 mm; green line) is better able to resolve high-frequency (sampled at 20 Hz) air temperature fluctuations as detected by the sonic anemometer (red line) than the larger diameter thermocouple (0.254 mm; blue line) (data collected from within a subalpine forest in Colorado at 6:20 MDT on July 22, 2015).
Source: Data and photograph provided by Sean P. Burns.

Since the thermometer itself emits longwave radiation, many infrared thermometers correct for this by measuring the cavity temperature of the instrument using a thermocouple or thermistor. It is obvious that a thermistor or thermocouple measures the temperature of the object it is in contact with; it is less clear with a radiometric-based instrument. The area detected by the infrared thermometer can be determined knowing the field of view of the sensor, the angle relative to the nadir (perpendicular to the surface would be 0°), and the sensor's height above the object (Figure 3.4). This method is applicable when the instrument's view does not include the horizon.

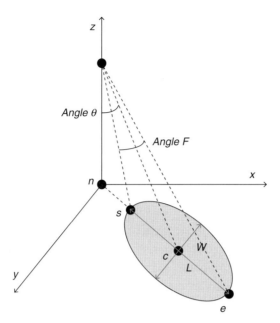

Figure 3.4. Sampling area of a radiometer such as viewed by an infrared thermometer.

For example, an infrared thermometer with a 40° degree field of view (F) placed 2 m above the surface (z) at an angle of 45° (θ) from the nadir would sample an area (A) of 6.31 m^2.

The distance (D) from the nadir (n) to the start (s) and end (e) of the elliptical viewing area:

$$D_{n:s} = z \tan(\theta - F/2) = 2\,\text{m} \tan(45° - 40°/2) = 0.93\,\text{m}$$

$$D_{n:e} = z \tan(\theta + F/2) = 2\,\text{m} \tan(45° + 40°/2) = 4.29\,\text{m}$$

Therefore the length of the ellipse sampled (L) is

$$L = 4.29\,\text{m} - 0.93\,\text{m} = 3.36\,\text{m}$$

and the distance from the nadir to the center of the ellipse ($D_{n:c}$) is

$$D_{n:c} = D_{n:s} + L/2 = 0.93\,\text{m} + 3.36\,\text{m}/2 = 2.61\,\text{m}.$$

The length of the hypotenuse from z to $D_{n:c}$ ($D_{H:c}$) is

$$D_{H:c} = \sqrt{z^2 + D_{n:c}^2} = \sqrt{2^2 + 2.61^2} = 3.29\,\text{m}$$

and the width of the ellipse (W) is

$$W = 2\left[D_{H:c} \tan(F/2)\right] = 2\left[3.29\,\text{m} \tan(40°/2)\right] = 2.39\,\text{m}.$$

Finally, the area sampled by the infrared thermometer, area of the ellipse, is

$$A = \pi \frac{L}{2}\frac{W}{2} = \pi \frac{3.36\,\text{m}}{2}\frac{2.39\,\text{m}}{2} = 6.31\,\text{m}^2.$$

This method can be used to estimate the area, or "footprint," only if the sensor does not "see" the horizon from an excessive tilt angle or if it is placed high above the surface. If the radiometer is placed level to the ground, then $\theta = 0°$.

D. Measuring Soil Temperature, Heat Flux, and Moisture

Soil temperature can be measured through the use of thermocouples, thermistors, or infrared thermometers. Placement of each sensor (depth or field of view) determines the location of the temperature measurement, with thermocouples or thermistors usually placed in a depth profile (see Figure 3.5). Since thermocouples and thermistors are essentially point measurements, care must be taken to place them properly in locations that are representative of the research area. For example, if the research area comprises bare ground and some vegetation, temperature sensors should be placed underneath the ground and then averaged by weighting each by the fraction of surface covered by each area. Care must also be taken when installing any sensor belowground, as disturbance must be kept to a minimum.

To measure the soil heat flux, G, a thermopile placed between two thin plates of material with a high thermal conductivity can be used. This is known as a soil heat flux plate. Ideally, the soil heat flux plate should be placed right at the surface since it is the heat flux across the atmosphere-soil boundary that needs to be measured. In practice, this cannot be done since the plate does not have the same thermal response as the soil, so the plate needs to be buried just below the surface (typically at a depth of 5 cm). Since the plate is now buried at some depth z, any energy stored or released in the soil above the plate (J_s) needs to be accounted for to calculate the "surface" G, G_o:

$$G_0 = G_z + J_s = G_z + C_S \frac{dT_S}{dt} dz \qquad\qquad 3.4$$

where C_S is the soil's volumetric heat capacity (J m^{-3} °C^{-1}), and dT_S/dt (°C s^{-1}) is the change is soil temperature (T_S) over the time interval between measurements (t; seconds). The soil's volumetric heat capacity, C_S, is the sum of the volumetric heat capacities of each of the soil's main constituents, air, solids, and water:

$$C_S = \rho_a c_a + \rho_S c_S + \rho_w c_w \theta \qquad\qquad 3.5$$

where ρ_a, ρ_S, and ρ_w are the densities of air, soil solids (the dry fraction only, known as the dry bulk density), and water, respectively, and c_a, c_S, c_w are the specific heat capacities of air, soil solids, and water, respectively. The product of the density and the specific heat capacity is the volumetric heat capacity, noted by uppercase C. As Table 3.1 shows, since $\rho_a c_a$ is small compared to the other terms, it is usually neglected, and the water and soil solid terms usually dominate the C_S calculation.

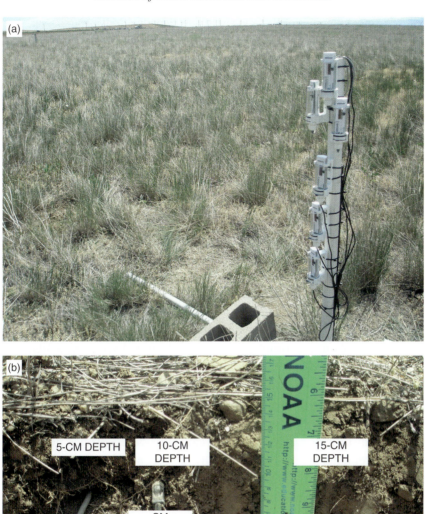

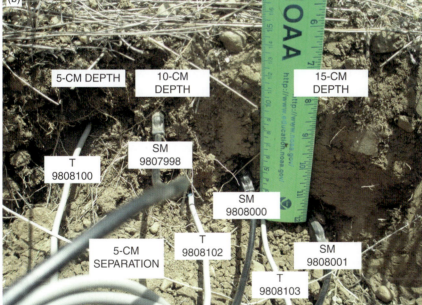

Figure 3.5. (A) A typical surface soil moisture and temperature measurement installation near Erie, Colorado. (B) TDR-based soil moisture probes ("SM"), and soil temperature probes ("T") are positioned to provide continuous in situ soil moisture measurements.

Source: Photographs provided by P.D. Blanken.

Table 3.1. Thermal Properties of Common Materials Found in Soils: Approximate Densities (ρ), Specific Heat Capacities (c), Volumetric Heat Capacities (C), and Thermal Conductivities (κ)

Material	Density (g cm^{-3})	Specific Heat Capacity (J g^{-1} °C^{-1})	Volumetric Heat Capacity (J cm^{-3} °C^{-1})	Thermal Conductivity (W m^{-1} °C^{-1})
soil mineral matter	2.65	0.733	1.94	2.93
soil organic matter	1.30	1.93	2.51	0.251
still water (20 °C)	1.00	4.19	4.19	0.595
still air (20 °C; sea level)	0.0012	1.00	0.0012	0.0260
Ice (−10 °C)	0.92	2.00	1.84	2.30

Source: Modified from Farouki, 1981.

To measure the surface soil heat flux properly, the soil temperature and moisture above the heat flux plate should be measured to account for heat storage above the plate.

Direct measurement of soil moisture is possible by the gravimetric method where soil samples are taken and weighed, before and after drying (typically drying is for 24 hours at 105 °C). This requires removal of a soil sample, thus is destructive and therefore cannot be reproduced at exactly the same site. The method was used in many historical data sets (Robock et al., 2000) and is still the preferred way to calibrate volumetric-based in situ measurements that are non-destructive, such as the time domain reflectometry (TDR) or capacitor-based probes described next. Indirect measurements can be made by TDR or soil capacitance (Figure 3.5). Both of these electromagnetic methods rely on the dependence of the dielectric permittivity of the soil on its moisture content. TDR-based sensors operate at higher frequencies and are therefore more accurate, but capacitance sensors are significantly cheaper.

Another instrument that measures the dielectric permittivity is the Hydra Probe, which has a measurement wavelength of 6 nm and uses a 50 MHz signal. Other indirect methods use neutron probes, electrical resistance, heat pulse, and gamma rays (see Vereecken et al., 2008). Merz et al. (2014) find good results for soil moisture profiles using nuclear magnetic resonance. They show distinctive shapes of soil moisture profile that characterize the first stage, where evaporation rate is controlled by atmospheric demand, and a second stage where evaporation rate is controlled by the porous medium. During stage one, an approximately uniform decrease of soil moisture over time was monitored, whereas during stage two, S-shaped moisture profiles developed, then receded progressively into the soil column.

Zreda et al. (2012) report on the COsmic-ray Soil Moisture Observing System (COSMOS), which uses a cosmic ray probe to measure neutrons emitted by soil moisture. There is an expanding global network of 95 COSMOS sites (two-thirds in the United States) that provide volumetric soil moisture for depths of a few hundred millimeters over an area of a few hectares. The data are either hourly or twice daily.

Near-global coverage of soil moisture in the upper few centimeters of the soil has been obtained by active and passive microwave remote sensing with a resolution of 30–50 km. These methods also make use of the dielectric permittivity. The European Space Agency (ESA) Soil Moisture and Oceanic Salinity (SMOS) sensor, launched in 2009, operates in the L-band (1.4 GHz, 21 cm), as will NASA's Soil Moisture Active-Passive (SMAP) mission, which was launched in 2015. The radar sensor operated only from April 21 to July 7. Dense vegetation cover and the shallow depth that is measured are major limitations. A further remote-sensing method uses ground penetrating radar.

A comparison of three global remote sensing products of soil moisture has been made by Rötzer et al. (2015). They analyzed (i) the SMOS Level 2 product, (ii) the MetOp-A Advanced Scatterometer (ASCAT) product retrieved with a change detection method from radar remote sensing backscattering coefficients, and (iii) the ERA Interim product from a weather forecast model reanalysis. The penetration depth for SMOS is 3–5 cm compared with only 0.5–2 cm for ASCAT. The spatial resolutions of SMOS and ASCAT are 30–50 km and 25–50 km, respectively. The ERA product used had 0.75° resolution and was for the layer 0–0.07 m. The comparisons were made for Köppen-Geiger climate classes. The ASCAT and ERA results are most similar while ERA and SMOS are least similar. The temporal patterns are reproduced well in all three products, while the absolute values are highly affected by acquisition and processing methods. The spatial patterns are influenced differently by the retrieval approach in different study areas and among the different products.

In a new field approach, Sayde et al. (2014) report on mapping soil water content and flux across 1–1000 m scales using the Actively Heated Fiber Optic (AHFO) method. This is shown to be capable of measuring soil water content several times per hour at 0.25 m spacing along cables multiple kilometers in length. AHFO is based on distributed temperature sensing (DTS) observation of the heating and cooling of a buried fiber-optic cable resulting from an electrical impulse of energy delivered from the steel cable jacket. Results were collected from 750 m of cable buried in three 240 m colocated transects at 30, 60, and 90 cm depths in an agricultural field under center pivot irrigation. The calibration curve relating soil water content to the thermal response of the soil to a heat pulse of 10 W m^{-2} for 1 minute duration was developed in the laboratory.

E. Measuring Radiation

Given that we are mainly concerned with shortwave (0.1–4.0 μm) and longwave (4.0–100 μm) in microclimatology, measuring these two distinct wavebands can be accomplished using a thermopile with an appropriate filter. Glass can be used effectively to filter out longwave radiation, allowing only shortwave radiation to reach the thermopile surface (painted black to minimize the albedo and maximize the absorption) of such a radiometer, known as a pyranometer. When a known energy source is used to illuminate the pyranometer's surface, the voltage output is measured to provide the instrument's calibration coefficient (sometimes referred to as the instrument's sensitivity). A typical pyranometer's sensitivity would be 14 μV produced by the thermopile for every 1 W m^{-2} received on its surface, so an output of 12000 μV (12 mV) is equal to 857 W m^{-2}:

$$\text{Output} = \frac{\text{Voltage}}{\text{Sensitiviy}} = \frac{12000\,\mu\text{V}}{14\,\mu\text{V} / \text{W}\,\text{m}^2} = 857\,\text{W}\,\text{m}^{-2}.$$

Most pyranometers have a field of view of a planar, hemisphere surface (i.e., $F = 180°$), and simple orientation or modifications can be used to measure variables of micrometeorological interest. A pyranometer facing upward would measure the downwelling (or incident) global shortwave radiation: the sum of all shortwave radiation including direct-beam, diffuse-beam, and downward-reflected ($S\uparrow$). A pyranometer facing downward would measure only the surface reflected and diffuse-beam shortwave radiation ($S\uparrow$). The direct-beam radiation component can be measured by a pyrheliometer that allows sunlight to enter a window where it is directed onto a thermopile. A mechanical solar tracker keeps the instrument pointing toward the Sun.

The diameter of the circular area sampled by a downward-facing radiometer with a 180° field of view is equal to roughly half the sensor's height (use this technique when the instrument does "see" the horizon). For example, a downward-facing pyranometer placed 2 m above the surface would sample a circle with a diameter of 1 m, a radius of 0.5 m, and a circular area of 0.79 m^2 ($A = \pi 0.5\text{m}^2$). Or a pyranometer with band ring designed to block out the Sun's sphere, thus shadowing the pyranometer and eliminating the direct-beam radiation, would measure the diffuse-beam solar radiation only.

As an alternative to thermopile-based pyranometers, silicon-based pyranometers are also available to measured global or PAR (400–700 nm) shortwave radiation. Many can even be used underwater. A silicon photodiode is used to produce a current that is proportional, when calibrated, to the incident shortwave radiation. Compared to the thermopile pyranometers, these radiometers are smaller, are cheaper, and do not require a glass dome.

Without the glass dome, a thermopile can be used to measure longwave radiation; this is known as a pyrgeometer. Whereas a glass dome is used to prevent longwave radiation from reaching the thermopile surface in a pyranometer, a silicon surface (not a photodiode) is used to prevent shortwave radiation from reaching the thermopile surface in a pyrgeometer. Just like a pyranometer, the pyrgeometer has a planar 180° field of view and can be mounted facing upward to measure the incident, downwelling longwave radiation ($L\downarrow$) or upward, emitted longwave radiation ($L\uparrow$). Since the body of the pyrgeometer also emits longwave radiation, this must be accounted for by subtracting the pyrgeometer's emitted longwave radiation by measuring the body temperature of the instrument (typically with a PT100 thermistor).

The net or all-wave radiation (R_n) is the sum of net short- and longwave radiation:

$$R_n = \left(S\downarrow - S\uparrow\right) + \left(L\downarrow - L\uparrow\right). \qquad 3.6$$

Each component of Eq. (3.6) can be measured individually using a four-component net radiometer. Or the voltage output from an integrated upward- and downward-facing thermopile can be used to measure R_n with a net radiometer. With a thermopile-based net radiometer, clear polyethylene domes over both the upper and lower surfaces are often used not to filter radiation (0.4–100 μm is required), but to keep surfaces clean and sheltered from wind.

F. Measuring Wind

Wind is a vector quantity, having both magnitude (velocity) and direction (a scalar quantity has magnitude only). Wind speed is measured with an anemometer, and the wind direction with a wind vane.

With the common description of wind speed and direction, it is implied that this is in the horizontal plane where the air is originating. This should not be confused with the direction to which the air is moving. For example, a wind speed of 5 m s^{-1} at 270° translates to the horizontal air moving from the west at a velocity of 5 m s^{-1}. Care should be taken when interpreting graphical descriptions of wind direction, since some show illustrations pointing toward the wind and some point away from the wind. Also, when installing wind vanes, the 360° (north) orientation could be aligned to true or magnetic north, and this should be specified when reporting data. Caution when averaging wind direction data is required; a north wind direction could erroneously be reported as a south wind direction when averaging directions close to the 0–360 boundary. Treating the wind as a vector, not a scalar, prevents this issue, as demonstrated in Box 3.2.

Box 3.2. Determining the Average Wind Speed and Direction Using Vectors

Since wind direction is given using a circular reference, from 0° to 360°, problems can arise when calculating the wind direction with north winds. Consider the following wind speed and direction observations made at 5-minute intervals:

Observation (5-min intervals)	Wind Speed (m s^{-1})	Wind direction (°N magnetic)
1	5	30
2	2	350

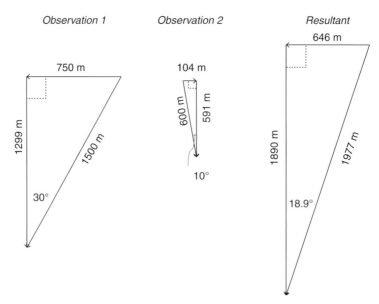

Figure 3.6. Resultant wind vectors calculated from Observation 1 and Observation 2.

The arithmetic mean wind speed and directions are 3.5 m s^{-1} and 190°, respectively. Although these are correct *scalar* means, both are incorrect *vector* means, and therefore incorrect as we know the average wind direction should be close to due north. The correct wind speed and direction are 3.33 m s^{-1} from 18.9 °N.

The proper wind speed and direction are determined by adding each wind vector from each observation, then determining the length (wind speed) and angle (wind direction) of the resultant vector. Each individual vector must be broken down into its west-east and south-north components to form right-angle triangles so that basic trigonometry can be used. Keep in mind that no angles can exceed 90° in a right-angle triangle, so the inside-triangle angle relative to the north-south direction must be used. The proper trigonometric function should be used depending on where the wind direction angle lies relative to the right angle (SOH-CAH-TOA).

First we express the observations as a distance: (5 m s^{-1})(300 s) = 1500 m for the first observation, and (2 m s^{-1})(300 s) = 600 m for the second observation.

As shown in Figure 3.6, the east-west and north-south components are the following:
Observation 1

East-west component = 1500 m × sin(30°) = 750 m; west direction

North-south component = 600 m × cos(30°) = 1299 m; south direction

Observation 2 (note the 350° wind direction, so 350 − 270 = 80, 90 − 80 = 10°)

East-west component = 600 × sin(10°) = 104 m; east direction

North-south component = 600 × cos(10°) = 591 m; south direction

The resultant vector has an east-west component of 750 m − 104 m = 646 m westward (subtraction because of the opposite directions), and a north-south component of

1299 m + 591 m = 1890 m southward (addition because of the same direction). The length of the resultant vector (the hypotenuse) is

$$\sqrt{646^2 + 1890^2} = 1997\,\text{m}$$

divided by the 600 seconds elapsed time gives an average wind speed of 3.33 m s^{-1}. The angle between the north-south component of the resultant vector and the hypotenuse (the proper wind direction) is

$$\tan^{-1}(646\,\text{m}/\,1890\,\text{m}) = 18.9°.$$

Wind direction can be measured mechanically with devices that are designed to point into the wind direction (e.g., a vane) or point away from the wind direction (e.g., a wind sock). If attached to a potentiometer, the voltage produced will vary as the wind vane rotates. With the advantage of not having any moving parts, a sonic anemometer can be used to measure wind direction using a minimum of two pairs of sonic transducers. The same trigonometry described in Box 3.2 is used to calculate the wind speed and direction of the resultant vector of the horizontal and cross-wind vectors.

With three pairs of transducers, a sonic anemometer can calculate the vertical component of the wind vector (w) in addition to the horizontal (u) and cross-wind (v) components. The vertical wind speed is especially relevant when measuring fluxes using the eddy covariance method (see Section I); therefore, sonic anemometers are commonly used when flux measurements are made, since air temperature and the three-component wind speed and wind direction can all be measured at a high sampling rate. The slower, mechanical propeller or cup-type anemometers can be used in general applications when slower responses and resolutions can be tolerated.

G. Measuring Precipitation

Wind speed can have a large effect on the accuracy of precipitation measurements. In theory, the task of measuring precipitation is simple: measure the depth of water collected in a vessel (a gauge). The location of the gauge is critical since the spatial variation in precipitation is large (hence the development of the Thiessen polygon technique), and the gauge must neither be placed in a location with overhead obstructions (not under a plant canopy or close to buildings) nor too close to the ground to allow rain to splash in or snow to blow into the gauge. The vast majority of surface meteorological observations including of precipitation are located where people are (cities, airports); thus observations are sparse over unpopulated regions such as the oceans or mountainous terrain.

There are around 100 different types of rain gauge design in use by national weather services in different countries around the world. These differ in diameter, orifice rim design, and elevation above the ground surface of the gauge rim. Hence, there can be considerable differences in the gauge catch. This is illustrated by the fact that mean precipitation isohyets on national maps may not meet correctly at the national boundaries, as is the case for Mexico, the United States, and Canada.

We are concerned with measuring the precipitation amount (volume converted to equivalent depth), duration (time), intensity (depth/time), and type (solid, liquid, diameter). Liquid precipitation (i.e., rain) can be measured with a standard rain gauge consisting of a 10.16 cm (4 inch) wide orifice that funnels rain into a cylinder with markings calibrated to the equivalent depth of 2.54 cm (1 inch) (volume/area of orifice). The graduated cylinder is placed within a larger one to catch any overflow. The advantages of this manual rain gauge are its low cost and simple design; the disadvantages are that it cannot be automated or discern types of precipitation (it is designed for rain only), and the rain intensity depends on the frequency of manual observations. A tipping-bucket rain gauge allows for automated data recording. As the name implies, under a funnel with a top diameter usually of 20.32 cm (8 inches is the standard diameter in the United States) a small bucket balance on a pin "tips" when filled with 0.254 mm (0.01 inch) of water. Each time the bucket tips, an electrical pulse is counted as a magnetic switch passes a magnetic contact. Therefore, both rainfall amount and intensity can be automatically recorded.

Solid precipitation, mainly in the form of snow, can represent the majority of the total annual precipitation in many high latitude and high altitude regions. For water resource planning and management, the amount of water contained in the snow, the snow water equivalent (SWE), is important. To measure this, the precipitation captured by a gauge can be melted either by applying heat, or, more commonly, by dissolving in an antifreeze solution. A pressure transducer can measure the increase in mass as rain or melted snow accumulates, thus providing a record of precipitation amount and intensity regardless of whether it is in the solid or liquid form. These are referred to as weighing precipitation gauges.

With any of these precipitation sensors that depend on the use of an orifice to capture the precipitation, wind and turbulence around the gauge can result in an underestimate of the actual precipitation. This undercatch error can be large, on the order of 2–10 percent for rain and 10–50 percent for snow (Plummer et al., 2003). Dozens of different types of wind shields and screens have been designed to reduce this undercatch error, with the current WMO standard the DFIR (Double Fence Intercomparion Reference), two octagonal vertical fences with 4 m and 12 m diameter circles, 3.0 and 3.5 m high, surrounding a Tretyakov precipitation gauge mount in the center at a height of 3.0 m (Goodison et al., 1998) (Figure 3.7). As reported by Rasmussen (2012), there is large variability in SWE observations depending on the type of precipitation gauge and windshield configurations (Figure 3.8).

(a)

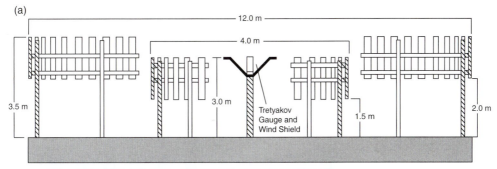

(b)

Figure 3.7. (A) The WMO accepted standard reference for shielding a precipitation gauge from wind, the Double Fence Intercomparison Reference (DFIR) (Goodison et al., 1998).
Source: World Meteorological Organization Report No. 67, p. 3, fig. 1.2.1.
(B) Photograph of a DFIR at Bratt's Lake, Saskatchewan, Canada (Rasmussen et al., 2012).
Source: Bulletin of the American Meteorological Society, 93(6), p. 815, fig. 3. Courtesy of the American Meteorology Society.

To avoid the undercatchment issue with orifice-based precipitation gauges, optical sensors or disdrometers can be used. By measuring the attenuation of light as precipitation passes through a laser beam, a disdrometer has the ability to discern the droplet size and velocity, and form of precipitation. The equivalent radar reflectivity can also be calculated by a disdrometer, and that can be compared to a radar-based estimate of precipitation. Using radar, the reflectivity (Z) is converted to rainfall rate (R) using the equation

$$R = aZ^b \qquad\qquad 3.7$$

where values of a and b were 0.017 and 0.714, respectively, in a study by Smith et al. (2010) conducted near Tulsa, Oklahoma. Although radar-based rainfall intensity tends to underestimate rain gauge observations significantly, radar estimates have the advantage of large spatial sampling.

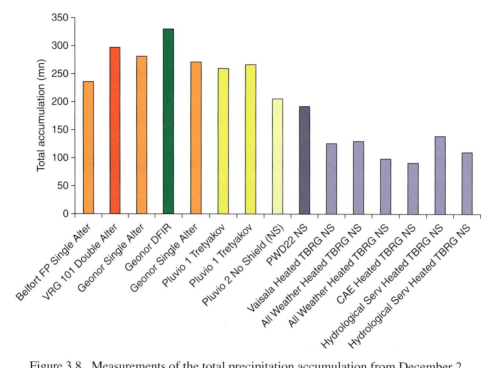

Figure 3.8. Measurements of the total precipitation accumulation from December 2, 2008, through April 15, 2009, near Egbert, Ontario, Canada, varies widely depending on the type of precipitation gauge and windshield configuration used (Rasmussen et al., 2012).
Source: *Bulletin of the American Meteorological Society*, 93(6), p. 812, fig. 1. Courtesy of the American Meteorology Society.

H. Measuring Total and Partial Atmospheric Pressures

Pressure (P) is a force (F) per area (A), with units of

$$P = \frac{F}{A} = \frac{ma}{A} = \frac{(\mathrm{kg})\left(\mathrm{m\,s}^{-2}\right)}{\mathrm{m}^2} = \frac{\mathrm{N}}{\mathrm{m}^2} = \mathrm{Pa} \qquad 3.8$$

where N is a newton and Pa is a pascal (note often units of millibars are used, where 1 mbar = 10 pascals; 10 mbar = 1 kPa). Equation (3.8) shows that if we assume that acceleration, in this case, equal to $g = 9.81$ m s^{-2}, is constant, then a measure of atmosphere pressure is essentially a measure of the weight of the atmosphere over a square meter surface area. On average, at sea level, this pressure is 101.3 kPa, representing the pressure exerted by all molecules in the air from sea level to the top of the atmosphere. Atmospheric pressure decreases exponentially with height above sea level (h; m) following the relationship

$$P_{\mathrm{h}} \approx P_0 e^{\left(-\frac{gMh}{RT_0}\right)} \qquad 3.9$$

where P_0 and T_0 are the standard pressure and temperature at sea level (101.3 kPa and 288.15 K, respectively), M is the molecular weight of dry air (28.9 g mol^{-1}), and R is the universal gas constant (8.31 J mol^{-1} K^{-1}). Denver, Colorado, known as the "Mile High City," has an elevation of 5280 feet above sea level (1609 m). Assuming a standard atmosphere, the pressure at this elevation would be

$$P_{1609} \approx P_0 e^{\left(-\frac{gMh}{RT_0}\right)} = 101.3\,\text{kPa}e^{\left(-\frac{9.81\text{m s}^{-2}\times 0.0289\text{kg mol}^{-1}\times 1609\text{m}}{8.31\text{J mol}^{-1}\text{K}^{-1}\times 288.15\text{K}}\right)} = 83.7\,\text{kPa}$$

The pressure exerted by the entire column of the atmosphere with all of its gases is referred to as barometric pressure and is measured with a barometer. Dalton's law states that the individual, or partial pressures, created by each of the gases can be added to give the total pressure of the gases combined ($P_{\text{total}} = P_1 + P_2 + P_3 + \cdots$). The first barometer created by Evangelista Torricelli (1608–1647) consistent of a small cylinder containing mercury placed inverted into a small dished filled with mercury. The height of the mercury column in the cylinder averages 760 mm at sea level, representing a balance between the atmospheric pressure on the exposed mercury fluid in the exposed dish and the pressure of the column of mercury in the inverted cylinder. The barometric pressure can therefore be read as the height of the column of fluid in the cylinder as it varies in response to changes in barometric pressure. At sea level, the atmosphere can support a column of water roughly 10 m high, far too tall for a practical barometer. Therefore, the denser fluid mercury was used, giving the well-known standard barometric pressure of 29.92 inches (760 mm) of mercury at sea level.

As with fluid thermometers, fluid barometers are seldom used since observations are hard to automate. Although the unit "inches of mercury" is still commonly used in the United States, electronic barometers now measure pressure through a calibrated pressure transducer.

Partial pressures of gases of interest to micrometeorology, namely, water vapor and CO_2, are much more difficult to measure because their concentrations in the atmosphere are so low (they are trace gases). Phase changes are often associated with pressure and temperature changes that occur as air samples are drawn through sample tubes, and adhesion to sample inlet surfaces makes water vapor even more difficult to measure. Although partial pressure is measured with units of pressure (kPa), the quantity of a gas is often expressed different ways relative to a volume of air (m^3): parts per million by volume; ppm$_V$), mass density (g m^{-3}), molar density (mol m^{-3}). The ideal gas law can be used to convert between units. Ratios are also often used; for example, the mixing ratio is mass of the gas divided by the total atmospheric mass. Humidity is a generic term for the amount of water vapor, and since humidity is such an important quantity, there are several ways to express it. The mixing ratio for water vapor is known as the specific humidity. Absolute humidity refers to the mixing ratio of water vapor in dry air (the pressure of water vapor is excluded in the denominator), and relative humidity refers to the water vapor pressure divided by the saturation vapor pressure (multiplied by 100 to give a percentage). The dew-point temperature is the temperature the air must be lowered to reach 100 percent relative humidity (commonly used in aviation since it is independent of changes in air temperature).

A hygrometer is the general term used for an instrument designed to measure water vapor. One of the earliest known hygrometer consisted of the hair of a horse's tail. Fixed at one end and attached to a movable pen and rotating cylinder at the other, the length of the hair changes in response to humidity (increases with increasing humidity). Perhaps the most accurate means of measuring the partial pressure of water (aka vapor pressure) is with a psychrometer, which consists of "dry" and "wet" bulb thermometer readings. The dry bulb is the ambient air temperature, properly shielded from solar radiation, and ventilated (the term "bulb" arose because traditional fluid thermometers were and still are used). The wet bulb is the ambient temperature measured by a thermometer enclosed by a wick that is constantly supplied with water. In dry air, water will evaporate from the wet wick. Since energy, the latent heat of vaporization, is required to evaporate water, evaporation from the wet wick will reduce the temperature. The difference between the dry- (T_a) and wet-bulb (T_w) temperatures (the wet-bulb depression) can then be used to calculate vapor pressure using the equation

$$e_a = e_s - \gamma P \left(T_a - T_w \right)$$

3.10

where e_s is the saturation vapor pressure at T_a, γ is the pseudopsychrometric "constant" of approximately 0.000644 $°C^{-1}$ ($\gamma = PC_p/\lambda\varepsilon_w$ but can be approximated by $\gamma = C_p/\lambda\varepsilon_w = 0.000644$ $°C^{-1}$ to facilitate the practical use of Eq. (3.10) when temperatures and barometric pressure is known). Note that the dew-point temperature is not the same as the wet-bulb temperature. For example, with a T_a and T_w of 32 °C and 19 °C, respectively, and $P = 102$ kPa, the vapor pressure can be calculated by first determining the saturation vapor pressure at the wet-bulb temperature (see Eq. (2.5)):

$$e_s = 0.61121e^{\left(\frac{17.502T_w}{240.97+T_w} \right)} = 6.1094e^{\left(\frac{17.502 \times 19\,°C}{240.97+19\,°C} \right)} = 2.20\,kPa$$

then using Eq. (3.10):

$$e_a = e_s - \gamma P \left(T_a - T_w \right) =$$

$$2.20\,kPa - 0.000644\,°C^{-1} \times 102\,kPa \times \left(32\,°C - 19\,°C \right) = 1.35\,kPa$$

The e_s at the ambient air temperature of 32 °C is 4.76 kPa, so relative humidity in this example is 1.35 kPa/4.76 kPa = 28.4 percent. Proper air circulation is required around the wet and dry bulbs (3–5 m s^{-1}) and can be achieved by physically swirling the psychrometer (a sling psychrometer) or by a mechanical fan (an aspirated psychrometer).

Whereas psychrometers have been frequently used to measures surface fluxes using the Bowen ratio-energy balance method (see Section I), high-frequency measurements of H_2O or CO_2 densities are required to measure fluxes using the eddy covariance method. Infrared gas analyzers (IRGAs) are commonly used for this purpose, since they can measure both H_2O and CO_2 densities at the minimum 10–20 Hz sampling frequency required for eddy covariance calculations. Both H_2O and CO_2 gases have strong absorption in the infrared wavelengths, so the greater the absorption of infrared radiation emitted from an infrared light source, the greater the concentration of infrared-absorbing gases. A broad-beam infrared light source emits radiation, and absorption bands centered at specific wavelengths (e.g., 2.59 μm and 4.26 μm for H_2O and CO_2, respectively; Burba, 2013) are used to measure the concentration in the air sampled either in the free atmosphere (open-paths IRGAs) or in air drawn through a sample line drawn into the IRGA (closed-path IRGAs). The ratio of the transmitted light in a chamber containing a reference gas compared to light transmitted in a chamber containing the sampled atmospheric gas is used to calculate the H_2O and CO_2 densities.

There are several other instruments available to measure the concentration of atmospheric gas, especially for water vapor (hygrometers), but psychrometers and IRGAs are the main ones in use today in micrometeorology.

I. Measuring Turbulent Fluxes

The measurement of the turbulent fluxes; the sensible heat flux (H); latent heat flux (λE); CO_2 flux (F_c); and momentum flux (τ) can be calculated using methods based on the measurement of profiles or gradients of scalars with height (Bowen ratio energy balance, BREB, or aerodynamic methods), or on measurement of the scalar in association with updrafts and downdrafts of air (eddy accumulation or eddy covariance methods).

Profile refers to the shape, expressed as the slope, of the plot of the mean (overbar) scalar quantity with height above the ground (z), theoretically expressed as a partial derivative, but practically expressed as a finite difference:

Wind profile: $\dfrac{\partial \bar{u}}{\partial z} \cong \dfrac{\Delta \bar{u}}{\Delta z}$

Air temperature profile: $\dfrac{\partial \bar{\theta}}{\partial z} \cong \dfrac{\Delta \bar{\theta}}{\Delta z}$

Water vapor density profile: $\dfrac{\partial \bar{\rho_v}}{\partial z} \cong \dfrac{\Delta \bar{\rho_v}}{\Delta z}$

For air temperature, θ is the potential temperature, the air temperature accounting for the dry adiabatic lapse rate of 0.0098 °C m^{-1}. The ability of the atmosphere to transport these scalars is termed the eddy diffusivity, K, which is unique for each scalar (momentum, K_M; heat, K_H; and water vapor, K_V). The mean gradients multiplied by the eddy diffusivity, with the inclusion of appropriate terms to express in units of an energy flux (W m^{-2}), gives the flux- gradient equations:

$$\text{Wind (momentum flux): } \tau = \rho K_M \frac{\partial \bar{u}}{\partial z} \qquad 3.11$$

$$\text{Air temperature (sensible heat flux)}: H = -\rho c_p K_H \frac{\partial \overline{\theta}}{\partial z} \qquad 3.12$$

$$\text{Water vapor (latent heat flux)}: \lambda E = -\lambda K_V \frac{\partial \overline{\rho_V}}{\partial z} \qquad 3.13$$

Standard micrometeorological flux sign convention indicates the direction of the fluxes relative to the surface. Positive indicated energy is directed away from the surface, negative toward it. For example: Air temperature decreasing with height would be a positive H (the atmosphere is gaining energy from the surface); a decrease in water vapor density with height would be a positive λE (evaporation is occurring from the surface); or a negative F_c would indicate that CO_2 is moving from the atmosphere to the surface (CO_2 uptake by vegetation). The momentum flux is always positive.

The Bowen ratio energy balance method employs the measurements of the Bowen ratio ($\beta = H/\lambda E$) combined with the surface energy balance equation. The Bowen ratio can be used to evaluate surface fluxes, or as a diagnostic measure to help classify the energy balance of a surface. For example, $\beta = 2$ would be a very dry surface; $\beta = 0.5$, a wet surface.

Appling the ideal gas law to Eq. (3.13) to replace vapor density with vapor pressure, finite vertical gradients of T_a and vapor pressure (e_a) for measurements taken at two heights (subscripts 1 and 2), and assuming $K_H = K_V$; the ratio of Eq. (3.12) and Eq. (3.13) can be written in terms of the Bowen ratio (β):

$$\beta = \frac{H}{\lambda E} = \frac{-\rho c_p K_H}{-\lambda K_V} \frac{\Delta \overline{\theta} / \Delta z}{\Delta \overline{\rho_V} / \Delta z} = \frac{P C_p}{\lambda \varepsilon_w} \frac{\left(T_{a1} - T_{a2} \right)}{\left(e_{a1} - e_{a2} \right)} \qquad 3.14$$

where ε_w is the ratio of the molecular weights of water to dry air (collectively, $P C_p / \lambda \varepsilon_w$ is the psychrometric constant, γ). The surface energy balance is written as

$$R_n - G = \lambda E + H \qquad 3.15$$

and when combined with Eq. (3.14) gives

$$\lambda E - \frac{R_n - G}{1 + \beta} \qquad 3.16$$

with H calculated as

$$H = \beta \lambda E \qquad 3.17$$

The BREB method assumes that the eddy diffusivities for scalars are equal for all atmospheric stability cases, and that there is no systematic bias in the gradient measurements. The routine swapping of sensors between heights by a mechanical system or by switching sample inlets between heights by means of a solenoid switch can eliminate this systematic bias. The BREB method still fails, however, when the gradients approach the resolution limits of the instruments, or when the available

Table 3.2. Typical Roughness Lengths (z_0) and Zero-plane Displacements (d) for Momentum

Surface	Roughness Length (m)	Zero Plane Displacement (m)
Calm lake or ocean	$0.1–10 \times 10^{-5}$	0.0002
Snow with no protruding obstacles	$0.5–10 \times 10^{-4}$	0.005
Short grass	0.003–0.01	0.03
Tall crops	0.04–0.20	0.25
Forests	1.0–6.0	1.0
Urban	5.0–15	> 2.0

Source: Modified from Oke, 1987.

energy, R_n-G, approaches 0 W m^{-2} at times such as sunrise and sunset. Perez et al. (1999) and Savage et al. (2009) provide quantitative methods to detect periods of suspect BREB data.

The aerodynamic method relies on measurements of the horizontal wind profile to calculate the turbulent fluxes. As with the BREB method, the aerodynamic method assumes that the eddy diffusivities for all scalars are equal. Starting with a neutral atmospheric stability, meaning the time-averaged horizontal wind at a height z can be predicted from the friction velocity u_*, following the (natural) logarithmic profile,

$$\overline{u}_z = \frac{u_*}{k} \ln\left(\frac{z}{z_0}\right)$$

3.18

where k is von Karman's constant (0.41) and z_0 is the roughness length for momentum. Momentum roughness lengths, conceptually equal to the height above the zero plane displacement (d), where the momentum flux is zero. Roughness elements, such as a plant canopy, elevate the imaginary height above the ground where the momentum flux is zero (d) by a small amount. For a completely flat, smooth surface, the momentum flux would be zero at $z = 0$ m, so d would also be zero. Typical values of z_0 and d are given in Table 3.2. As an initial approximation, $d = 2/3h$ where h is the canopy height, and $z_0 = 0.10d$.

The friction velocity is related to the wind profile by

$$u_* = kz\left(\frac{\partial \overline{u}}{\partial z}\right)$$

3.19

and u_* is related to the momentum flux τ by

$$u_* = \sqrt{\tau/\rho}$$

3.20

Combining (3.19) with (3.20) gives

$$\tau = \rho k^2 z^2 \left(\frac{\Delta \bar{u}}{\Delta z} \right)^2 \qquad 3.21$$

where ρ is air density and the slopes of the wind and air temperature profiles have been approximated with finite differences. Similar to the BREB approach, we can take the ratio of two fluxes, momentum (Eq. (3.21)) and one of the other fluxes; with gradients approximated with finite differences (Eq. (3.12)–(3.13)) if the all scalars are measured at the same height intervals, assuming the eddy diffusivities are equal gives the aerodynamic flux equations

$$\frac{H}{\tau} = \frac{-\rho c_p K_H \frac{\Delta \bar{\theta}}{\Delta z}}{\rho K_M \frac{\Delta \bar{u}}{\Delta z}};$$

$$H = \left[\rho k^2 z^2 \left(\frac{\Delta \bar{u}}{\Delta z} \right)^2 \right] \left[\frac{-\rho c_p K_H \frac{\Delta \bar{\theta}}{\Delta z}}{\rho K_M \frac{\Delta \bar{u}}{\Delta z}} \right] = -\rho c_p k^2 z^2 \left(\frac{\Delta \bar{u}}{\Delta z} \right) \left(\frac{\Delta \bar{\theta}}{\Delta z} \right) \qquad 3.22$$

$$\frac{\lambda E}{\tau} = \frac{-\lambda K_V \frac{\overline{\Delta \rho_V}}{\Delta z}}{\rho K_M \frac{\Delta \bar{u}}{\Delta z}};$$

$$\lambda E = \left[\rho k^2 z^2 \left(\frac{\Delta \bar{u}}{\Delta z} \right)^2 \right] \left[\frac{-\lambda K_V \frac{\overline{\Delta \rho_V}}{\Delta z}}{\rho K_M \frac{\Delta \bar{u}}{\Delta z}} \right] = -\lambda k^2 z^2 \left(\frac{\Delta \bar{u}}{\Delta z} \right) \left(\frac{\overline{\Delta \rho_V}}{\Delta z} \right) \qquad 3.23$$

Equations (3.22) and (3.23) apply to neutral atmospheric conditions only, so the following stability functions (non-dimensional factors) must be used when under stable or unstable conditions:

Stable conditions:

$$(\Phi_M \Phi_x)^{-1} = (1 - 5 Ri)^2 \qquad 3.24$$

Unstable conditions:

$$(\Phi_M \Phi_x)^{-1} = (1 - 16 Ri)^{3/4} \qquad 3.25$$

where Φ_M is the dimensionless stability function for momentum, and Φ_x is the dimensionless stability function for the scalar x (e.g., heat, water vapor, or CO_2). These stability functions are meant to account for changes in the curvature of the neutral wind profile due to unstable (i.e., vertical stretching of eddies) or stable (i.e., horizontal stretching of eddies) conditions.

Classification of atmospheric stability is determined through the calculation of the Richardson number (Ri) or the Obukov stability length (L) (Obukhov, 1971) when divided into the observation height (z/L). The aerodynamic flux equations for H (Eq. (3.22)) and λE (Eq. (3.23)) are simply multiplied by the stability functions depending on whether Ri or z/L is positive, indicating stable atmospheric conditions (use Eq. (3.24)), or whether Ri or z/L is negative, indicating unstable atmospheric conditions (use Eq. (3.25)). Neutral stability occurs when Ri or z/L approaches 0, generally from $-0.01 < Ri < 0.01$. The Richardson number is calculated as

$$Ri = \frac{g}{\overline{T}} \frac{\left(\Delta \overline{T} / \Delta z\right)}{\left(\Delta \overline{u} / \Delta z\right)^2} \qquad 3.26$$

where $\overline{T}$ is the mean air temperature in Kelvin in the layer Δz. The Obukov stability length (L), with the absolute value equal to the height where turbulent and shear production of kinetic energy are equal, is given by

$$L = -\frac{u_*^3 \, \overline{\theta_v}}{kg\left(\overline{w' \theta_v'}\right)} \qquad 3.27$$

where θ_v is the mean virtual potential temperature (virtual temperature, T_v, includes the effects of humidity on temperature; $T_v \approx T_a (^\circ C)\left[1 + 0.61 r_v\right]$ where r_v is the mixing ratio in g kg^{-1}; the potential temperature, θ, accounts for the dry adiabatic lapse rate), and $\left(\overline{w' \theta_v'}\right)$ is the potential temperature flux. The relationship between the virtual potential temperature flux and the potential temperature flux is given by

$$\overline{w' \theta_v'} = \overline{w' \theta'} + 0.61 \overline{T_a} \, \overline{w' q'} \qquad 3.28$$

where the mean T_a is in Kelvin and q is the specific humidity.

The dimensionless index z/L provides the stability parameter to classify stability, and z/L is roughly equal to Ri in most situations.

The $\left(\overline{w' \theta_v'}\right)$ term in Eq. (3.28) highlights the use of the eddy covariance method for calculated turbulent fluxes. Deviations are noted by the primes, and the overbars note a temporal mean, typically 30 minutes. Over an averaging period of 30 minutes, for example, the mean values of continuous measurements of the vertical wind speed w and the scalar quantity x sampled at a high rate (typically 10–20 Hz to capture the small, fast turbulent eddies) are calculated. The time series of w and x usually have any trend first removed for each 30-minute period using a detrending technique, so that the deviations from the detrended means (noted by the primes), when multiplied together, are proportional to a flux (Figure 3.9). The vertical wind speed must also be mathematically rotated so that the mean value is zero.

With a three-axis sonic anemometer to measure w, u, and T_a (T_v is actually measured by a sonic anemometer, so sometimes a fine-wire thermocouple is used to measure the actual T_a), the momentum and sensible heat fluxes can be measured:

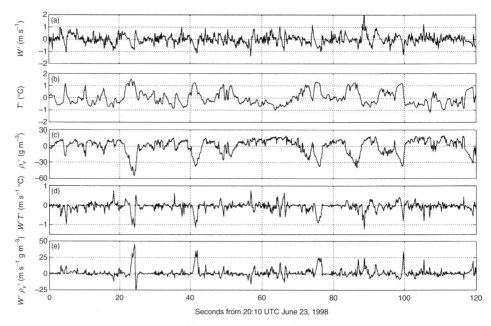

Figure 3.9. Time series of 10-Hz measurements of the fluctuations (primes) of vertical wind speed (w), air temperature (T), water vapor density (ρ_v), and the products wT and $w\rho_v$, as measured above Great Slave Lake, North West Territories, Canada (Blanken et al., 2003).
Source: *Journal of Hydrometeorology*, 4(4), p. 687, fig. 10. Courtesy of the American Meteorology Society.

$$\tau = -\rho\left(\overline{w'u'}\right) \qquad\qquad 3.29$$

$$H = \rho c_p \left(\overline{w'T_a'}\right) \qquad\qquad 3.30$$

If the sonic anemometer is colocated with an IRGA that is capable of simultaneous high-frequency measurements of the H_2O and CO_2 densities, then λE and the CO_2 flux, F_c, can also be measured:

$$\lambda E = \lambda\left(\overline{w'\rho_v'}\right) \qquad\qquad 3.31$$

$$F_c = \left(\overline{w'\rho_c'}\right) \qquad\qquad 3.32$$

The eddy covariance method has the advantage of its theoretical simplicity, independence from stability corrections, and potential biases involved with fluxes calculated from profile measurements. Figure 3.10 shows an example of a typical eddy covariance installation on Lake Huron where conditions preclude the profile measurements. There are many corrections concerning assumptions embedded in the eddy covariance approach that should be addressed when using this method (see Aubinet et al., 2012; Burba, 2013). This method has become the standard for flux measurements in large networks of sites, including the AmeriFlux Network in the United States and the EUROFLUX Network in Europe.

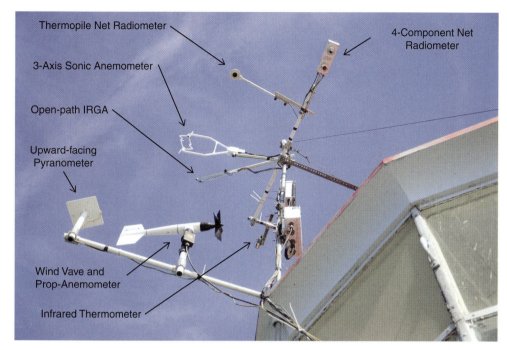

Figure 3.10. Example of eddy covariance and ancillary meteorological instruments mounted at a height of 30 m above the surface of Lake Huron atop the Spectacle Reef Lighthouse.
Source: Photographs provided by P.D. Blanken.

These three flux measurement techniques, BREB, the aerodynamic method, and eddy covariance, offer the advantage of sampling over a fairly large up-wind area by relying on the atmosphere to mix and integrate the concentration of the scalar. A general rule of thumb is that the fetch, or upwind area contributing to the flux measurement, is 1:100, or for every 1 m in height above the effective source height, 100 m upwind is sampled. This "flux footprint" varies with sensor height, surface roughness, wind direction, turbulence, and atmospheric stability, but can be estimated using a variety of empirical and statistical techniques (see Leclerc and Foken, 2014).

References

Aubinet, M., Vesala, T., and Papale, D. (eds.) 2012. *Eddy covariance: A practical guide to measurement and data analysis*. Dordrecht: Springer.

Blanken, P. D., Rouse, W. R., and Schertzer, W. M. 2003. Enhancement of evaporation from a large northern lake by the entrainment of warm, dry air. *J. Hydromet.* 4(4), 680–93.

Burba, G. 2013. *Eddy covariance method*. Lincoln, NE: LI-COR.

Farouki, O. T. 1981. Thermal properties of soils. Hanover, NH: *CRREL Monograph* 81-1.

Goodison, B. E., Louie P. Y. T., and Yang, D. 1998. WMO solid precipitation measurement intercomparison. Final report no. 67. WMO/TD – No. 872.

Leclerc, M. Y. and Foken, T. 2014. *Footprints in micrometeorology and ecology*. Berlin: Springer-Verlag.

Merz, S. et al. 2014. Moisture profiles of the upper soil layer during evaporation monitored by NMR. *Water Resour. Res.* doi: 10.1002/2013WR014809.

Mudelsee, M. 2014. *Climate time series analysis*. Netherlands: Springer.

Navarra, A., and Simoncini, V. 2010. *A guide to empirical orthogonal functions for climate data analysis*. Netherlands: Springer.

Obukhov, A. M. 1971. Turbulence in an atmosphere with a non-uniform temperature, *Boundary-Layer Met.* 2, 7–29.

Perez, P. J., Castellvi, F., Ibañez, M., and Rosell, J. I. 1999. Assessment of reliability of Bowen ratio method for partitioning fluxes, *Agric. Forest Met.* 97, 141–50.

Plummer, N., Allsopp, T., and Lopez, J. A. 2003. Guidelines on climate observation networks and systems. World Meteorological Organization, WMO/TD no. 1185.

Rasmussen, R et al. 2012. How well are we measuring snow? *Bull. Amer. Met. Soc.* 93(6), 811–29. doi: 10.1175/BAMS-D-11-00052.1.

Robock, A. et al., 2000. The global soil moisture data bank. *Bull. Amer. Met. Soc.* 81(6), 1281–99.

Rötzer, K., Montzka, C. and Vereecken, H. 2015. Spatio-temporal variability of global soil moisture products. *J. Hydrol.* 522, 187–202.

Savage, M. J., Everson, C. S., and Metelerkamp, B. R. 2009. Bowen ratio evaporation measurement in a remote montane grassland: Data integrity and fluxes. *J. Hydrol.* 376, 249–60.

Sayde, C. et al. 2014. Mapping variability of soil water content and flux across 1–1000 m scales using the Actively Heated Fiber Optic method, *Water Resour. Res.* 50. doi:10.1002/2013WR014983.

Smith, J. A., Seo, D. J., Baeck, M. L., and Hudlow, M. D. 2010. An intercomparison study of NEXRAD precipitation estimates. *Water Resour. Res.* 32(7), 2035–45.

Storch, H., von, and Navarra, A (eds.) 1999. *Analysis of Climate Variability*, 2nd Ed. Berlin: Springer-Verlag.

Storch, H., von, and Zwiers, F. W. 2002. *Statistical analysis in climate research*. Cambridge: Cambridge University Press.

Vereecken, H. et al. 2008. On the value of soil moisture measurements in vadose zone hydrology: a review. *Water Resour. Res.* 44, W00D06.

Zreda, M. et al. 2012. COSMOS: the cosmic-ray Soil Moisture Observing System. *Hydrol. Earth System Sci.* 16(11).

4

Radiation

This chapter surveys the characteristics of electromagnetic radiation. Solar radiation is described at the top of the atmosphere and in its transmission to the surface, detailing the effects of atmospheric gases, aerosols, and clouds. Radiation received on a horizontal and inclined surface is discussed, together with the surface reflectivity. Infrared or longwave radiation emitted by the Earth's surface and its absorption and reemission in the atmosphere, particularly by greenhouse gases, are examined. The chapter concludes with a discussion of net all-wavelength radiation.

A. Solar Radiation

The Sun is a gaseous body almost 1.4 million km in diameter, composed mainly of hydrogen and helium. Radiation is constantly emitted by the Sun as electromagnetic energy that is transmitted through space at 3×10^8 m s^{-1}, the speed of light. The sun is essentially a blackbody, meaning that it absorbs and radiates the maximum possible radiation at all wavelengths; the amount of energy emitted (E) by a black body was shown by Stefan and Boltzmann to be proportional to the fourth power of its absolute surface temperature (Kelvin):

$$E = \sigma T_s^4 \qquad\qquad 4.1$$

where $\sigma = 5.67 \times 10^{-8}$ W m^{-2} K^{-4} is the Stefan-Boltzmann constant. Hence, the Sun emits vast quantities of energy amounting to 74 million W m^{-2} (see the notes for conversion of energy units).[1] Almost all objects are not blackbodies, but are "graybodies," meaning that less than the maximum possible radiation at some or all wavelengths is radiated. This is described by the emissivity (ε) of an object, which is 1 for a blackbody and less than 1 for a graybody. Equation (4.1) written to include ε is

$$E = \varepsilon \sigma T_s^4 \qquad\qquad 4.2$$

Objects with a high water content tend to have an ε close to 1 (e.g., 0.95–0.99) since water is such a good absorber and emitter of radiation. For example, a lake with a surface temperature of 18 °C (18 °C + 273.15 = 291.15 K) and an emissivity of 0.99 would emit 403 W m^{-2} of longwave radiation directed upward to the sky:

$$E = (0.99)(5.67 \times 10^{-8}\ \text{W m}^{-2}\text{K}^{-4})(291.15\,\text{K})^4$$
$$= 403.35\ \text{W m}^{-2}$$

This example shows that as an object warms, it will dramatically lose energy to the fourth power of its surface temperature.

Solar radiation has a wide spectrum from ultra-short gamma rays to long radio waves, but here we are concerned only with three ranges – ultraviolet, visible, and infrared radiation. Their respective wavelength ranges are 0.1–0.4 × 10^{-6} m, 0.4–0.76 × 10^{-6} m, and 0.76–100 × 10^{-6} m (10^{-6} is written as μ). The Sun's photosphere has a temperature of around 5800 K, which determines the wavelength of the maximum energy emission (λ_{max}) by virtue of Wien's displacement law:

$$\lambda_{max}\ (\mu m) = \frac{2897\,\mu m\,\text{K}^{-1}}{T_s\,\text{K}} \qquad\qquad 4.3$$

where T_s is in Kelvin (K). Hence, solar radiation has a maximum energy emission at about 0.5 μm, in the 0.4–0.7 μm visible range:

$$\lambda_{max} = \frac{2897\,\mu m\,\text{K}^{-1}}{5800\,\text{K}} \simeq 0.50\,\mu m$$

By contrast, the infrared radiation (IR) emitted by the Earth's surface at around 290 K has a maximum near 10 μm:

$$\lambda_{max} = \frac{2897\,\mu m\,\text{K}^{-1}}{290\,\text{K}} \simeq 10\,\mu m$$

This difference is illustrated for two blackbodies in Figure 4.1.

Short ultraviolet (UV) radiation (< 0.3 μm) is mostly absorbed by ozone in the stratosphere and UV comprises about 7 percent of the total solar radiation incoming at the surface. By comparison, visible radiation accounts for 43 percent and infrared radiation 50 percent of the total (Figure 4.2).

Solar Radiation in the Atmosphere and at the Surface

The mean Sun-Earth distance of ~1.5 × 10^8 km implies that only a minute fraction of the emitted solar energy reaches the Earth. At the top of the atmosphere, the extraterrestrial solar radiation perpendicular to a surface is ~1361.5 W m^{-2}, indicated by recent data from NASA's Solar Radiation and Climate Experiment (SORCE). This amount is known as the "solar constant" (S_o) although it fluctuates over the ~11-year sunspot cycle by about 1.6 W m^{-2} (Kopp and Lean, 2011). Sunspots slightly increase the emission as a result of the brighter faculae surrounding the darker spots outweighing the emission from the cooler spots. Averaged annually over the globe, the incoming

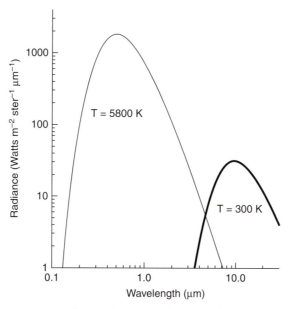

Figure 4.1. Emission from blackbodies at 6000 K and 300 K.

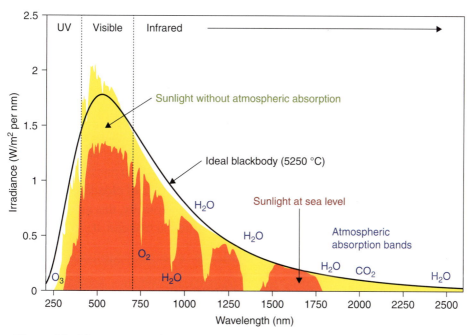

Figure 4.2. The spectrum of solar radiation and atmospheric absorption bands. nm = 1×10^{-9} m) (from Barry and Chorley, 2010).

Source. Atmosphere, Weather and Climate, 9th ed., p. 42, fig. 3.10 London: Routledge.

Table 4.1. The Energy (Percentage) in the Spectrum Emitted by the Sun

Wavelength (μm)	Energy (Percentage)
0–0.20	0.7
0.20–0.28 (UV-C)	0.5
0.28–0.32 (UV-B)	1.5
0.32–0.40 (UV-A)	6.3
0.40–0.70 (Visible, PAR)	39.8
0.70–1.50 (Near-infrared)	38.8
1.50–1000 (Thermal infrared)	12.2

radiation is one quarter of the solar constant, or ~340 W m^{-2}. This is because the surface area of the sphere is $4\,\pi r^2$, whereas the area of the circular disc of the Earth is πr^2.

The distribution of energy in the various spectral bands emitted by the Sun is summarized in Table 4.1. Ultraviolet radiation shorter than 0.32 μm is almost all absorbed by ozone in the stratosphere.

Atmospheric gases absorb a small fraction (8 percent by air molecules, 6 percent by water vapor, and 3 percent by stratospheric ozone) of the incoming solar radiation, but most passes through the atmosphere when it is cloud-free. The main contributor to atmospheric turbidity (its relative clarity) is aerosols – particles and/or liquid water droplets and gases. They are made up of sulfates, nitrates, mineral dust, and black carbon. Aerosols scatter solar radiation (both forward to the surface – 7 percent – and backward to space – 4 percent) and absorb it (about 2 percent). The total clear-sky absorbing effect of the atmosphere is, therefore, about 19 percent. Most of the absorption takes place in the near-infrared wavelengths between about 0.9 and 2 μm.

Clouds have a variable effect on absorption depending on cloud thickness, droplet (or ice crystal) concentration, and cloud fraction, as well as the solar elevation angle. A typical range is of the order of 10–20 percent. Clouds have a much greater effect on reflection of radiation, as discussed later. The main effect of clouds is in transforming the incoming direct beam into diffuse radiation. Typically, this averages between about 30 and 50 percent of the total radiation. Under an overcast sky all of the radiation is diffuse.

Astronomical Effects

The Earth's orbit about the Sun is slightly elliptical so the Sun-Earth distance varies between 152.1 million km on July 4 (apogee) and 147.3 million km on January 3 (perigee). Therefore, currently, the Earth as a whole receives 3 percent more radiation in January than July. The axis of rotation is tilted 23.4 degrees so that the North (South) Pole is toward the Sun in the boreal summer (austral summer). At noon on the solstices the Sun is overhead at the Tropic of Cancer (23.4° N) on June 21 and the

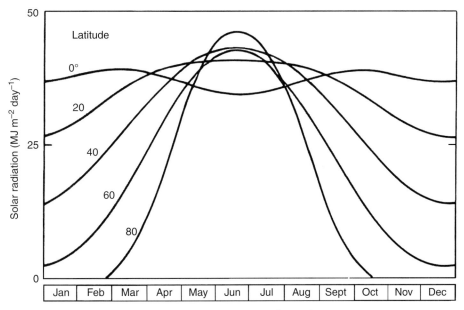

Figure 4.3. The amount of solar radiation (MJ m^{-2} day^{-1}) received over the year at different latitudes (adapted after D. M. Gates, 1962, p.8 fig. 2).

Tropic of Capricorn (23.4° S) on December 22. The Sun is overhead at the equator at noon at the equinoxes on March 20 and September 23.

The solar declination is the angular distance of the Sun from the equatorial plane, and it ranges from +23.4 to −23.4 degrees over the year. The amount of solar radiation received at the surface depends on the time of year, which determines the declination, and the time of day, which determines the solar elevation angle above the horizon. Figure 4.3 illustrates the effects of season and latitude on incoming solar radiation.

Atmospheric Effects

When the Sun is directly overhead at noon, the optical path length (also called the optical air mass or optical depth) of the solar beam through the atmosphere is at its absolute minimum. The daily minimum occurs when the Sun is at its zenith at local solar noon. The path length increases as the solar elevation angle decreases. The Beers-Bouguer-Lambert law relates the surface radiation (S) to the solar constant, S_o, the atmospheric depth (x; km), and the extinction coefficient (k; km^{-1}):

$$S = S_o e^{-kx} \qquad 4.4$$

The extinction coefficient is determined by gaseous absorption and molecular and aerosol scattering and ranges from about 0.01 km^{-1} for clear skies to 0.03–0.05 km^{-1} in turbid atmospheres.

Measurements of S and the calculation of S_o from astronomical data can be used to estimate the value of k. At noon on the summer solstice in Boulder, Colorado,

S is typically around 1000 W m^{-2} for a cloud-free condition, and S_o is 1268 W m^{-2}. Using Eq. (4.4) when solved for k over an atmospheric depth of 150 km for all wavelengths gives

$$k = -\frac{\ln\left(\dfrac{1000\,\mathrm{Wm}^{-2}}{1268\,\mathrm{Wm}^{-2}}\right)}{150\,\mathrm{km}} = 0.0016\,\mathrm{km}^{-1}.$$

Absorption of solar radiation involves, first, the ozone layer around 20–30 km altitude in the lower stratosphere, which absorbs almost all short wavelength UV ($< 0.3\ \mu$m), amounting to 1.5–3.0 percent of the total incoming. Second, water vapor in the lower troposphere mainly below 5 km, which has absorption bands at 0.9, 1.1, 1.4, and 1.9 μm, absorbs between about 6 and 13 percent of the extraterrestrial radiation in different latitudes and seasons (Tarasova and Fomin, 2000). Scattering comprises, first, Rayleigh scattering where the wavelength of the radiation is greater than the diameter of the air molecules; it is proportional to λ^{-4} so blue light (0.4 μm) is scattered 10 times more than red light (0.7 μm), making the sky appear blue. Rayleigh scattering is more or less isotropic (equal in all directions), but it is mainly responsible for the reduction in incoming visible radiation compared with that at the top of the atmosphere (TOA). Second, Mie scattering occurs where the wavelength of the radiation and diameter of the particles (e.g., cloud droplets, dust, and smoke) are comparable. It is proportional to λ^{-1} and in this case the scattering is forward (aniso-tropic). As all wavelengths are affected equally, clouds appear white, for example.

Aerosols are airborne particles and/or liquid droplets and gases (see Chapter 2, Section G). The major types from a radiative viewpoint are mineral dust blown from deserts and dry soil surfaces and soot (black carbon) particles from biomass burning, forest fires, and industrial activity. However, they also include sea salts, sulfates, nitrates, particulate organic matter, and volcanic ash. There are three size classes – Aitken nuclei with radii from 10^{-3} to 10^{-1} μm, large particles of 0.1–1 μm radius, and giant particles in the 1–100 μm range. Concentrations range from ~3000 cm^{-3} in clean continental air to 50,000 cm^{-3} in polluted continental air. Approximately 20–25 percent of the optical depth is contributed each by sulfates, sea salts, dust, and particulate organic matter, with a small percentage from black carbon.

The photosynthetic active radiation (PAR) is required for photosynthesis. Essentially, it is defined as the radiation in the same wavelengths as the visible spec-trum (0.4–0.7 μm) and is about 48 percent of the total energy emitted by the Sun since the λ_{max} (from Wien's law; Eq. (4.3)) is ~ 0.5 μm. The molecule chlorophyll is highly efficient in capturing red and blue light, but reflects green light, giving leaves their characteristic green color. Other pigments such as carotenes absorb red, orange, and yellow light, but these colors are masked by the chlorophyll except in autumn when many leaves turn red and yellow. Most of the red color in fall is a result of the accumulation of anthocyanins in the leaf vacuoles. Anthocyanins play a role in photo-inhibition and serve as antioxidants (Lee and Gould, 2002). The theoretical maximum storage of energy in photosynthesis is about 10 percent, but agricultural photosynthetic efficiencies are typically only between about 0.1 and 1.0 percent (Rosenberg et al., 1983, p. 312).

The total solar radiation incoming at the surface, also called global radiation, comprises the direct (beam) radiation and the diffuse (sky) radiation. The latter is received from all of the sky and hence it can penetrate more readily into plant canopies. Diffuse radiation is caused by the scattering effects of air molecules, water drops, and aerosols, discussed previously. For a clean, cloudless atmosphere, the diffuse radiation component comprises about 0.10–0.15 of the global irradiance for low zenith angles. Monteith (1962) shows that it is about 20 percent for zenith angles of 30°–60°, but increases to 60–80 percent for a zenith angle of 85°. The diffuse fraction increases with increasing cloud amount and reaches unity when the sky is totally overcast and the Sun is obscured. At Kew, United Kingdom, the diffuse radiation averaged for 1956–60 on cloudless days was 33 percent in January and 13 percent in July (Monteith, 1962). Corresponding mean monthly values were 68 and 56 percent, respectively.

A formula to estimate diffuse radiation was proposed by Bristow et al. (1985). The diffuse atmospheric transmission coefficient ($T_d = S_d/S_{b,o}$) is given by

$$T_d = T_T \left[1 - e^{\frac{0.6\left(1 - \frac{B}{T_T}\right)}{(B - 0.4)}} \right] \qquad 4.5$$

where T_T is the total atmospheric transmission coefficient ($S_t/S_{b,o}$), B is the maximum clear-sky transmissivity of the atmosphere (taken as 0.76), S_t is the global solar transmission, and $S_{b,o}$ is the extraterrestrial solar radiation incident on a horizontal surface, 1362 W m^{-2}). The diffuse radiation S_d is calculated by multiplying the maximum radiation that would reach the Earth's surface if there was no atmospheric attenuation, by T_d.

The distribution of diffuse radiation over the clear-sky hemisphere has been calculated by Stevens (1977). Figure 4.4 illustrated this for solar zenith angles of 35° and 55°. It is apparent that there is a circumsolar concentration of incoming diffuse radiation at moderate zenith angles that changes as the zenith angle increases. Limb brightening becomes relatively more important as a result of the increased atmospheric scattering caused by the longer path length.

Box 4.1 summarizes the calculation of incident solar radiation on a horizontal and an inclined surface. Brock (1981) provided a survey of the calculation of solar radiation for ecological studies. Details of the various algorithms for solar radiation (spectral and broadband) under cloudless and cloudy skies are given by Iqbal (1983).

Box 4.1 Calculation of Incident Solar Radiation

(a) On a Horizontal Surface

The spectral transmission at normal incidence integrated between the top of the atmosphere and a horizontal surface at the ground is given by Beer's law:

$$X = \exp(-\tau_\lambda/\mu)$$

where τ_λ is the atmospheric optical depth at a given wavelength λ, and μ is the cosine of the solar zenith angle (Davies and Hay, 1980). The absorbing gases (ozone and water vapor) and the Rayleigh scattering air molecules and Mie scattering aerosols that compose the optical depth have been described previously.

The direct beam solar radiation incident on a horizontal surface ($S\downarrow$) is expressed as

$$S\downarrow = S_o\,\mu[T_0(u_0\,m)\,T_R\,(m) - a_w(u_w m)]\,T_a(m)$$

where S_o is the extraterrestrial radiation; T_o, T_R, and T_a are the transmittances after absorption by ozone, Rayleigh scattering, and extinction by aerosol; u_o and u_w are the optical path lengths for ozone and water vapor; and m is the optical air mass. Values of m are shown in Table 4.2. Up to 60° solar zenith angle the values are identical for $m = 1/\mu$.

Table 4.2. Optical Air Mass Values (m) for Different Solar Zenith Angles (θ)

θ	0°	20°	40°	60°	80°
m	1	1.064	1.305	1.998	5.685

Source: From Rogers, 1967.

The diffuse radiation, assuming that absorption occurs before scattering and that T_a is the product of components due to absorption, T_{aa}, and to scattering, T_{as}, can be written

$$D\downarrow o = So\,\mu.T\,(u_o m)\,Tw(u_w m)\,T_{aa}(m)\,[1 - T_R(m)\,T_{aa}(m)]F$$

where F = the ratio of forward to total scatter. F ranges from 92 at zenith angle (θ) = 0, to 78 at $\theta = 60°$, to 60 at $\theta = 79°$ (Robinson, 1962).

Cloudiness has substantial impacts on incident solar radiation. The type and height of cloud layers are important as well as multiple reflections between the clouds and the surface, especially if it is snow covered. Over vegetated surfaces the enhancement by multiple reflection for cloud amounts >0.5 is 6 percent, but for snow cover this can reach 30 percent (Loewe, 1963). In the Antarctic, multiple reflections between overcast skies and a snow-covered ice pack increased the incoming solar radiation by 85 percent compared to low-albedo water surfaces (Wendler et al., 2004).

The solar irradiance ($S\downarrow$) beneath n cloud layers is

$$\prod_{i=1}^{n}\left[(1-C_i)+t_i C_i\right]/\left(1-\alpha_s\alpha_c C_T\right).$$

where $\prod$ is the Cartesian product, $1 - C_i$ defines the cloudless portion of the sky, $t_i C_i$ is the transmittance of the cloudy portion, and α_s and α_c are the albedos of the surface and cloud. C_T is the total cloud amount

$$C_T = [a/\,S_0\,m]\,\exp(-\,bm)$$

where *a* and *b* are parameters depending on cloud type after Haurwitz (1948) as shown in Table 4.3. They were determined from observations at Blue Hill Observatory, Massachusetts.

Table 4.3. Values of Cloud Transmittance Parameters

Cloud Type	a (W m^2)	b
Fog	179	0.028
Nimbostratus	130	−0.167
Stratus	277	−0.159
Stratocumulus	404	0.104
Altostratus	453	0.063
Altocumulus	611	0.112
Cirrostratus	101	0.148
Cirrus	956	0.079

Source: From Haurwitz, 1948.

(b) On an Inclined Surface

Calculation of the radiation incident on a sloping surface is discussed in detail by Hay and Davies (1980). In the basic equation the zenith angle (θ) is replaced by the angle between the Sun and the normal to the slope (ι).

The radiation on the slope $S\!\downarrow_s$ is given by

$$S\!\downarrow_s = S\!\downarrow \cos \iota / \mu.$$

$$\cos \iota = \cos a \sin h + \sin a \cos h \cos (\psi - \psi')$$

where a = the slope angle from the horizontal, h = solar elevation angle, ψ = azimuth of the Sun measured from the south in the Northern Hemisphere, and ψ' = the slope azimuth.

The effect of slope and azimuth on direct beam radiation is greatest in high latitudes where the solar elevation angle remains low. In winter when the fraction of radiation that is diffuse is larger, as a result of the low Sun angle, the effects of slope and azimuth angle are much reduced.

Garnier and Ohmura (1968) and Williams et al. (1972) use this approach to map direct beam solar radiation as a function of slope angle and azimuth and incorporate topographic screening. Dozier et al. (1991) developed a faster algorithm in the program TOPORAD. More recently, GIS-based models have been developed for GRASS v. 5.4 open source software. Šuri and Hofierka (2004) provide the code for r.sun, which provides raster maps of direct, diffuse, and reflected solar radiation for given day, latitude, surface, and atmospheric conditions. Time specific or daily totals can be determined and the shadow effect of topography can be optionally derived. Linke turbidity factors (referring to the overall spectrally-integrated attenuation by water vapor and aerosol), which model the atmospheric absorption and scattering of solar radiation under clear skies for an optical air mass of 2, are used. Linke turbidity is about 3 for Europe.

Diffuse radiation on a slope comprises radiation from the sector of the sky seen by the slope and that reflected onto the slope from the neighboring ground surface. It was originally assumed that the radiation is isotropic, but anisotropic corrections have been developed (Hay and Davies, 1980). The diffuse radiation on a slope is given by the difference in radiation from a full hemisphere incident on a horizontal surface and that emanating from the fraction of the hemisphere that is obscured by the slope.

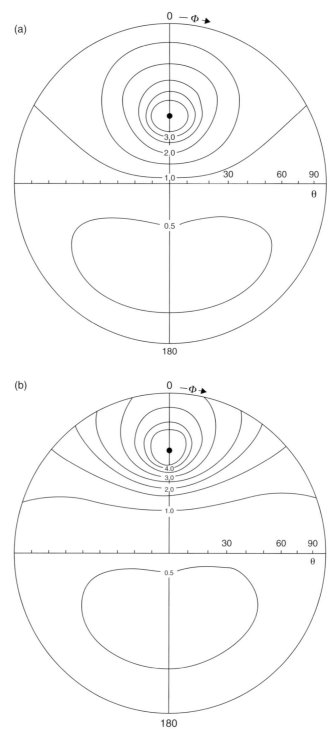

Figure 4.4. Standardized distribution of normalized sky radiance for a zenith angle (θ) of (a) 35° and (b) 55°. Isopleths are sr^{-1} (steradian, sr, is the SI unit of solid angle; there are 2π sr in a hemisphere). The isopleths are the ratio of "equivalent flux density" to horizontal diffuse irradiance. Since the plot is a Lambert projection, equal solid angles are represented by equal areas. The radial scale is proportional to $(1 - \cos \theta)^{0.5}$ (M. D. Stevens 1977).

Source: Quarterly Journal of the Royal Meteorological Society, 103, p. 460, fig. 1.

Cloud Effects

Global average cloud cover is 68 percent, 58 percent over land and 72 percent over the oceans (Rossow and Schiffer, 1999). Cloud effects on solar radiation depend on the cloud fraction and type (Twomey, 1976; Twomey and Bohren, 1980). The primary effect of clouds is in reflecting the solar beam (the cloud albedo ranges from 0.2 to 0.6), but cloud droplets and ice crystals also scatter and absorb radiation. High-level cirrus cloud, which is composed of ice crystals, has the least effect. Deep cumulonimbus with high liquid water content has a major effect. Thick nimbostratus and cumulonimbus reflect 80–90 percent of the solar radiation incident upon them according to Liou (1976). Thin stratus with a thickness of 0.1 km reflects about 45–72 percent while a 0.6 km thick altostratus reflects about 57–72 percent. Overall, clouds absorb a small fraction of incoming near-infrared radiation. Thick nimbostratus and cumulonimbus absorb 10–20 percent of the solar radiation incident upon them. Thin stratus, 0.1 km thick, absorbs about 1–6 percent, while a 0.6 km thick altostratus absorbs 8–15 percent. Aircraft observations reveal that clouds may absorb as much as 30–40 percent of the incident solar flux.

Shortwave cloud forcing is defined as the difference of the net solar irradiances at the top of the atmosphere between all-sky and cloudless conditions. Globally, it is about -45 W m^{-2}, with a maximum in the summer hemisphere around 60° latitude of -120 W m^{-2}. Hence global cloudiness has a net cooling effect on the climate system.

As a result of cloud effects, the radiation spectrum of diffuse radiation differs from that under cloudless skies. There is a secondary maximum in the near-infrared at 0.70–0.72 μm in the presence of large water droplets (0.1 μm), implying a "reddening" of the spectrum (Kondratyev, 1969, p. 366).

Absorbed shortwave radiation at the surface has maximum values over 300 W m^{-2} at the equator and minimum values of 80 W m^{-2} at the poles in summer. The global average is approximately 150 W m^{-2}, corresponding to 44 percent of that incoming at the top of the atmosphere.

Radiation at the Surface

The amount of radiation that is absorbed at the surface depends primarily on its reflectivity or albedo (α), which is the ratio of the reflected solar radiation ($S\uparrow$) divided by the incident solar radiation ($S\downarrow$), expressed as a fraction or percentage when multiplied by 100:

$$\alpha = \frac{S\uparrow}{S\downarrow} \qquad\qquad 4.6$$

For example, the albedo for a surface that reflects 200 W m^{-2} while receiving 800 W m^{-2} is 0.25 (25 percent). Alternatively, if a surface has a known α of 16 percent, and $S\downarrow$ is 750 W m^{-2}, then $S\uparrow$ is (0.16)(750 W m^{-2}) = 120 W m^{-2}.

Table 4.4. Albedo Values for a Variety of Natural Surfaces

Surface Type	Albedo
Ocean	0.06–0.10
Coniferous forest	0.08–0.15
Grass	0.25
Crops	0.20
Bare soil	0.17–0.25
Desert sand	0.35
Fresh snow cover	0.80–0.90
Mean global	0.31

The albedo (also known as the reflection coefficient) can be measured by using a pair of pyranometers, one facing up and the other down, as described in Chapter 3, Section E. Table 4.4 provides typical albedo values for a variety of surface types.

Water surfaces and coniferous forest are dark, compared with grass and desert sand, since both water and chlorophyll are effective absorbers in the visible (same as PAR) wavelengths. Snow cover is the brightest surface, although values decline as it ages, or when it is covered by dust. For most natural surfaces, the albedo also varies with the angle of incidence between the Sun's rays and the surface. For example, water has a very high albedo at large solar zenith angles, and in windy (wave) conditions, as both act to increase the angle of incidence (see Figure 8.1). Vegetation height affects albedo; taller vegetation tends to have a lower albedo than short vegetation because of the greater absorption of PAR.

Most surfaces display bidirectional reflectance (BDR), which is the variation in reflectivity that results from the location of the sensor in relation to the ground target and the position of the Sun (Asner et al., 1998). A detailed overview of the concept is given by Schaepman-Srub et al. (2006). Typically, the maximum reflectivity is at the antisolar point. This is the position where the sensor is located directly between the Sun and the ground target. The reflectivity of vegetation is strongly anisotropic because of the complexity of plant geometry. The BDR is influenced by both the landscape structure (vegetation type, canopy dimensions, and crown spacing) and canopy structure (leaf area index, leaf and stem angle distribution, and leaf properties).

The bi-directional reflectance function (BDRF) is the ratio of the quantity of reflected light in direction α_o to the amount of light that reaches the surface from direction α_i. It can be expressed as

$$\text{BDRF} = L_o/[L_i \cos\theta_i \, d\alpha_i] \qquad\qquad 4.7$$

where L_i is the light source, L_o is the outgoing reflected light, and θ is the solar zenith angle. It has units of sr^{-1}.

Diffuse radiation does not satisfy the single direction geometric condition of the BDRF, and hence this needs to be modeled. Walthall et al. (1985) and Liang and

Strahler (1994) proposed a statistical model for BDRF. For clear-sky conditions, the reflectance, r, at a given zenith and azimuth angle is

$$r = a\theta 2 + b\theta \cos(\varphi - \varphi i) + c \qquad 4.8$$

where a, b, and c are parameters depending on the incidence angle θi, φ is the view azimuth angle, and φi is the solar azimuth angle. The first term on the right controls the general surface curvature and the second provides a linear dependence on view azimuth. The third term is the nadir reflectance. The equation can be integrated over the hemisphere of view angles to give the surface hemispheric reflectance, α_{hem}

$$\alpha_{hem} = \frac{2.305a}{\pi} + c \qquad 4.9$$

Note that the second term does not contribute to the hemispheric reflectance. Walthall et al. (1985) point out that reflectance generally increases (i) with increasing zenith view angle for all azimuth angles; (ii) in the principal plane because of back-scatter in the direction of the Sun; and (iii) with increasing solar zenith angles.

Solar radiation decreases through reflection and absorption as light penetrates into a forest canopy. The amount reaching the forest floor depends on the canopy architecture and the leaf area index (LAI). In multistory tropical rain forests, light levels are very low and the solar radiation may be only a small percentage of that at the top of the canopy. This is also true of deciduous forests in full leaf in summer. Sun flecks can play a major role in the energy received by the understory and at the ground surface. The leaf area index is the main control of light penetration, but it is important to note that leaves tend to be clumped rather than randomly distributed.

In more general terms, the transmissivity of direct radiation for each canopy layer can be calculated according to Goudriaan (1988):

$$\tau b, i = \exp\left\{ -\sum_{j=1}^{np} K_j L_i j \right\} \qquad 4.10$$

where L_i, j, and K_j are leaf area index and extinction coefficient for plant species j of the canopy layer and n_p is the number of plant species in the canopy layer. The extinction coefficient is dependent on the orientation of the plant leaves and the angle of incident radiation. For leaves with random orientation,

$$K_j = 1/2\sin \beta \qquad 4.11$$

where β is the angle of the solar beam on the surface (which may be sloping).

As a result of absorption and reflection within a canopy, there is a change in light quality (Ross, 1975). Gates (1965) showed that for a single poplar leaf, 85–90 percent of the visible radiation is absorbed, 5–10 percent is reflected, and 5 percent transmitted. Absorption of the near-IR is low from 0.7 to 1. 4 μm, but it increases from 1.7 to 2.0 μm; reflection and transmission vary oppositely to absorption. Hence, light in canopies tends to be enriched in the green and near-IR wavelengths and depleted in other wavelengths. Under clear skies, the diffuse radiation may represent 15 percent

Table 4.5. Changes in the Extinction Coefficient, per LAI, for a Deciduous Boreal Aspen Forest under Leafless (LAI = 0 m² m⁻²) and Full-leaf (LAI = 2.3 m² m⁻²) Conditions

Radiation Stream	Leafless Aspen Canopy	Full-leaf Aspen Canopy	Decrease (%)
Net radiation	1.22	0.49	60
PAR	0.88	0.44	50
Solar	0.90	0.54	40

Source: Blanken et al. 2001.

of the incident radiation, but within a forest canopy this proportion can increase to 45 percent (of a much smaller total). The PAR intercepted by a forest is greater than the intercepted solar radiation. For example, extinction coefficients in an oak-hickory forest were 0.65 for PAR compared with 0.51 for $S\!\downarrow$ (Hutchison and Baldocchi, 1989). Blanken et al. (2001) measured extinction coefficients in a southern boreal deciduous aspen forest, and compared values with and without leaves (Table 4.5). They found significant decreases in the extinction coefficients for net radiation (~ 0.4–100 μm), PAR (0.4–0.7 μm), and solar (0.4–4.0 μm) wavelengths as the canopy leaf conditions changed.

There are numerous models available to calculate canopy radiation. Lemeur and Blad (1974) distinguish geometrical and statistical models. In the geometrical approach plant shapes are simulated by various geometrical forms. They can be divided into two classes – those that consider individual shapes and those that consider an arrangement of shapes. In the statistical approach the location of plant elements is parameterized by various distributions. Here the type of leaf dispersion in space – random, clumped, regular, or variable – is the most important consideration.

B. Infrared (Longwave or Terrestrial) Radiation

Every object with a temperature above absolute zero emits radiation proportional to the fourth power of its absolute temperature. For non-blackbodies this is scaled by the emissivity (ε), as described in Eq. (4.2). The emissivity is wavelength-dependent, and for the infrared radiation wavelength 8–14 μm, ε ranges from 0.90 to 0.94 for desert, 0.91 to 0.93 for soil, 0.95 to 0.96 for water, 0.98 for rough ice, to 0.99 for melting snow. The emission from the Earth's surface, with a temperature of around 290 K, has a maximum energy emission at about 10 μm; the atmosphere having an effective temperature of about 255 K has a maximum at a slightly longer wavelength, but all of this emission is in the infrared (IR) wavelengths between about 5 and 50 μm. About 90 percent of the IR emitted by the surface is absorbed by water vapor (5–8 μm) and carbon dioxide (~15 μm) in the lowest kilometer of the atmosphere. There is an "atmospheric window" between 8 and 12 μm where virtually all IR radiation exits to space. As a result, nighttime air temperatures on a clear night tend to be much lower compared to a cloudy night. The IR that is absorbed in the atmosphere is reemitted both back to the surface and up to the next atmospheric layer, where it is again absorbed. Water drops in cloud layers also absorb IR radiation.

Longwave radiation is measured with a pygeometer (see Chapter 3, Section E), but this is a difficult measurement to make and few meteorological stations have pygeometers. Therefore, several attempts have been made to estimate the incident longwave radiation ($L\downarrow$) empirically from routinely measured air temperature measurements. For example, Swinbank (1963) performed a linear regression of measured $L\downarrow$ and air temperature (T_a in K) under clear-sky conditions and derived the relationship

$$L\downarrow = 5.31 \times 10^{-13} \left(\text{Wm}^{-2}\,\text{K}^{-6}\right) \times T_a^6 (\text{K}) \qquad 4.12$$

Alternatively, $L\uparrow$ can be estimated from the Stefan-Boltzmann law using surface temperature measured with an infrared thermometer (see Chapter 3, Section C).

The net longwave radiation, $L^* = L\downarrow - L\uparrow$, at the surface depends on the vapor pressure (e in kilopascals) and cloud conditions according to Brunt (1934):

$$L^* = \sigma T^4 \,(a - b\sqrt{e})\,(1 - cC) \qquad 4.13$$

where a, b, and c are constants (c depends on cloud type) and C is the cloud cover in tenths. It makes intuitive sense that clouds affect the net longwave radiation since water in its vapor and liquid forms is a strong emitter of longwave radiation. For most surfaces, on average, the net longwave radiation tends to be close to zero since both $L\downarrow$ and $L\uparrow$ are roughly around 350–400 W m^{-2}.

C. Net Radiation

The all-wavelength, or net, radiation at the surface (R_n) is given by the sum of the net shortwave ($S_n = S\downarrow - S\uparrow$) and net longwave radiation ($L_n = L\downarrow - L\uparrow$):

$$R_n = S_n + L_n = \left(S\downarrow - S\uparrow\right) + \left(L\downarrow - L\uparrow\right) \qquad 4.14$$

Substitution of albedo (Eq. (4.5)) into Eq. (4.14) yields

$$R_n = S\downarrow (1 - \alpha) + \left(L\downarrow - L\uparrow\right) \qquad 4.15$$

An illustration of the diurnal variation of net solar radiation (S_n), net infrared radiation (L_n), and net all-wave radiation (R_n) over Lake Ontario on a cloudless day in August is shown in Figure 4.5. The small net infrared radiation value is the sum of large up- and downward fluxes whereas the large net solar radiation is due to the small albedo value of the water surface. The difference results in a large net radiation term.

When positive, the net radiation can be thought of as the energy available at the surface to do work, such as through convection (sensible heat flux), conduction (soil heat flux), phase changes (such as evaporation), or energy storage (an increase in

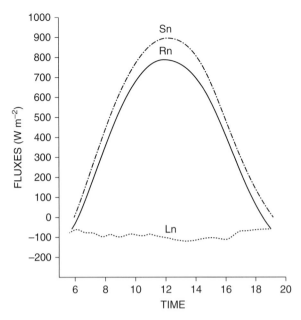

Figure 4.5. The diurnal variation of net solar, net infrared, and net all-wave radiation on a cloudless day over Lake Ontario in August (adapted from A. Henderson-Sellers and P. Robinson, 1986).

temperature over time). When negative, the net radiation can be thought of as an energy deficit, meaning that the surface will likely lose energy and cool. For many terrestrial surfaces, the net radiation peaks around local noon and at the summer solstice. It is usually negative at night, when there is no solar radiation, and in winter in high latitudes because of the polar night. In middle latitudes, R_n is typically positive from April to October and negative from November to February/March. Canopy interception can greatly diminish the net radiation reaching the ground. For example, Denmead et al. (1962) reported a 75 percent loss in a 1-m-high corn canopy. Blanken et al. (2001) reported that the ratio of R_n measured below/above a 21.5-m-tall boreal aspen forest varied with the aspen LAI (a_l) following:

$$\frac{R_n\left(\text{below}\right)}{R_n\left(\text{above}\right)} = 0.4682e^{-0.3038a_l} \qquad 4.16$$

There are many empirical formulae in the literature relating net and solar radiation. However, these have limited geographical applicability. According to Stanhill et al. (1966), the ratio of net radiation to solar radiation varies from about 0.58 over open water to 0.25 over a desert surface. Monteith and Szeicz (1962) report ratios of 0.37 for bare soil and 0.46 for a tall crop. Given the relative low cost and availability of net radiometers, and the increase in the use of remote sensing, measurements of R_n over many surfaces are now widely available.

Note

1 Conversion of energy units:

1 MJ m^{-2} day^{-1} = 11.6 W m^{-2} ~ 0.412 mm day^{-1} of equivalent evaporation.
1 W m^{-2} = 0.0864 MJ m^{-2} day^{-1} ~ 0.0355 mm day^{-1} of equivalent evaporation.
1 mm of evaporation over 1 hour requires ~ 680 W m^{-2}.

References

Asner, G. P. et al. 1998. Ecological research needs from multi-angle remote sensing data. *Remote Sensing Environment* 63, 155–65.

Barry, R.G. and Chorley, R. J. 2010. *Atmosphere, weather and climate*, 9th edn. London, Routledge.

Blanken, P. D. et al. 2001. The seasonal energy and water exchange above and within a boreal aspen forest. *J. Hydrology* 245(1–4), 118–36.

Bristow, K. L. et al. 1985. An equation for separating daily solar irradiation into direct and diffuse components. *Agric. For. Met.* 35, 123–31.

Brock, T. D. 1981. The calculation of solar radiation for ecological studies. *Ecol. Modelling* 14, 1–19.

Brunt, D. 1934. *Physical and dynamical meteorology*. London: Cambridge University Press.

Davies, J. A., and Hay, J. E. 1980. Calculation of the solar radiation incident on a horizontal surface. In J. E. Hay and T. K. Won (eds.), *Proceedings: First Canadian radiation data workshop*. Ottawa, Canada: Minister of Supply and Services. pp. 32–58.

Denmead, O,T., Fritschen, I. J., and Shaw, R. H. 1962. Spatial distribution of net radiation in a cornfield. *Agron. J*, 54, 505–10.

Dozier, J., Bruno, J., and Downey, P. 1991. A faster solution to the horizon problem. *Computers Geosci.* 7, 145–51.

Garnier, B. J., and Ohmura, A. 1968. A method of calculating the direct shortwave radiation income of slopes. *Solar Energy* 7, 796–800.

Gates, D. M. 1962. *Energy exchange in the biosphere*. New York: Harper & Row.

Gates, D. M. 1965. Radiant energy, its reception and disposal. In P. E. Waggonner (ed.), *Agricultural meteorology. Met. Monogr.* 6. Boston: American MeteorlogicalSociety, pp. 1–26.

Goudrian, J. 1988. The bare bones of leaf-angle distribution in radiation models for canopy photosynthesis and energy exchange. *Agric. Forest Met.* 43, 155–69.

Haurwitz, B. 1948. Insolation in relation to cloud type. *J. Met.* 5, 110–13.

Hay, J. E., and Davies, J. A. 1980. Calculation of the solar radiation incident on an inclined surface. In J. E. Hay and T. K. Won (eds.) *Proceedings: First Canadian radiation data workshop*. Ottawa, Canada: Minister of Supply and Services. pp. 59–72.

Henderson-Sellers, A. and Robinson, P. 1986. *Contemporary climatology*, London, Longman.

Hutchison, B. A., and Baldocchi, D. D. 1989. Analysis of biogeochemical cycling processes in Walker Branch watershed. In D. W. Johnson and R. I. Hook (eds.), *Forest meteorology*. New York: Springer Verlag, pp. 22–95.

Iqbal, M. 1983. *An introduction to solar radiation*. New York: Academic Press.

Kondratyev, K. Ya. 1969. *Radiation in the atmosphere*. New York: Academic Press.

Kopp, G., and Lean, J. L. 2011. A new lower value of total solar irradiance L evidence and climate significance. *Geophys. Res. Lett.* 38, L01706

Lee D. W., and Gould, K. S. 2002. Why leaves turn red. *Amer. Scientist* 990, 524–30.

Lemeur, R., and Blad, B. L. 1974. A critical review of light models for estimating the short-wave radiation regime of plant canopies. *Agric. Met.* 14, 255–86.

Liang, Sh-L, and Strahler, A. H. 1994. Retrieval of surface BDRF from multiangle remotely sensed data. *Rem. Sens. Environ.* 50, 18–30.

Liou, K-N. 1976. On the absorption, reflection and transmission of solar radiation in cloudy atmospheres. *J. Atmos. Sci.* 33, 798–805.

Loewe,F. 1963. On the radiation economy, particularly in ice- and snow-covered regions. *Gerlands Beiträge Geophysik* 72, 371–6.

Monteith, J. L. 1962. Attenuation of solar radiation – a climatological study, *Quart. J. Roy. Met. Soc.* 88, 508–21.

Monteith, J. L., and Szeicz, G.1962. Radiative temperature in the heat balance of natural surfaces. *Quart. J. Roy. Met. Soc.* 88, 496–507.

Robinson, G. D. 1962. Absorption of solar radiation by atmospheric aerosol, as revealed by measurements from the ground. *Archiv. Met. Geophys. Biokl.* B12, 19–40.

Rogers, C. D. 1967. The radiative heat balance of the troposphere and lower stratosphere. Planetary Circulation Project, Report A2. Cambridge, MA: MIT, Dept. of Meteorology.

Rosenberg N. J. et al. 1983. *Microclimate: The biological environment.* New York, Wiley.

Ross, J. 1975. Radiative transfer in plant communities. In Monteith, J. L (ed.), *Vegetation and atmosphere.* Vol. 1. *Principles.* London: Academic Press. pp. 13–55.

Rossow, W. B., and Schiffer, R. A. 1999. Advances in understanding clouds from ISCCP. *Bull. Amer. Met. Soc.* 80, 2261–87.

Schaepman-Strub, G. et al. 2006. Reflectance quantities in optical remote sensing – definitions and case studies. *Remote Sens. Environ.* 103, 27–42.

Stevens, M. D. 1977. Standard distribution of clear sky radiance. *Quart. J. Roy. Met. Soc.* 103, 457–65.

Stanhill, G., Hofstede, G. J., and Kalma, J. D. 1966. Radiation balance of natural and agricultural vegetation. *Quart. J. Roy. Met. Soc.* 92, 128–40.

Šuri, M., and Hofierka, J. 2004. A new GIS-based solar radiation model and its application to photovoltaic assessments. *Transactions in GIS.* 8, 175–90.

Swinbank, W. C. 1963. Long-wave radiation from clear skies. *Quart. J. Roy. Met. Soc.* 89, 339–48.

Tarasova, T. A., and Fomin, B. A. 2000. Solar radiation absorption due to water vapor: Advanced broadband parameterizations. *J. Appl. Met.* 39, 1946–51.

Twomey, S. 1976. Computations of the absorption of solar radiation by clouds. *J. Atmos. Sci.* 33, 1087–91.

Twomey, S., and Bohren, C. F. 1980. Simple approximations for calculations of absorption in clouds. *J. Atmos. Sci.* 37, 2086–95.

Walthall, C. L. et al. 1985. Simple equation to approximate the bidirectional reflectance from vegetation canopies and bare soil surfaces. *Appl. Opt.* 24, 383–7.

Wendler, G., Moore, B., Hartmann,B., Stuefer, M., and Flint, R. 2004. Effects of multiple reflection and albedo on the net radiation in the pack ice zones of Antarctica. *J. Geophys. Res.* 109, D06113. doi:10.1029/2003JD003927.

Williams. L. D., Barry, R. G., and Andrews, J. T. 1972. Application of computed global radiation for areas of high relief. *J. Appl. Met.* 11, 526–33.

5

The Energy Balance

A. Introduction

In Chapter 2 we introduced the surface radiation balance. Here we consider the disposition of the net radiation, R_n. A positive R_n, which is typical during the daytime, indicates an energy surplus that must be partitioned at the Earth's surface through some combination of conduction, convection, a phase change(s), or energy storage. Conduction occurs through the soil heat flux term (G), convection through the sensible heat flux term (H), and phase changes for water through the latent heat flux term (λE, where λ is the latent heat of vaporization and E is the evaporation rate):

$$R_n = G + H + \lambda E \qquad 5.1$$

with positive values of G, H, and λE representing the direction of the flux away from the surface, negative toward the surface. A positive λE is evaporation from the surface; negative is condensation on the surface. A typical diurnal breakdown of the four components is illustrated in Figure 5.1 for a grass surface in summer at Davis, California.

As with the radiation balance terms, the energy balance terms are measured in units of energy per time per area, J s^{-1} m^{-2} (W m^{-2}); this is the energy exchange across an imaginary plane parallel to the surface at the height where the measurements are being made. Objects in the climate system are three-dimensional, having volume and heat capacity, so these energy exchanges across the planar surface originate from and enter into a volume of air, water, plants, soil, and so forth. The extent to which the energy balance "balances" or not then determines the energy storage in the volume beneath the plane where the aboveground measurements are being made. Energy storage is represented by the rate of change of temperature of the volume integrated with depth (dz) (or height when aboveground) beneath the imaginary plane where the radiation and energy exchange is occurring (Figure 5.2), so Eq. (5.1) should include this energy storage term:

$$R_n = G + H + \lambda E + \rho c_p \frac{\Delta T}{\Delta t} dz \qquad 5.2$$

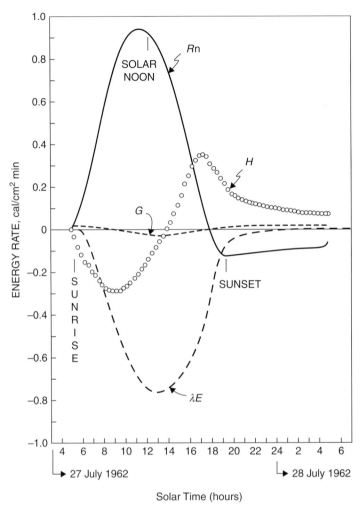

Figure 5.1. Example of the diurnal variation of the energy budget components on July 27, 1962, over a short grass surface (1 cal/cm^2 min^{-1} = 697.8 W m^{-2}) (from F. A. Brooks et al. 1963).

Source: Investigation of energy and mass transfers near the ground including the influences of the soil-plant-atmosphere system. Davis: Department of Agricultural Engineering and Department of Irrigation, University of California. Contract DA- 36-039- SC-80334 U.S. Army Electronic Proving Ground Fort Huachuca, AZ, p. 7, Fig. 1.4.

where ρ and c_p are the density and specific heat capacity, respectively, and $\Delta T/\Delta t$ is the rate of temperature change. This energy storage term is not a "heat flux" because there is no temperature gradient, just a heating or cooling of the entire volume up to depth (height) z. Formally, the last term in Eq. (5.2) should be measured and then summed for all objects in the volume beneath the measurement plane, but typically it is calculated only for the dominant objects if it is even calculated at all.

Imagine a situation where H, λE, and G are all negligible yet R_n is large and positive. This must result in a large increase in energy storage in order for energy to be conserved, thus an increase in the temperature of the volume. This situation is exactly

Table 5.1. General Relationship between the Energy Balance across a Plane and Energy Storage in a Volume When λE, H, and G Are Negligible

Energy Balance	Net Radiation	Energy Storage	Temperature Change
Imbalances: Inputs > outputs	Positive: daytime	Positive	Increase: Object warms
Imbalances: Inputs < outputs	Negative: nighttime	Negative	Decrease: Object cools
Balances: Inputs = outputs	Zero: sunrise or sunset	zero	Zero: Temperate not changing but not necessarily 0 °C

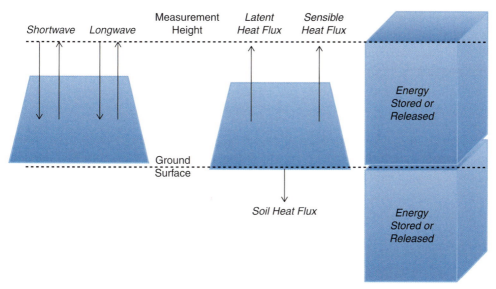

Figure 5.2. Illustration of the radiation and energy balance components across a planar surface where the aboveground measurements are made, and the energy storage in the volume beneath that plane.

what is observed in large water bodies such as the North American Great Lakes in the summer; R_n is large yet λE and H are small and often negative (directed downward) because the air is more humid and warmer that the air at the cold water surface; therefore, all the R_n is partitioned as energy storage as an increase in water temperature (see Figure 8.4). Table 5.1 presents a summary of the energy storage in relation to the energy balance.

There may be additional terms depending on the surface that is being examined. For example, a surface that has a snowpack needs to account for processes within the snowpack such as the energy required for melting; a vegetated surface needs to account for the energy used by photosynthesis; an urban surface needs to account for anthropogenic heat generated by industry and transportation. It should be noted that the geothermal heat flow due to radioactive decay in the Earth's interior amounts to

$0.5-2$ W m^{-2} and this small contribution to the surface energy balance is normally ignored.

After discussing the soil heat flux, we consider momentum and mass exchange in the atmosphere. Then we examine the sensible heat flux and latent heat fluxes, including the effects of interception and advection.

B. Soil Heat Flux

Heat is transferred into the soil by molecular conduction. It is transferred into the soil by day and back to the atmosphere at night, and correspondingly on an annual basis in the warm and cold seasons (thus the sign convection is positive when G is directed away from the surface, as during the day, and negative when G is directed toward the surface, as during the night). The soil heat flux (G) is given by the vertical temperature gradient in the soil ($\partial T_s/\partial z$, approximated by using a finite difference $\Delta T_s/\Delta z$) and the soil's thermal conductivity, K_s, which is the heat flowing in unit time through unit cross section of the soil (W m^{-1} K^{-1}):

$$G = -K_s \frac{\partial T_s}{\partial z} \cong -K_s \frac{\Delta T_s}{\Delta z} \qquad\qquad 5.3$$

The thermal conductivity depends on soil mineral composition, soil porosity, moisture content, and the content of organic matter. Sands have a high thermal conductivity as they contain a lot of quartz, whereas clay minerals have low conductivities. Organic soils have a very low thermal conductivity – about a quarter to a third that of mineral soils. Porosity (or void fraction) is the volume of soil voids that can be filled with air and/or water. It is defined as (1 – bulk density/particle density) and is typically in the range 0.3 (mineral soils)–0.6 (organic soils) of the total soil volume. A large air-filled pore space corresponds to a low thermal conductivity since still air has such a low thermal conductivity ($K_a = 0.0260$ W m^{-1} K^{-1}). If these pore spaces are filled with water, however, the thermal conductivity can increase dramatically since water has such a high thermal conductivity ($K_w = 0.595$ W m^{-1} K^{-1}). The thermal properties of key soil constituents can be found in Chapter 3, Table 3.1. The moisture content has a much larger effect on thermal conductivity than bulk density or grain size. A sandy soil with 40 percent porosity has a thermal conductivity of 0.3 W m^{-1} K^{-1} when dry, versus 2.2 when saturated (Monteith and Unsworth, 1990, p. 224). In general, thermal conductivity increases rapidly with moisture content up to a certain degree of wetness and then shows little further increase (Rider, 1957). Soil organic matter typically ranges between 1 and 6 percent for upland soils and 12–18 percent for organic soils.

The soil thermal diffusivity, $\kappa_s = K_s/(\rho_s c_s)$ in units of m^2 s^{-1}, is the ratio of the thermal conductivity, K_s, to the heat capacity (the amount of heat required (released) to raise (lower) the temperature of a unit of soil by 1 °C), where ρ_s is the soil density and c_s is the specific heat of soil. The specific heat is the ratio of the heat capacity

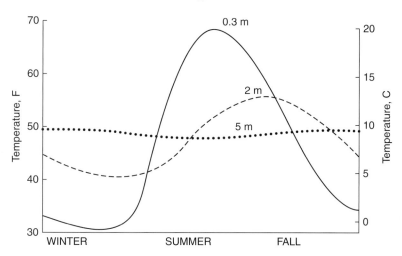

Figure 5.3. The variation of soil temperature with depth (0.3, 2, and 5 m) at Ottawa, Canada (after G. Williams and L. Gold, 1976, figure 2).
Source: *Canadian Building Digest*, 1976. Ottawa: National Research Council of Canada, Institute for Research in Construction No. 180. http://irc.nrc-cnrc.gc.ca/cbd/cbd180e.html, N. 180.

of the material to that of water at the same temperature and pressure. The specific heat of soils is about 1 MJ m^{-3} K^{-1} compared with 4.186 MJ m^{-3} K^{-1} for water (see Table 3.1). For this reason, dry soils warm up and cool down rapidly, whereas water bodies warm and cool much more slowly. The diffusivity is the primary control of soil temperature. Typical values of κ are in the range 0.2×10^{-6} for clay, 0.5×10^{-6} for silt, to 1×10^{-6} m^2 s^{-1} for sand (Al Nakshabandi and Kohnke, 1965). For unfrozen soil, thermal diffusivity reaches a maximum at a relatively low value of moisture content and then decreases with increasing wetness (Farouki, 1981, p. 49). The hydraulic conductivity and diffusivity have been shown to be related to the particle size distribution (percentage sand, silt, and clay) as defined by the textural classes in the United States Department of Agriculture soil classification (Cosby et al., 1984).

Non-conductive heat transfer can result from the infiltration of rainwater or snowmelt into seasonally frozen soils; freezing of this water releasing latent heat of fusion is an important mechanism for the acceleration of warming in surficial soils in the spring, according to Kane et al. (2001).

Temperature variations in soil are greatest at the surface, and these variations are transmitted to the subsurface layers with decreasing amplitude and with a time lag (Figure 5.3). Typically, the diurnal wave penetrates to ~35 cm in sands and 20 cm in clays. Corresponding values for the diurnal and annual wave are shown in Table 5.2, which illustrates the effect of moisture content.

The damping depth, D, is the depth at which the amplitude of the temperature wave is 0.37 time that at the ground surface. It is given by

$$D = (2 \kappa/\omega)^{0.5} \qquad\qquad 5.4$$

Table 5.2. Penetration Depth (m) of Diurnal and
Annual Temperature Waves

Material	Day (m)	Year (m)
Rock	1.10	20.5
Wet clay	0.95	18.0
Wet sand	0.80	14.5
Dry clay	0.40	6.5
Dry sand	0.30	4.5

Source: From Williams and Gold, 1976.

where the angular frequency for the annual variation $\omega = 2\pi/365$ day. At a depth $z = \pi D$, the temperature wave is exactly out of phase with that at the surface. Hence, when the surface temperature is at its maximum, that at πD is at its minimum, and vice versa. The time lag in the daily cycle is about 8–12 hours and in the monthly cycle it is about 6 months. At a depth of 8–10 m the temperature is approximately the same as the annual mean value of air temperature.

The annual and daily cycles of soil temperature can be closely approximated by a finite number of sine and cosine terms in a Fourier analysis (Carson, 1963; see Conrad and Pollak, 1950). This provides an objective measure of phase and amplitude at different depths.

Soil temperatures are strongly influenced by the presence of leaf litter and especially by snow cover, both of which greatly dampen the penetration of the temperature wave (see Table 5.3). Goodrich (1982) shows in a simple model that the buildup of a 50 cm cover displaces the mean annual ground temperature by 2.3 °C. The thermal offset is the difference between the mean annual temperature at the base of the active layer (the seasonal thaw layer) and that at the ground surface. Nearly half of the frost penetration depth is accomplished in the brief freezing period before snow cover inception. In the permafrost case, the mean annual ground surface is increased substantially with snow cover, with little effect on the thermal offset. The seasonal amplitude of temperature at depth is greater with snow cover than without.

Seasonally Frozen and Permafrost Soils

Freezing has major effects on the conduction of heat in soils. In general, there is a reduction in thermal conductivity in frozen compared with unfrozen soils. For example, the thermal conductivity of clay drops from 1.7 W m^{-1} K^{-1} to 1.0 W m^{-1} K^{-1} when frozen (Gold et al., 1972). Goodrich (1982) shows comparable decreases for thawed soils compared with their frozen state. With increasing ice content, thermal diffusivity in frozen soil slows but does not reach a maximum (Farouki, 1981, p. 49). The thermal conductivity of ice itself is about 2.3 W m^{-1} K^{-1} at −10 °C (Farouki, 1981, p. 30), whereas for water, K ranges from about 0.56 to 0.61 W m^{-1} K^{-1}. In Alaskan permafrost the thermal conductivity is 3.4 W m^{-1} K^{-1}.

Table 5.3. Summary Model Results for Seasonal Frost and Permafrost Cases

Soil Type	Snow cover	Max. Frost (Thaw) depth m	Min. GST °C	MAGST Offset °C	Thermal sfc °C	Ground freezing index °C day
Fine soil	snow free	1.40	−10.0	+5.0	−2.3	931
Fine soil	0.5 m snow cover	0.18	−1.12	7.35	−0.1	48
Fine soil	0.25 m snow cover	0.43	−1.53	7.08	−0.4	153
Permafrost fine soil	snow free	(0.74)	−30.0	−10.0	−1.1	4458
Permafrost fine soil	0.5 m snow cover	(0.84)	−12.25	−3.42	−1.4	2027

GST = ground surface temperature; MAGST = mean annual ground surface temperature.
Source: From Goodrich, 1982. *Canadian Geotechnical Journal* 19: 424, part of table 2.

Permanently frozen ground (permafrost) is defined as ground that remains frozen through two consecutive summers. It occupies about 23 percent of the surface land area of the Northern Hemisphere and seasonally frozen ground extends over 50 percent (Barry and Gan, 2011). There are four latitudinal zones – continuous permafrost (>90 percent of the surface is underlain by permafrost), extensive discontinuous (50–90 percent permafrost), discontinuous (20–50 percent permafrost), and isolated (<10 percent). The equatorward limit of continuous (discontinuous) permafrost corresponds approximately to a mean annual air temperature of – 6 to −8 °C (−1 to 1 °C). On the Tibet Plateau the lower altitudinal limit is about 4200 m in the northern part, increasing to 4800 m in the south (French, 2007). Recently, the mean annual ground temperature (MAGT) has been mapped at 1-km resolution for the lands surrounding the North Atlantic. Westermann et al. (2015) used MODIS land surface temperature for 2003–12 fused with ERA-Interim derived temperatures to calculate freezing and thawing degree-days. These values are employed in CryoGrid (Gisnås et al., 2013), an equilibrium ground temperature model, with adjustments relating to snow cover and vegetation type. In ice-free northern Greenland, MAGT ranges from −10 to −15 °C while south of Disko Bay they exceed −5 °C. Permafrost may consist only of frozen rock or it may contain ice. It is defined solely by temperature.

Not all soil water freezes at 0 °C because of dissolved mineral salts that depress the freezing point in the pore water. When water freezes it undergoes a volume expansion of about 9 percent. This expansion gives rise to frost heave of soil particles. There are four major types of ground ice: pore (or interstitial) ice, segregated ice, intrusive ice, and vein (or wedge) ice (Mackay, 1972). The "ice content" is defined as the weight of ice to dry soil, expressed as a percentage. Fine-grained soils may have ice contents of 50–150 percent while low ice content soils have values below 50 percent.

The volumetric content of ice in permafrost is very poorly known. In a borehole at Prudhoe Bay, Alaska, Lachenbruch et al. (2012) calculated a mean ice content of 39 percent. Maximum thicknesses of permafrost in Arctic North America and Siberia exceed 600 m with more than 1000 m in central northern Siberia. Near the southern boundary it may be only 25–50 m thick.

The upper soil layer thaws in summer, forming an active layer that may be 0.5–4.0 m deep. The active layer is the layer above the permafrost that is frozen in winter and thaws in summer. Streletsky et al. (2008) show that in the Alaska foothills the active later thickness (ALT) ranges from 41 cm in moist acidic tundra to 56 cm in moist non-acidic tundra, and on the coastal plain from 40 cm on moist nonacidic tundra to 63 cm on wet graminoid tundra. Zhang et al. (2005) showed that average ALT is about 1.87 m in the Ob basin, 1.67 m in the Yenisei, and 1.69 m in the Lena basin.

The seasonal freezing and thawing of the ground, respectively, release and consume large amounts of heat as a result of the phase change of water to ice, and vice versa. At 0 °C, 334 MJ is involved in the phase change of 1 kg of water – the latent heat of fusion. The seasonal freeze-up releases latent heat, which slows the freeze-up process by keeping the soil temperature at 0 °C (see Box 5.1).

Box 5.1. Phase Changes and Temperature

Energy is required, or released, when a material undergoes a phase change from a liquid, solid, or gas. The amount of energy depends on the substance, temperature, and pressure, and between which phases the substance is transitioning. A phase diagram shows the phase of the substance with pressure on the y-axis and temperature on the x-axis. Water is the only substance on Earth that exists naturally in all three phases, an important fact for the climate system and life. The latent heats in kJ kg^{-1} between the different phases of water are shown in Figure 5.4.

Since energy must be conserved, the energy required to evaporate 1 kg of water (the latent heat of fusion) is released when 1 kg of water is condensed (the latent heat of condensation). This release of heat when water vapor condenses is the reason thunderstorms and hurricanes have so much energy and can transport energy from the surface where the water evaporated to another place and time when the water condenses.

Energy is required to melt ice, so an equal amount of energy must be released when water freezes. When frozen soil warms, the soil temperature will remain at 0 °C until the ice is melted with the energy being used in the phase change, not a temperature increase. The release of this energy upon freezing results in a small but important temperature increase that has biological importance. When freezing is expected, citrus growers will often spray the crop with water. When the water surrounding the citrus freezes, the release of heat protects the crop from short-term frost damage.

The most energy is required (released) when water transitions directly between the solid and gas phases. In alpine regions where the snowpack is exposed to intense solar radiation, the snow can sublimate and thus never melt and supply water to the watershed. Sublimation losses can be large and represent a significant loss of water in these regions, which depend on the winter snowpack for the summer water supply.

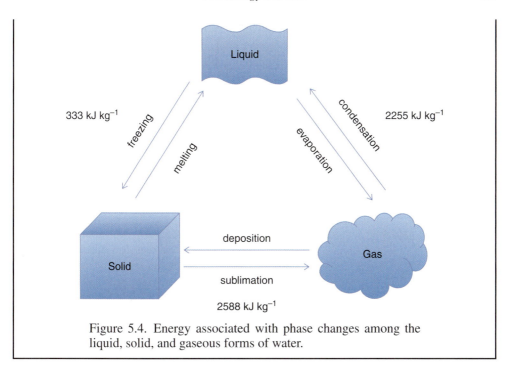

Figure 5.4. Energy associated with phase changes among the liquid, solid, and gaseous forms of water.

C. Momentum and Mass Exchange

The frictional retardation of the horizontal wind by the ground surface gives rise to a downward flux of momentum that is driven by turbulent motion. To explore this we look first at the vertical wind profile. A linear plot of wind speed with height above the ground shows that the speed increases rapidly at first and then more slowly. The wind velocity profile was first studied by Hellmann (1915) near Potsdam at heights between 5 and 200 cm (Yoshino, 1975, p. 60). Hellmann showed that on a logarithmic plot of height, the wind speed increases linearly when the atmosphere is adiabatic with "neutral" stability. Figure 2.3 illustrates this for hourly wind profiles over short grass up to 10 m height at Davis, California (Brooks et al., 1963). Later, Hellmann (1917) developed a power law relationship

$$u = u_r \, (z/z_r)^a \qquad\qquad 5.5$$

where u_r is the wind at a reference height, z_r, and a is an empirically derived coefficient that depends on atmospheric stability. For neutral stability, a is approximately 1/7 or 0.143.

The logarithmic relation can be expressed according to Thom (1975) as

$$u_z = A \, \ln \, (z/z_o) \qquad\qquad 5.6$$

where z_o is the roughness length for momentum, the theoretical height at which the wind speed becomes zero close to the surface. It is about a tenth of the height of the surface roughness elements. Typical values are shown in Chapter 3, Table 3.2, and in Box 5.2.

Box 5.2. The Classification of Effective Terrain Roughness for Momentum (after Davenport et al., 2000)

Class	Roughness length (m)	Landscape description
Sea	0.0002	Open water, snow-covered flat plain, featureless desert, tarmac, and concrete, with a free fetch of several kilometers
Smooth	0.005	Featureless landscape and little if any vegetation (marsh, snow-covered or fallow open country)
Open	0.03	Level country with low vegetation and isolated obstacles (e.g., grass, tundra)
Roughly open	0.10	Low crops or plant cover; moderately open country with occasional obstacles (isolated trees, low buildings) separated by 20 obstacle heights
Rough	0.25	High crops, or crops of varying height; scattered obstacles separated by 8 to 15 obstacle heights (e.g., buildings, tree belts)
Very rough	0.5	Intensely cultivated with large farms and forest clumps separated by 8 obstacle heights; bushland, orchards. Urban areas with low buildings interspaced by 3 to 7 building heights; no high trees
Skimming	1.0	Covered with large, similar-height obstacles, separated by 1 obstacle height (e.g., mature forests). Dense urban areas without significant building-height variation
Chaotic	≥2	Irregularly distributed large obstacles (dense urban areas with mix of low- and high-rise buildings, large forest with many clearings)

In Eq. (5.6), $A = u_*/(kz)$ where u_* is the friction velocity and k is von Karman's constant ~0.4. This constant describes the logarithmic velocity profile of a turbulent fluid flow near a boundary. The friction velocity is a scaling value given by the square root of the surface stress divided by the air density $(\tau/\rho_a)^{0.5}$; the ratio u_*/u is about 0.05–0.1, where u is the wind speed at the standard measurement height of 10 m (see end note, Chapter 2). However, corrections need to be made for atmospheric stability and there is no simple procedure (see Chapter 3, Section I).

The drag force per unit area of the surface is called the shear stress, τ. Its physical dimensions are mass × acceleration × area. It is also dimensionally equivalent to the product of air density, ρ_a, and the square of the friction velocity, u_*

$$\tau = \rho_a u_*^2 \qquad\qquad 5.7$$

The height variation of u can be expressed as

$$\delta u/\delta z = u_*/kz \qquad\qquad 5.8$$

where $kz = l$ is the effective eddy size, or "mixing length," at level z.

In the case of a plant community, the eddy size is related not to the height above the surface, but to the distance above a canopy-related measure known as the zero plane displacement, d, that is a major fraction of the plant height h ($d \sim 2/3\ h$). Hence, for plant canopies

$$u_z = u_*/k \ln [(z - d)/z_o] \qquad\qquad 5.9$$

and

$$\delta u/\delta z = u_*/k(z - d). \qquad\qquad 5.10$$

The eddy diffusivity of momentum, K_M, also known as the eddy viscosity, is defined by the ratio of momentum flux, τ, to the gradient of its concentration, $\delta(\rho u)/\delta z$:

$$\tau = \rho_a K_M\ \delta u/\delta z \qquad\qquad 5.11$$

$$\text{and } K_M = ku_* (z - d) = lu_* \qquad\qquad 5.12$$

where $z \geq h$; l is the mixing length, kz.

Atmospheric stability affects both the wind profile and the eddy structure. In a neutrally stable atmosphere the wind profile is logarithmic and the eddy structure comprises a set of circular eddies with diameters increasing with height, given by the mixing length, rotating at a uniform tangential speed equal to the friction velocity (Thom, 1975) (see Figure 5.5a). In unstable (stable) air the vertical motion in the eddies is enhanced (diminished) by buoyancy (damping effect).

Atmospheric stability is an important characteristic of the lower atmosphere since it determines how effectively energy and mass are transported. With a stable atmosphere, mixing and turbulent transport are suppressed. With an unstable atmosphere, mixing and turbulent transport are enhanced. Two methods to quantify atmospheric stability, the Richardson number and the Obukov stability length, are described in Chapter 3, Section I.

Convection is the upward transfer of thermal energy by air motion. It involves the combined effects of molecular conduction and mass exchange. There are two basic types of convection – free and forced. Free convection occurs when the motion is caused by local buoyancy forces that result from density differences due to temperature contrasts within the air mass. A heated surface creates air of lower density that rises, displacing overlying cooler air, which sinks to replace it. Forced convection results when the air is stable and the flow over the surface is perturbed by inertial forces due to surface irregularities forcing it to rise (Priestley, 1955: Monteith and Unsworth, 2014). Forced convection is generally more effective in heat transfer than free convection.

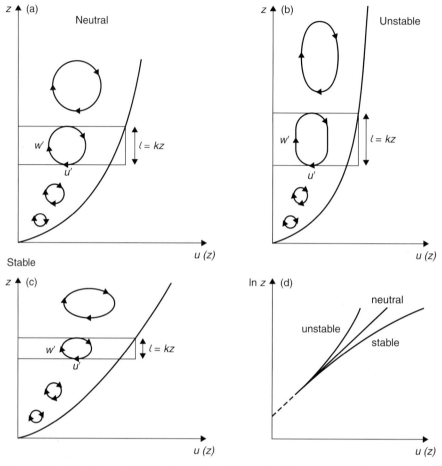

Figure 5.5. (a), (b), and (c) Wind speed profiles and eddy structures characteristic of three basic stability states in airflow near the ground. Panel (d) shows the wind speed as a function of ln(z) for the three states.
Source: Modified after A. S. Thom, 1975 in J. L. Monteith, Vegetation and atmosphere, p. 80. fig, 6, Academic Press, London.

Approximations for free convection (Cf) and forced convection (Cw) over a narrow flat leaf in W m^{-2} are given by Gates and Janke (1966) as

$$Cf = 4.18\ (\Delta T/D)^{0.25}\ \Delta T \qquad\qquad 5.13$$

and

$$Cw = 3.97\ (V/D)^{0.5}\ \Delta T \qquad\qquad 5.14$$

where D is the leaf dimension in the flow direction, V is wind speed (cm s^{-1}), ΔT is the difference between leaf and air temperature.

The Nusselt number (Nu) is the non-dimensional ratio of the convective to conductive heat transfer normal to the surface. Nu is typically in the range 100–1000 for active convection.

D. Sensible Heat Flux

When the surface is warmed, it transfers heat by molecular diffusion to a thin (approx-imately a millimeter) layer adjacent to it, known as the laminar boundary layer. This heat is then transferred into the overlying atmosphere by convection through rising

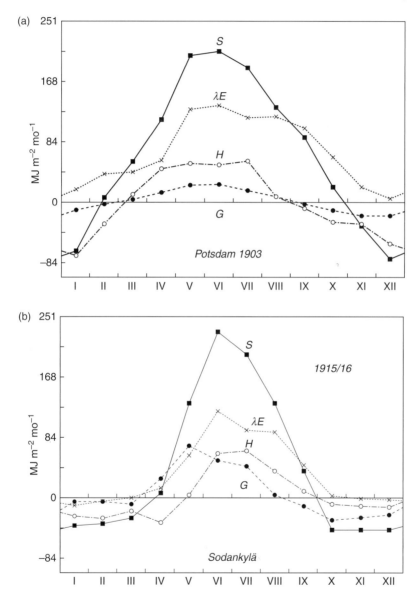

Figure 5.6. The diurnal variation of monthly energy budget components in different climatic regimes (a) Potsdam, Germany; (b) Sodankylä, Sweden, Irkutsk, Siberia; (d) Ikengüng, Gobi Desert; (e) Batavia, Java; (f) Greenland (Eismitte). *S* = net radia-tion (adapted after F. Albrecht, 1940).

Source: *Untersuchung über den Wärmehaushalt der Erdoberfläche in den verschie-denen Klimagebieten. Reichsamt f. Wetterdienst (Berlin), Wissenschaft. Abhand.* 8(2), p. 44, tafel III.

(*continued*)

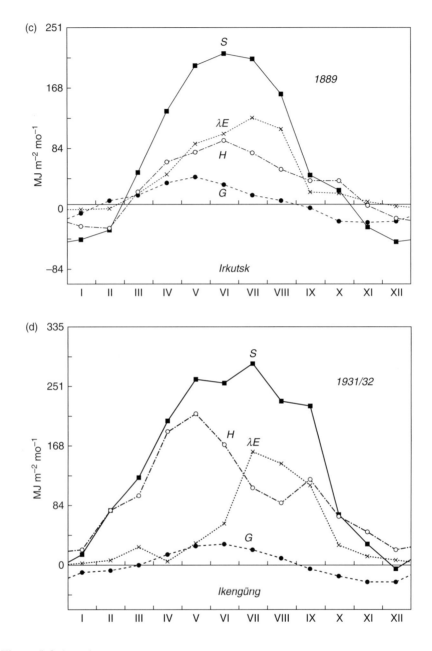

Figure 5.6 *(cont.)*

plumes of warm air, largely through turbulent transport, through the convective boundary layer. The transfer of sensible heat (or enthalpy), *H*, is determined by the vertical temperature gradient in the air ($\delta T_a/\delta z$), an eddy transfer coefficient for heat (K_H) as described in Chapter 3, Section I.

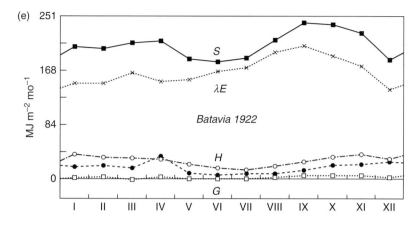

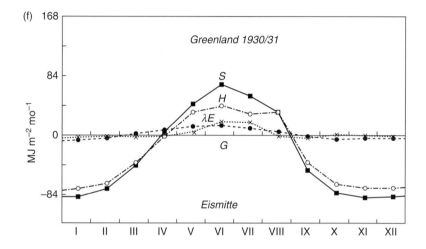

Figure 5.6 (*cont.*)

On average, H accounts for about 40 percent of net radiation. Its annual variation is illustrated in Figure 5.6, showing that H tends to peak in summer in midlatitudes, but is strongest in spring over a desert surface (Figure 5.6d). In Batavia, Java, it is consistently small in all months.

Over a snow or ice surface, particularly at night and/or in winter, there may be a temperature inversion near the surface. Hence, the coldest air is at the surface because of infrared radiational cooling, and the temperature increases with height to the top of the inversion layer. In this case the sensible heat flux is directed downward toward the surface to help offset the negative R_n.

Inversions may also develop during the day when there is strong evaporative cooling from a crop canopy lowering its temperature relative to the air flowing over it, or within a forest canopy at night when the canopy is warmer than the air beneath.

E. Latent Heat Flux

The transfer of latent heat from the surface into the atmosphere involves evaporation of moisture from water surfaces, the soil surface, and foliage and transpiration from plants. These contributions are usually jointly referred to as evapotranspiration (ET). This process involves three factors: (i) an energy source, (ii) a moisture source, and (iii) air motion to remove the moisture from the surface. The energy source is solar radiation, or more strictly the net radiation, which is essentially independent of the other terms. When net radiation is zero, the latent heat needed is supplied by the near-surface air and ground. The latent heat of vaporization (λ) is approximately 2500 kJ kg^{-1}, but varies slightly with temperature. This energy is returned to the atmosphere when water vapor condenses as clouds form. The availability of water to evaporate depends on the soil moisture status and on the relative dryness of the air (the vapor pressure deficit, D), and the surface. The ability of the air to transfer water vapor away from the surface depends on the vertical gradient of atmospheric water vapor, and the efficiency of turbulent mixing in the boundary layer. Where there is a surface boundary – land/water, grassland/forest, or desert/oasis, for example – the heat and moisture fluxes have to adjust to new equilibria in response to the horizontal advection of upwind conditions. Hence, most studies are undertaken where there is an extensive surface of relatively homogeneous properties. The factors that may change across a boundary include temperature, moisture content, and wind profile as well as albedo and surface roughness. Advective effects are treated in Chapter 5, Section F.

The measurement of evaporation using the Bowen ratio energy balance, aerodynamic, and eddy covariance method were described in Chapter 3, Section I. Two relatively simple and straightforward means of measuring evaporation, evaporation pans and lysimeters, are discussed here.

An evaporation pan is simply a small pan filled with water on the ground surface. The Class A pan used by the U.S. National Weather Service is a cylinder with a diameter of 120.7 cm and a depth of 25 cm. It is filled with water to within 5 cm of the top of the pan. Each 24 hours, the amount of water needed to refill the pan to the same level is measured. Rainfall amounts are recorded separately and the pan reading is adjusted. It cannot be used on days with >30 mm rainfall unless more frequent readings are taken. Given the simplicity of pan evaporation measurements, they have been used to analyze long-term, global trends in potential or reference evaporation worldwide (Brutsaert and Parlange, 1998); however, care must be taken to account for errors due to artificial heating and alteration of wind. Evaporation pans should not be used to infer evaporation from lakes and reservoirs because the pans are too shallow to represent temperature structure and mixing. And for large bodies of water, offshore conditions are not the same as those onshore.

A second method is the weighing lysimeter. This is a tank sunk into the ground with the soil and vegetation or plant cover on it. The drainage is retained and measured together with the precipitation. The block is weighed hourly or daily to determine the amount of water that a grass cover, natural vegetation, or crops use, by detecting losses of soil moisture. Accuracy is increased as the size of the block increases, but then weighing it and ensuring there are no leaks can become a problem. A lysimeter 1.6 m long, 1.1 m wide, and 0.6 m deep was able to resolve mass changes of 0.1 mm. A lysimeter encompassing a tree might have dimensions of $3 \times 3 \times 2.15$ m.

Given the limitations in both evaporation pans and lysimeters, they are seldom the method of choice to measure evaporation. Today, the eddy covariance method is widely used as part of a suite of micrometeorological instruments to measure the full surface radiation and energy balance, in addition to the net ecosystem exchange of CO_2.

Given that the widespread use of the eddy covariance method spans only the past 20–30 years and requires significant costs, expertise, and site characteristics, there is still a need to estimate evaporation from using temperature-based formulae. The semiempirical temperature-based methods are summarized for convenience and completeness in Box 5.3.

Box 5.3. Temperature-based Formulae for Estimating Evapotranspiration

For average conditions, climatological estimates of ET can be derived using air temperature data alone (see also McMahon et al., 2013). There are a number of such formulae that are summarized here.

Thornthwaite (1948) developed a method for estimating monthly "potential" evapotranspiration (PET) in millimeters. The term "potential" originally referred to the evaporation from a surface continually supplied with water; however, the term has subsequently been abandoned owing to the ambiguity involved in its definition. Instead the term ET_o – the reference evaporation – is used especially for water loss from irrigated crops.

$$ET_o \text{ (mm)} = 0.533 \, N \, (L_d/12)(10 \, T_a/I)^a \qquad 5.15$$

where L_d is day length (h), N is number of days in the time interval, T_a is the mean monthly air temperature (°C), and a is given by

$$a = 0.0675 \, I^3 - 7.71 \, I^2 + 1792 \, I + 49239 \times 10^{-5} \qquad 5.16$$

where I is a heat index derived from the sum of the 12 monthly index values of i

$$i = (T_a/5)^{1.514}. \qquad 5.17$$

Thornthwaite and Mather (1955) provide convenient tables and nomogram to simplify the calculations. Palmer and Havens (1958) describe a graphical method.

Blaney and Criddle (1950) provided a method to estimate actual evaporation. On a monthly basis, ET is given by

$$ET = k_m f \qquad 5.18$$

where k_m is an empirical coefficient related to the type of crop, f is a monthly evaporation factor given by

$$f = 0.01(1.81 \, T_a + 32)p \qquad 5.19$$

where p is the monthly percentage of total annual daylight hours. The total evaporation for the season is

$$ET = \sum_1^{12} k_m f \qquad 5.20$$

Box 5.3 (*cont.*)

Linacre (1977) developed a method that relies on temperature alone. It requires information on the latitude, elevation, daily maximum and minimum temperature, and mean dew-point temperature (T_d) of the location.

$$\text{ET}(\text{mm}) = \left(\frac{700\, T_m / (100 - l) + 15(T_a - T_d)}{(80 - T_a)} \right) \qquad 5.21$$

where $T_m = T_a + 0.006\, z$, z is the elevation (m), T_a is the mean air temperature (°C), and l is the latitude (degrees).

On the basis of water balance data from around the world, Turc (1954) proposed the relationship

$$E/P = 1/[1 + (P/ET_o)^n]^{1/n} \qquad 5.22$$

where n is an exponent; E and P (precipitation) are in millimeters. Turc graphically looked for the most appropriate value for n and concluded that the best fit was $n = 3$ or possibly 2 (Turc, 1954, p. 563).

Hargreaves and Allen (2003) evaluated a formula proposed by Hargreaves in 1985 and concluded that it was satisfactory without additional terms. Their equation is

$$ET_o = 0.0023\, S_{etr}\, (T + 17.81)\, TR^{0.5} \qquad 5.23$$

where S_{etr} is the extraterrestrial radiation (converted to evaporation units), T is the air temperature (°C), and TR is the daily temperature range.

Haslinger and Bartsch (2015) apply Hargreaves' method, with the constant 0.0023 recalibrated to the Penman-Monteith equation, to determine daily and mean ET for Austria. Surface elevation is used as a predictor to estimate the recalibrated Hargreaves parameter in space. The Hargreaves method is applied to the temperature fields, yielding reference evapotranspiration for the entire grid and period from 1961 to 2013.

As discussed so far, evaporation requires energy, the latent heat of vaporization, usually dominantly provided by a positive net radiation as described by the radiation balance equation. Available liquid water is also required. The evaporation from a surface with an unlimited water supply calculated from the available energy is referred to as the potential evaporation (see Box 5.4). Early investigations of evaporation, however, found that the evaporation rate may exceed what is predicted by the available radiative energy. It was found that evaporation rates in excess of the available radiative energy could be accounted for when an advective evaporation term was included. Warm, dry air passing over a wet surface can enhance evaporation.

A method combining the aerodynamic and energy budget approaches was developed by Howard Penman (1948) and has been widely used and subsequently modified (Thom and Oliver, 1977) to describe the radiative and advective evaporation terms explicitly:

$$\lambda E = \frac{\Delta(R_n - G)}{\Delta + \gamma} + \frac{\gamma K_w (q_a - q)}{\Delta + \gamma} \qquad 5.24$$

Table 5.4. Average ET_o for Different Agroclimatic Regions (mm day^{-1})

Climatic Regions	Mean Daily Temperature (°C)		
	10 °C	20 °C	> 30 °C
Tropics and subtropics			
– humid and subhumid	2–3	3–5	5–7
– arid and semi-arid	2–4	4–5	6–8
Temperate regions			
– humid and subhumid	1–2	2–4	4–7
– arid and semi-arid	1–3	4–7	6–9

Source: From Allen et al., 1998, table 2.

where $\gamma = c_p/L$ (the psychrometric constant approximately equal to 66 Pa K^{-1}), Δ is the rate of change of the saturated vapor pressure at air temperature (T_a), $\delta q_s/\delta T$ (kPa K^{-1}), q_a = specific humidity at T_a, and K_w is an eddy exchange coefficient for water vapor. By linearizing the saturated vapor pressure curve by the slope, Δ, of the curve determined at T_a, we can eliminate the surface temperature between the equations of H and λE and the energy balance equation. Slatyer and McIlroy (1961) developed a combination method similar to Penman's except that it uses the wet-bulb temperature depression with respect to the dry-bulb reading, instead of the vapor pressure deficit. Priestley and Taylor (1972) found that over extensive, well-watered surfaces, evaporation calculated from the first term in Eq. (5.24) needed to be multiplied by 1.26 (the Priestley-Taylor α). In other words, advection increased evaporation by 26 percent.

For agricultural purposes, such as irrigation estimation, the United Nations Food and Agricultural Organization (FAO) developed a version of the Penman equation with specific crop factors (Dorenbos and Pruitt, 1977) that was widely used by water managers. This was revised by Allen et al. (1998) using the Penman-Monteith equation to obtain what is termed reference evaporation (ET_o). It utilizes a reference surface that is a hypothetical grass crop with an assumed crop height of 0.12 m, a fixed surface resistance of 70 s m^{-1}, and an albedo of 0.23. The reference surface closely resembles an extensive surface of green, well-watered grass of uniform height, actively growing and completely shading the ground. The fixed surface resistance of 70 s m^{-1} implies a moderately dry soil surface resulting from about a weekly irrigation frequency. The effects of a crop are obtained by a crop coefficient, K_c. This coefficient accounts for differences in the crop canopy and aerodynamic resistance relative to the reference crop. K_c can be split into two factors that separately describe the evaporation and transpiration components or treated as a single combined value.

Values of average ET_o for different agroclimatic regions in millimeters per day are shown in Table 5.4 for different temperature ranges.

Box 5.4. Definitions of Potential Evaporation

Three primary definitions are recognized by Granger (1989): (i) the "equilibrium evaporation," which is defined as the evaporation rate that would occur from a saturated surface
with a constant energy supply to the surface; this represents the lower limit of actual evaporation from a wet surface; (ii) the evaporation rate that would occur from a saturated surface with constant energy supply to and constant atmospheric conditions over the surface;
this is equivalent to the Penman evaporation from a free-water surface; (iii) the upper limit
of evaporation, defined as the evaporation rate that would occur from a saturated surface
with constant atmospheric conditions and constant surface temperature.

The reference evapotranspiration (ET_o ; mm day^{-1}) is expressed as

$$ET_o = \frac{0.408 \, \Delta\left(R_n - G\right) + \gamma\left\{900 / \left(T + 273\right)\right\} u_2 \left(e_s - e_a\right)}{\Delta \, + \, \gamma \left(1 \, + \, 0.34 \, u_2\right)} \qquad 5.25$$

where R_n = net radiation at the crop surface [MJ m^{-2} day^{-1}],

  G = soil heat flux density [MJ m^{-2} day^{-1}],

  T = mean daily air temperature at 2 m height [°C],

  u_2 = wind speed at 2 m height [m s^{-1}],

  e_s = saturation vapor pressure [kPa],

  e_a = actual vapor pressure [kPa],

  $e_s - e_a$ = saturation vapor pressure deficit [kPa],

  Δ = slope of the vapor pressure curve [kPa °C^{-1}],

  γ = psychrometric constant [kPa °C^{-1}].

K_c at midseason ranges between about 0.3 (pineapples) and 1.2 (maize, cotton, and
vegetables). In the initial stages of growth K_c ranges from about 0.3 for infrequent
wetting to 1.1 for frequent wetting of the crop. Allen et al. (1998) provide detailed
tables of crop coefficients for numerous crops.

  In all vegetated surfaces, especially for crops where bare soil is often exposed
between the rows of vegetation, the practical issue of knowing a crop's water requirement requires an understanding of both transpiration and evaporation from the soil.
The issue of evaporation from drying soils was treated theoretically by Philip (1957).
He shows that there are three phases: (i) as long as the soil is sufficiently moist, the
evaporation rate is constant and indistinguishable from that of a saturated surface
(E_o); (ii) at intermediate soil water content, the evaporation rate is independent of E_o
and depends on the soil moisture distribution; (iii) when the soil surface is sufficiently
dry, the evaporation is sensitive to the heat flux in the soil. Phases (i) and (ii) corre-
spond to the constant and falling rate phases of the "isothermal" drying of an initially
saturated soil. In phase (i), evaporation is governed by the atmospheric conditions,
while in phase (ii) it is governed by conditions in the medium.

To account for the role of vegetation in evaporation, Monteith (1965) modified the original Penman equation by incorporating the surface's canopy resistance (1/conductance) term, resulting in the well-known Penman-Monteith equation:

$$\lambda E = \frac{\Delta A + \left(\rho_a c_p D \right) / g_a}{\Delta + \gamma \left(1 + g_s / g_a \right)} \qquad 5.26$$

where A is the energy available for evaporation $(R_n - G)$, ρ_a is the air density, D is the vapor pressure deficit (kPa), g_a is the aerodynamic conductance (the inverse of the aerodynamic resistance, r_a), and g_s is the bulk surface conductance.

When λE is measured using the Bowen ratio energy balance or eddy covariance method, then g_c can be solved as a residual of Eq. (5.26), as has been the case in many studies (e.g., Blanken and Black, 2004):

$$\frac{1}{g_c} = \left[\left(\frac{S}{\gamma} \right) \beta - 1 \right] \left(1 / g_a \right) + \frac{\rho_a c_p}{\gamma} \frac{D}{\lambda E} \qquad 5.27$$

The aerodynamic conductance can be calculated as the sum of the eddy diffusive (g_e) and leaf boundary layer (g_b) conductances (i.e., resistors acting in series):

$$\frac{1}{g_a} = \frac{1}{g_e} + \frac{1}{g_b} = \frac{u}{u_*^2} + \frac{B^{-1}}{u_*} \qquad 5.28$$

where B^{-1} is the dimensionless sublayer Stanton number (e.g. $B^{-1} = 2.50$ for an aspen canopy with an LAI of 2.3 m^2 m^{-2}; Blanken et al., 1997).

Alternatively, g_c can be measured at the leaf level using a porometer (a cuvette that attaches to the leaf to measure transpiration and stomatal resistance directly), and then scaled to the canopy level knowing the stomatal resistance and LAI of all species (e.g. Blanken and Rouse, 1995). Or, g_c can be modeled using a photosynthesis-based model such as the Ball-Berry model (Collatz et al, 1991).

Brutsaert (1982) points out that the physical meaning of canopy resistance is difficult to comprehend. Lhomme (1991) notes that the plant canopy cannot be treated as a single effective source-sink height for H and λE (a condition for the Penman-Monteith equation) as they have different temperatures. The zero plane displacements for heat and water vapor have erratic behavior. Lhomme develops a multilayer approach to the estimation of resistance. His combination equation is

$$\lambda E = \frac{\Delta \left(R_n - G \right) + \rho c_p (D + \Delta \left(Te_V - Te_H \right) / r_{c,a}}{\Delta + \gamma \left(1 + r_{c,s} / r_{c,a} \right)} \qquad 5.29$$

where Te_V and Te_H are the weighted means of leaf and soil temperatures, and $r_{c,a}$ is the bulk aerodynamic resistance, the sum of the aerodynamic resistance of the airstream above the canopy and the bulk resistance of the canopy to H, combining the diffusive resistances and leaf boundary resistances of the canopy. The $r_{c,s}$ term is the bulk stomatal (or surface) resistance, which is the difference between the bulk canopy resistance to vapor transfer and the bulk aerodynamic resistance to transfer of H. The algorithm is given by Lhomme (1988). The values of $r_{c,s}$ in relation to solar radiation range from 34.9 s m^{-1} at 400 W m^{-2}, to 24.1 at 600 W m^{-2}, and 19.7 at 800 Wm^{-2}. Using the combination equation, Lhomme determines rates of λE as 393 W m^{-2} for a 1 m s^{-1} wind at a reference height of 3 m, increasing to 444 at 3 m s^{-1} and 477 at 5 m s^{-1}.

Shuttleworth (2007) made a further modification to the Penman-Monteith equations by using measured net radiation and writing the equation in SI units. He gives an extensive review of the development of evaporation theory from which only a few salient points can be summarized here. A major issue is the drying of vegetation canopies after rainfall, which changes the surface resistance. Overall, it is preferable to separate the evaporation of intercepted water on leaves from the direct transpiration by the vegetation cover. In wet climates considerable energy is required to evaporate the intercepted rainfall, whereas in drier climates, vegetative transpiration is more important. Leaf wetness sensors can be used to detect when the canopy has intercepted precipitation.

The interception of rainfall by vegetation was first studied almost 100 years ago by Horton (1919). Gash (1979) determined that interception can be estimated from the forest structure, the rainfall pattern, and the mean rates of rainfall and of evaporation during rainfall. The principal controls of rainfall interception are the direct throughfall proportion and the canopy storage capacity per unit leaf area. The latter is the amount of water left on the canopy in zero evaporation conditions when rainfall and throughflow have ceased. The model assumes that the canopy dries out between rain events. Evaporation from water on the trunk is also determined. A revised version of the model has been developed for sparse canopies by Gash et al. (1995).

In the Amazon rain forest, Lloyd et al. (1988) reported measured interception of 9 percent of total rainfall. For a 60 m tall Douglas fir – western hemlock forest in southwestern Washington State, Link et al. (2004) report total interception of 22.8 percent for April 8–November 8, 1999, and 25.0 percent for March 30–December 3, 2000. Seasonal values had a great range; in spring, summer, and autumn of 2000, the net interception values were 28.0, 69.4, and 18.0 percent, respectively. For similar forest in Vancouver Island they cite McMinn's (1960) results: interception of 20–40 percent annually over a 5.5-year period, with a range of 30–57 percent in the summer. In a global survey using multiple satellite data and Gash's analytical model, Miralies et al. (2010) estimated that interception loss is responsible for approximately 13 percent of the total incoming rainfall over broadleaf evergreen forests, 19 percent in broadleaf deciduous forests, and 22 percent in needle-leaf forests. The product is sensitive to the volume of rainfall, rain intensity, and forest cover.

In absolute terms, interception storage capacity of forests is between 0.5 and 2 mm for rain and 2–6 mm for snow according to Zinke (1967). Rutter (1975, table 1) noted that there is no separation between forests (ranging from 0.4 to 2.1 mm) and herbaceous communities (0.5 to 2.8 mm).

F. Advective Effects

The effects of horizontal advection on turbulent fluxes downwind of a surface boundary were briefly noted earlier. Now the effects of surface heterogeneity are examined in more detail. Garratt (1990) provides a review of the internal boundary layer (IBL) defined for neutral, small-scale flow, as the layer in which there is a significant change from upstream conditions in wind velocity and shear stress with height, downstream of a change in surface roughness. Its top is the height at which there is a significant discontinuity in $\delta u/\delta z$. Above the IBL, the flow field corresponds to that upstream, except for a small vertical displacement of the streamlines of the outer flow. For a smooth-to-rough transition, the stress is initially increased to about twice its final (large fetch) value, whereas for a rough-to-smooth transition, it initially decreases to about half of its final value. Garratt cites fetch/height ratios of 10 for convective cases (usually cold air off the sea over warmer land) and 2000 for stable cases. The neutral IBL grows at a rate of approximately $x^{0.8}$ (where x is downstream distance) for a smooth-to-rough transition and slightly less in the opposite case. Very unstable conditions give a growth rate dependence of $x^{1.4}$.

The changes in the profiles of wind velocity and shear stress are accompanied by responses in the turbulent fluxes of sensible and latent heat. McNaughton (1976A, B) introduced the concept of an advective enhancement term, which depends on the surface resistance and the wind speed, as well as on the efficiency of downward vertical mixing. It diminishes to zero at some distance downwind of the boundary. The boundary layer is modified downwind of a boundary by vertical diffusion. Thus, it may deepen if the transition is from a dry land surface to a lake or irrigated area. The increase in evaporation leads to greater convectively available potential energy (CAPE). Baldocchi and Rao (1995) measured H and λE over an irrigated potato field growing beside a desert patch with 800 m diameter dimensions. The turbulent fluxes measured at 4 m height exhibited marked variations with downwind distance over the field. Only after the fetch to height ratio exceeded 75:1 did λE and H become invariant with downwind distance. Contrary to McNaughton, the advection of hot, dry air did not enhance surface evaporation rates near the upwind edge of the irrigated field, as a result of negative feedbacks among stomatal conductance, humidity deficits, and λE.

Dyer (1963) discussed the response of dry air moving downwind from the leading edge of an irrigated plot, or a stream channel in semi-arid terrain (a process known as the "oasis" effect). He showed that the downwind distance to the point where the moisture flux and wind profile are 90 percent adjusted (for near-neutral stability) is ~100–1000 m and the time interval ~1–15 minutes for a wind speed of 1.9 m s^{-1} at a reference height of 1 cm. At 1 m height the downwind fetch for 90 percent adjustment is 170 m; at 5 m height it is 1350 m, and at 10 m it is 3330 m. Fetch/height ratios range from 170 at 1 m to 330 at 10 m. The corresponding times required for 90 percent adjustment are 1 minute, 6 minutes, and 14 minutes, respectively. Where there is a small pond or an evaporation pan surrounded by a dry surface, the evaporation is greatly increased as dry air flows over it. A similar case occurs where the vegetation height is locally higher than its surroundings. This gives rise to what is termed the "clothesline" effect, and it represents a biased measurement that is difficult to interpret.

References

Allen, R. G. et al. 1998. Crop evapotranspiration. Guidelines for computing water requirements. FAO Irrigation and Drainage Paper 56. Rome: UN Food and Agriculture Organization.

Al Nakshabandi, G. and Kohnke, H. 1965. Thermal conductivity and diffusivity of soils as related to moisture tension and other physical properties. *Agric. Met.* 2, 271–9.

Baldocchi, D. D. and Rao, K. S. 1995. Intra-field variability of scalar flux densities across a transition between a desert and an irrigated potato field. *Boundary-Layer Met.* 76, 109–356.

Barry, R.G. and Gan, T.Y. 2011. *The global cryosphetre; Past, presentand future.* Cambridge Cambridge University Press.

Blaney, H.F, and Criddle, W.D. 1950 Determining consumptive use and irrigation water requirements. *Tech. Bull.,* 1275. Washington, DC US Dept. of Agriculture.

Blanken, P. D., and Rouse, W. R. 1995. Modelling evaporation from a high subarctic willow-birch forest. *Int. J. Climatl.* 15(1), 97–106.

Blanken, P. D. et al. 1997. Energy balance and canopy conductance of a boreal aspen forest: Partitioning overstory and understory components. *J. Geophys. Res.* 102(D24), 28915–27.

Blanken, P. D., and Black, T. A. 2004. Canopy conductance of a boreal aspen forest, Prince Albert National Park, Canada. *Hydrol. Proces.,* 8, 1561–78.

Bowen, I. S. 1926. The ratio of heat losses by conduction and by evaporation from any water surface. *Phys. Rev.* 27, 779–87.

Brooks, F. A., Pruitt, W. O., and Nielsen, D. R. et al. 1963. Investigation of energy and mass transfers near the ground including the influences of the soil-plant-atmosphere system. Davis, CA: Department of Agricultural Engineering and Department of Irrigation, University of California. Contract DA- 36-039-SC-80334 U.S. Army Electronic Proving Ground Fort Huachuca, AZ,

Brutsaert, W. H. 1982. *Evaporation into the atmosphere.* New York: Reidel.

Brutsaert, W. H., and Parlange, M. B. 1998. Hydrologic cycle explains the evaporation paradox. *Nature* 396(30). doi: 10.1038/23845.

Cai, D-L. et al. 2014. Climate and vegetation: An ERA-Interim and GIMMS NDVI analysis. *J. Climate* 27, 5111–18.

Carson, J. E. 1963. Analysis of soil and air temperatures by Fourier technique. *J. Geophys. Res.* 68, 2217–32.

Collatz, G. J. et al. 1991. Physiological and environmental regulation of stomatal conductance, photosynthesis and transpiration: a model that includes a laminar boundary layer. *Agric. For. Met.* 54, 107–36.

Conrad, V., and Pollak, L. W. 1950. *Methods in climatology,* 2nd ed. Cambridge, MA: Harvard University Press.

Cosby, B. J. et al. 1984. A statistical exploration of the relationships of soil moisture characteristics to the physical properties of soils. *Water Resour. Res.* 20, 682–90.

Davenport, A. G. et al. 2000: Estimating the roughness of cities and sheltered country. *Preprints, 12th Conf. on Applied Climatology, Asheville, NC.* Boston, MA: Amer. Meteor. Soc., pp. 96–9.

Dirks, B. O. M., and Hensen, A. 1999. Surface conductance and energy exchange in an intensively managed peat pasture. *Climate Res.* 12, 29–37. doi:10.3354/cr012029.

Dorenbos, J., and Pruitt, W. O. 1977. Guidelines for predicting cropwater requirements. *FAO Irrigation and Drainage Paper,* no. 24. 2nd ed. Rome: UN Food and Agriculture Organization.

Dyer, A. J. 1963. The adjustment of profiles and eddy fluxes. *Quart. J. Roy. Met. Soc.* 89, 276–80.

Farouki, O. T. 1981. Thermal properties of soils. *CRREL Monograph* 81-1. Hanover, NH: U.S. Army Corps of Engineers, Cold Regions Research and Engineering Laboratory.

French, H. M. 2007. *The periglacial environment.* New York: Wiley.

Garratt, J. R. 1990. The internal boundary layer: A review. *Boundary-Layer Met.* 59L, 171–203.

Gash, J. H. C. 1979. An *analytical* model of *rainfall interception* by forests. *Quart. J. Roy. Met. Soc.* 105, 45–53.

Gash, J. H. C., Lloyd, C. R., and Lachaud, G. 1995. Estimating sparse forest rainfall interception with an analytical model. *J Hydrol.* 170, 79–86.

Gates, D.M. and Janke, R. 1966. The energy environment of the alpine tundra. *Oecol. Plant,* 1, 39–62.

Gold, L. W. et al. 1972. Thermal effects in permafrost. *Proceedings, Canadian Northern Pipeline Research Conference. Tech. Mem.* No.104. Ottawa: National Research Council, Canada. pp. 25–39.

Goodrich, L. E. 1982. The influence of snow cover on the ground thermal regime. *Canad. Geotech. J.* 19, 421–32.

Granger, R. J. 1989. An examination of the concept of potential evaporation. *J. Hydrol.* 111, 9–19.

Grisnås, K. et al. 2013. CryoGRID 1.0: Permafrost distribution in Norway estimated by a spatial numerical model. *Permafrost Periglac. Proc.* 24, 2–19.

Hargreaves, G. H., and Allen, R. G. 2003. History and evaluation of Hargreaves evapotranspiration equation. *J. Irrig. Drain. Eng.* 129, 53–63.

Haslinger, K., and Bartsch, A. 2015. Creating long term gridded fields of reference evapotranspiration in Alpine terrain based on a re-calibrated Hargreaves method. *Hydrol. Earth System Sci., Disc.* 12, 5055–82.

Hellmann, G. 1915; 1917. Über die Bewgung der Luft in den untersten Schichten der Atmosphäre. *Met. Zeit.* 32, 1–16. 34, 273–85.

Horton, R. E. 1919. Rainfall interception. *MWR* 47, 603–23.

Kane, D. L. et al. 2001. Non-conductive heat transfer associated with frozen soils. *Global Planet. Change* 29, 275–92

Lachenbruch, A. H. et al. 2012. Permafrost, heat flow, and the geothermal regime at Prudhoe Bay, Alaska. *J. Geophys. Res., Solid Earth.* B11(87), 9301–16.

Linacre, E. T. 1977. A simple formula for estimating evapotranspiration rates in various climates, using temperature data alone. *Agric Met.* 18, 409–24.

Link, T. E., Unsworth, M., and Marks, D. 2004. The dynamics of rainfall interception by a seasonal temperate rainforest. *Agric. Forest Met.* 124, 171–91.

Lhomme, J. P. 1988. A generalized combination equation derived from a multi-layer micro-meteorological model. *Boundary-Layer Met.* 45, 103–15.

Lhomme, J. P. 1991. The concept of canopy resistance: Historical survey and comparison of different approaches. *Agric. Forest Met.* 54, 227–40.

Lloyd. C. R, Gash, J. H. C., and Shuttleworth, W. J. 1988. The measurement and modelling of rainfall interception by Amazonian rain forest. *Agric. Forest Met.* 43, 277–94.

Mackay, J. R. 1972. The world of underground ice. *Ann. Assoc. Amer. Geog.* 62, 1–22.

McMahon, T. A. et al. 2013. Estimating actual, potential, reference crop and pan evaporation using standard meteorological data: A pragmatic synthesis. *Hydrol. Earth System Sci.* 17, 1331–63.

McMinn, R. G, 1960. Water relations and forest distribution in the Douglas-fir region of Vancouver Island. Publ. 1091, Div. For. Biol., Dept. Agric. Ottawa, Canada.

McNaughton, K. G. 1976A. Evaporation and advection. I. Evaporation from extensive homogeneous surfaces. *Quart. J. Roy. Met. Soc.* 102, 181–91.

McNaughton, K. G. 1976B. Evaporation and advection. II: Evaporation downwind of a boundary separating regions having different surface resistances and available energies. *Quart. J. Roy.Met Soc.* 102, 193–202.

Miralies, D. G. et al. 2010. Global canopy interception from satellite observations. *J. Geophys. Res., Atmos.* 115 (D16122). doi:10.1029/2009JD013530

Monteith, J. L. 1965. Evaporation and environment. *Symposia of the Society for Experimental Biology* 19: 205–24.

Monteith, J.J. and Unsworth, M.H. 1990. *Principles of environmental physics,* 2nd edn. London . Arnold.

Monteith, J. L., and Unsworth, M. H. 2014. *Principles of environmental physics. Plants, animals and the atmosphere*, 4th ed. Amsterdam: Elsevier.

Palmer, W. C., and Havens, A. V. 1958. A graphical technique for determining evapotranspirayion by the Thornthwaite method. *MWR* 86, 123–8.

Penman, H. L. 1948: Natural evaporation from open water, bare soil and grass. *Proc. Roy. Soc. London A* 193, 120–45.

Philip, J. R. 1957. Evaporation and moisture and heat fields in the soils. *J. Met.* 14, 354–66.

Priestley, C. H. B. 1955. Free and forced convection in the atmosphere near the ground. *Quart. J. Roy. Met. Soc.* 81, 139–43.

Priestley, C. H. B. Taylor, R. J. 1972. On the assessment of surface heat flux and evaporation using large-scale parameters. *MWR* 100, 81–92.

Rider, N. E. 1957. A note on the physics of air temperature. *Weather* 12(8), 236–66.

Rutter, A. J. 1975. The hydrological cycle in vegetation. In Monteith, J. L. (ed.) *Vegetation and atmosphere.* Vol. 1. *Principles.* London: Academic Press. pp. 111–54.

Shuttleworth, W. J. 1989. Micrometeorology of tropical and temperate forest. *Phil. Trans. Roy. Soc., London* B233, 321–46.

Shuttleworth, W. J. 2007. Putting the vap into evaporation. *Hydrol. Earth System Sci.* 11, 210–44.

Slatyer, R. O., and McIlroy, I. C. 1961. *Practical microclimatology.* Melbourne, Australia: CSIRO.

Streletskiy D. A. et al. 2008. 13 years of observations at Alaskan CALM sites: Long-term active layer and ground surface temperature trends. In D. L. Kane and K. M. Hinkel (eds.), *Ninth International Conference on Permafrost.* Fairbanks: Institute of Northern Engineering, University of Alaska Fairbanks, pp. 1727–32.

Thom, A. S. 1975. Momentum, mass and heat exchange of plant communities. In J. L. Monteith (ed.), *Vegetation and atmosphere.* Vol. 1. *Principles.* London: Academic Press, pp. 57–109.

Thom, A. S., and Oliver, H. R. 1977. On Penman's equation for estimating regional evaporation. *Quart. J. Roy. Met. Soc.* 103, 345–57.

Thornthwaite, C. W. 1948. An approach toward a rational classification of climate. *Geog. Rev.* 38, L 55–94.

Thornthwaite, C. W., and Mather, J. R. 1955. *The water balance.* Publ. Climatology Centerton, NJ: Laboratory of Climatology, 8: pp. 1–104.

Turc, L., 1954. Le bilan d'eau des sols: relation entre les précipitations, l'évaporation et l'écoulement. *Annal. Agronom.* Série A(5), 491–595.

Westermann, S. et al. 2015. A ground temperature map of the North Atlantic permafrost region based on remote sensing and reanalysis data. *Cryosphere Discuss.* 8, 753–90.

Williams, G., and Gold, L. 1976. Canadian Building Digest, 1976. Ottawa: National Research Council of Canada, Institute for Research in Construction No. 180. http://irc.nrc-cnrc.gc.ca/cbd/cbd180e.html, N. 180.

Yoshino, M. M. 1975. *Climate in a small area: An introduction to local meteorology.* Tokyo: Tokyo University Press.

Zhang, T-J. et al. 2005. Spatial and temporal variability in active layer thickness over the Russian Arctic drainage basin. *J. Geophys. Res., Atmos.* 110(D16). doi: 10.1029/2004JD005642

Zinke, P. J. 1967. Forest interception studies in the United States. In W. A. Sopper and H. W. Lull (eds.), *Forest hydrology.* Oxford: Pergamon Press.

6

Monitoring Radiation, Energy, and Moisture Balance via Remote Sensing and Modeling with Land Surface Models

The previous discussion has focused on the measurement of components of the radiation and energy balances at a small scale, such as a field. However, these data do not provide a basis for scaling up to large geographical areas such as watersheds or biomes, or globally. Over the last three decades this problem has increasingly been addressed by the application of multispectral remote sensing techniques. Here, we describe these approaches. This is followed by discussion of land surface models leading on to the topic of downscaling model outputs to small-scale conditions.

Microclimatic models have evolved over the last 20–30 years from empirical, statistical correlation-based approaches to physical models that are incorporated into landscape models and urban structure models. Here we briefly introduce the principal empirical and physical models that are in use.

A. Approaches

Microclimatic research was for most of its history carried out at a few sites for limited time intervals. However, the scaling up of results from field studies has become a focus of attention in the last two to three decades. This has been made possible by the application of high-resolution remote sensing data and by numerical modeling.

Courault et al. (2005) identify three general categories of approach that incorporate remote sensing data: (i) direct empirical methods, (ii) combined empirical and physically based methods, and (iii) soil-vegetation-atmosphere-transfer (SVAT) schemes.

Empirical Direct Methods

One of the most basic approaches is the remote sensing of leaf area index (LAI). It is defined as the one-sided green leaf area per unit ground surface area (square meters per square meter) in broadleaf canopies, while in conifer canopies, three different definitions for LAI have been used: half of the total needle surface area per unit ground surface area, projected (or one-sided) needle area per unit ground area, and total needle surface area per unit ground area. LAI ranges from 0 over bare ground to more than 10 for dense conifer forests. LAI is a significant indicator of energy and mass exchange by plant canopies and is useful to scale from the leaf to canopy-level measurements. Its importance was recognized by the National Research Council (1986, p. 102) as "the key to studying vegetation biomass and productivity from space."

One of the first studies of LAI via remote sensing was that of Curran (1983). He showed that the Normalized Difference Vegetation Index (NDVI – see later discussion) is linearly related to LAI up to some threshold above which it becomes asymptotic; the threshold is dependent on vegetation and crop type.

A global synthesis of LAI data for 15 biomes is provided by Asner et al. (2003). Values ranged from 1.3 for deserts to 8.7 for tree plantations, with temperate evergreen forests (needle leaf and broadleaf) displaying the highest average LAI (5.1–6.7) among the natural vegetation classes. Wetlands averaged 6.3, crops averaged 3.6, and grasslands 1.7. The global mean LAI, after statistical removal of outliers, was 4.5 (terrestrial surfaces only). A further global climatology (GEOCLIM) of LAI has been carried out by Verger et al. (2015) using the SPOT VEGETATION program.

Interannual average values from the GEOV1 Copernicus Global Land time series of biophysical products at 1-km resolution and 10-day frequency were computed for 1999–2010. Maps of maximum and mean annual LAI are shown.

Recently, Tesemma et al. (2014) used remotely sensed LAI data from the Moderate Resolution Imaging Spectroradiometer (MODIS) and gridded climatic data from the Australian Water Availability Project to relate seasonal and annual LAI of three different land cover types (trees, pasture, and crops) with climatic variables. The analysis was made for the period 2000–2009 in the Goulburn–Broken catchment, Australia. Good relationships were obtained between annual LAI of the three cover types and annual precipitation (R^2 = 0.70, 0.65, and 0.82, respectively). The monthly LAI of each cover type also showed a stronger relationship ($R^2 \geq 0.92$) with the difference between precipitation and reference crop evapotranspiration (see Chapter 5, Section E) for crop, pasture, and trees. Independent model calibration and validation showed good agreement with remotely sensed LAI from MODIS.

The basis of these methods is the relationship between surface temperature (T_s) or ($T_s - T_a$) and λE where H/R_n is assumed to be constant and G is assumed to be 0 W m^{-2}. T_s is determined by airborne or satellite thermal infrared measurements corrected for the atmospheric effects of aerosols, water vapor, and clouds, and surface emissivity. For example, Lagouarde (1991) uses AVHRR data to estimate evaporation. The coefficients (A and B) in the relationship

$$\text{ET} = R_n + A - B(T_s - T_a) \qquad\qquad 6.1$$

may be parameterized in terms of wind speed, roughness, and atmospheric stability. R_n can be obtained via satellite measurements of solar and infrared radiation.

In a study for the northern monsoon region of Mali, West Africa, Bateni et al. (2014) apply remotely sensed land surface temperature (LST) and fraction of photosynthetically active radiation absorbed by vegetation (FPAR) to estimate surface energy fluxes and vegetation dynamics. The region is covered with C4 annual grasses. They show that spaceborne LST observations can be used to constrain the surface energy balance (SEB) equation and obtain its key two unknown parameters – the neutral bulk heat transfer coefficient and evaporative fraction. The spatial patterns of these two parameters resemble, respectively, independently observed vegetation index and soil moisture. The framework is a variational data assimilation (VDA) scheme that yields surface heat fluxes, LAI, and specific leaf area estimates that are constrained by remotely sensed LST and FPAR data. It couples the SEB equation and the VDA via the linkage between transpiration and photosynthesis. The SEB scheme provides estimates of transpiration, which is used as the key environmental input variable to the vegetation dynamics model (VDM). The VDM predicts the seasonal LAI variation, which is then utilized by the SEB scheme for the partitioning of available energy among heat fluxes. FPAR data are assimilated in order to constrain the key unknown variable of the VDM, which is the specific leaf area (c_g); this captures seasonal phenology and vegetation dynamics. c_g is a measure of leaf thickness, calculated by dividing the area of a portion of a leaf by the dry matter weight (DM) of that same portion. It ranges from about 0.01 to 0.025 m^2 g DM^{-1} for grassland sites.

Figure 6.1 shows the energy balance components at two stations in the region (Bamba, 17.1 °N, and Agoufou, 15.3 °N) over the periods June, July, August, and September. λE at Bamba is only significant during the monsoon in August. At Agoufou, the first two months are dry and H dominates, while in July and August λE is large. Note that the retrieved and observed fluxes are in good agreement.

Pan-Arctic estimation of ET has been carried out by Mu et al. (2009) using the Penman-Monteith formulation driven by MODIS-derived vegetation indices and daily surface meteorological data (incoming solar radiation, air temperature, and vapor pressure deficit (D)). Three different sources of meteorological data were compared: (i) site weather station data, (ii) air temperature and D derived from AMSR-E passive microwave data, and (iii) Global Modeling and Assimilation Office (GMAO) reanalysis meteorology-based surface air temperature, humidity, and solar radiation data. Vapor pressure deficit and minimum air temperature are used to constraint stomatal conductance, and LAI is used to scale leaf-level stomatal conductance to canopy conductance. The Enhanced Vegetation Index (EVI) is used to calculate the vegetation cover fraction. The surface conductance is estimated from the Normalized Difference Vegetation Index (NDVI) and LAI. The NDVI is based on the difference in reflectivity (α) between the near-IR (NIR) and visible wavelength (VIS) bands, with the latter representing the photosynthetic-active radiation or PAR. It is determined from

$$\text{NDVI} = \frac{\alpha_{NIR} - \alpha_{VIS}}{\alpha_{NIR} + \alpha_{VIS}} \qquad\qquad 6.2$$

The EVI is calculated from

$$\text{EVI} = 2.6(\alpha_{NIR} - \alpha_{red})/(\alpha_{NIR} + 6\,\alpha_{red} + 7.5\,\alpha_{blue} + 1.0) \qquad 6.3$$

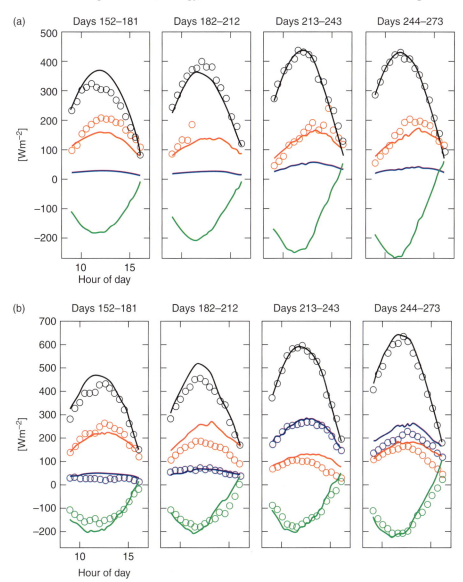

Figure 6.1. Monthly mean diurnal cycles of estimated surface energy balance components, June to September, at (top) Bamba and (bottom) Agoufou. Thick lines and open circles indicate retrieved and measured heat fluxes, respectively. Colors show different components of the surface energy balance: net radiation (black), latent heat (blue), sensible heat (red), and ground heat plotted as negative values (green) (S. M. Bateni et al., 2014). *Source: Water Resources Research* 50, p. 8636, fig, 13. Courtesy American Geophysical Union.

where α is albedo for the NIR, red, and blue wavelength bands, respectively. EVI has a higher correlation with ET than the Normalized Difference Vegetation Index (NDVI).

Model performance was assessed via a North American latitudinal transect of six eddy covariance flux towers representing northern temperate grassland, boreal forest, and wet-sedge tundra. They found good agreement ($r > 0.7$) between the model results and observations. However, the assumption of negligible G in the algorithm used is a problem in tundra areas, where G can be a significant fraction of R_n early in the growing season.

Estimates of monthly global land evaporation at $1°$ resolution were made for 1986–1993 by Fisher et al. (2008) using AVHRR and International Satellite Land Surface Climatology Project (ISLSCP)-II data. For the Priestley-Taylor based model, five inputs are required: net radiation, NDVI, soil adjusted vegetation index, maximum air temperature, and vapor pressure. The results were validated with measurements at 16 FLUXNET sites. An improved algorithm for estimating global terrestrial evapotranspiration using MODIS was subsequently developed by Mu et al. (2011). The algorithm calculates ET as the sum of daytime and nighttime components; includes soil heat fluxes; improves the estimation of stomatal conductance and aerodynamic and boundary layer resistance; separates dry canopy surface from wet; and divides the soil surface into saturated wet and moist surfaces. The algorithm was validated with data from 46 flux towers in North America. The contribution of average annual nighttime ET to annual total ET was 9.5 percent. The calculated global annual total ET over the vegetated land surface for 2000–2006 was 62.8×10^3 km^3 (or 432 mm yr^{-1}). This represented 58 percent of the precipitation over the land surface.

NDVI data were used by Cai et al. (2014) in an analysis of global geobotanic data applying M. I. Budyko's approach of the dryness ratio of net radiation/precipitation (Budyko, 1956). This is described in Box 6.1.

Combined Empirical and Physically Based Methods

There are numerous approaches under this heading. Most employ the spatial variability of remote sensing data directly to estimate input parameters and ET. One uses a vegetation index (VI) measured at frequent intervals by remote sensing and assumes a relationship between foliage density or LAI and unstressed ET. A crop coefficient is assigned to each vegetation class, which is a fraction of the reference crop evaporation – defined as the evapotranspiration from an extensive green grass surface of uniform height (8–15 cm tall), actively growing, completely shading the ground, and not short of water.

Nagler et al. (2005) used the Enhanced Vegetation Index from MODIS satellite data combined with eddy covariance and Bowen ratio data from flux towers for western U.S. river basins. A scaled EVI from 0 to 1 was found to improve the correlations further. The predictive equation for ET incorporates air temperature from the flux towers. Huang et al. (2014) estimated soil respiration (R_s) at a deciduous broadleaf forest site in the Midwest of the United States. They derived land surface temperature (LST) and spectral vegetation index from MODIS, and root zone soil moisture from the assimilation of AMSR-E and a land surface model. The models based on mean LST (i.e., averaged nighttime and daytime LST) and root zone soil moisture

Box 6.1. Climate and Geobotanical Vegetation Categories

The concept of water demand (mean annual net radiation, R_n) versus water/energy limitation (the dryness ratio D of net radiation/annual precipitation) was developed by M. I. Budyko in the Soviet Union in 1956 to relate climate to geobotanical vegetation categories (Budyko, 1956). The dryness ratio is divided in to five ranges: >3 = desert, 2–3 = semi-desert, 1–2 = steppe or prairie, 0.33–1 = forest, and <0.33 = tundra, as illustrated in Figure 6.2b.

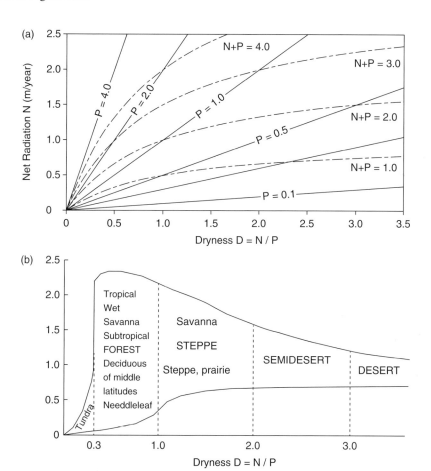

Figure 6.2. Continental surface climate state space spanned by the mean net radiation R_n and dryness ratio D (all units in meters per year[1] water equivalent). (a) Precipitation $R_n = PD$ corresponds to the slope of straight lines through the origin; the curve of total supply of energy and water satisfies $R_n = (R_n + P)[D/(D + 1)]$. (b) Geobotanic zonality types are adapted from Budyko (1956); the boundary (full line) includes all area units of (R_n, D) pairs observed on the global land surface; and D classes are separated by vertical lines (dashed) with additional subclasses depending on the magnitude of net radiation (D-L. Cai et al., 2014).

Source: *Journal of Climate* 27, p. 5114, figure 2,Courtesy of American Meteorological Society.

explained 82 percent and 72 percent of seasonal variations in R_s for spring and winter dormant periods, respectively. In the growing season, the models depending on mean LST, root zone soil moisture, and photosynthesis-related enhanced vegetation index showed comparable accuracy with the models entirely based on in situ measured data, except for the midgrowing period. Drought stress led to a somewhat lower explanation (76 percent) for the R_s model based on spatial data products during the midgrowing period.

Soil-Vegetation-Atmosphere-Transfer (SVAT) Schemes

These models describe the exchanges and interactions among the plants, soil, and atmosphere on the basis of the physical processes at work. The vegetation layer may be described by a single "big leaf" with one surface conductance value, or by a multilayer model where radiative and energy budgets are determined for each layer (see discussion by Raupach and Finnigan, 1988).

Courault et al. (2005) note that remote sensing data can be applied in three different ways:

- forcing the model input directly,
- correcting the state variables in the model at each time remote sensing data are available (sequential assimilation),
- reinitializing or changing unknown parameters using data sets acquired over temporal windows of several days/weeks (variational assimilation).

B. Land Surface Models

SVATs have subsequently been greatly expanded in land surface models (LSMs) that are a component of global climate models (GCMs). A review by Pitman (2003) details the evolution of these models. The Project for Intercomparison of Land Surface Parameterization Schemes (PILPS) was conducted in the 1990s, beginning in 1992 (Henderson-Sellers et al., 1995). Twenty-seven models contributed to PILPS. Initially, the models were run off-line and then later coupled with GCMs. These models were developed by both individuals and modeling groups; the increased complexity of the schemes has necessitated the collaboration of large numbers of scientists with a wide range of expertise (meteorologists, biologists, hydrologists, biogeochemists, soil scientists, and computer programmers). Comparisons were made in PILPS with observational data for short grass on a deep saturated soil at Cabauw in the Netherlands, the Konza tall grass prairie in Kansas under the First ISLSCP Field Experiment (FIFE), the Hydrological Atmospheric Pilot Experiment (HAPEX) in the pine forest of the Landes in southwestern France, and the Amazon Rainforest Meteorological Experiment (ARME) in Brazil, among others.

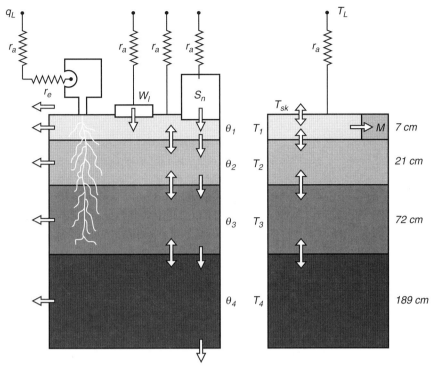

Figure 6.3. Model structure of the ECMWF land surface parameterization scheme (from Viterbo and Beljaars, 1995).

Double arrows are diffusivity processes, single arrows are drainage-type terms (soil drainage, snow melt, throughfall), horizontal arrows show surface and subsurface runoff. In the heat transfer panel the horizontal arrow represents heat exchange due to melting. θ is the volumetric soil water content, W is the interception water, r_a is the aerodynamic resistance and r_c is the canopy resistance, and q_L is the specific humidity at the lowest level (Viterbo and Beljaars, 1995).

Source: *Journal of Climate*, 8, p. 2718, fig.1.

Courtesy: American Meteorological Society.

The state of the art in the mid-1990s is presented by Viterbo and Beljaars (1995) for the land surface parameterization adopted in the ECMWF model. The model structure is illustrated in Figure 6.3. There are four soil layers each with a root fraction, hydraulic 3.conductivity and diffusivity are functions of soil water content, and snow cover and intercepted water are treated.

There are now many third generation LSMs. One is the Joint UK Land Environment Simulator (JULES) of the UK Met Office (Best et al., 2011). The model structure is shown in Figure 6.4. It is a 1° tiled model of sub-grid heterogeneity with separate surface temperatures, shortwave and longwave radiative fluxes, sensible and latent heat fluxes, ground heat fluxes, canopy moisture contents, snow masses, and snow melt rates computed for each surface type in a grid box. Nine surface types are normally used:

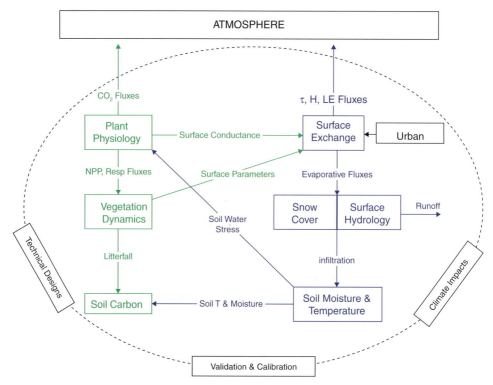

Figure 6.4. Model structure for JULES. The boxes show each of the physics modules while the lines between the boxes show the physical processes that connect these modules. The surrounding three boxes show the cross-cutting themes (from Best et al., 2011).
Source: *Geosci. Model Dev.*, 4, p. 680, fig. 1. Meteorological Office, FitzRoy Road, Exeter, EX1 3PB, UK.

- Five plant functional types
  - Broadleaf trees
  - Needleleaf trees
  - C3 (temperate) grass
  - C4 (tropical) grass
  - Shrubs

- Four non-vegetation types
  - Urban
  - Inland water
  - Bare soil
  - Land ice

Except for those classified as land ice, a land grid box can be made up from any mixture of the other surface types. Air temperature, humidity, and wind speed above the surface and soil temperatures and moisture contents below the surface are treated as homogeneous across a grid-box.

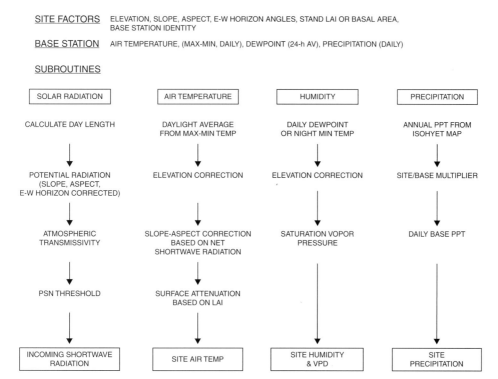

Figure 6.5. MTCLIM model for estimating daily microclimate conditions in mountainous terrain. The site factors and base station variables are required inputs for the model (from Hungerford, R. D. et al. 1989. MTCLIM: A mountain microclimate simulation model. Ogden, UT: Inter Mountain Research Station, Res. Pap. INT 414. U.S. Dept. of Agriculture, Forest Service).

C. Empirical Models

The MTCLIM (Mountain Climate) model was developed for the western United States by Hungerford et al. (1989) to predict daily solar radiation, air temperature, relative humidity, and precipitation for mountainous sites by extrapolating data measured at National Weather Service stations (Figure 6.5). The model may be used to generate data for use in ecological models, fire models, insect and disease models or in developing silvicultural practices. MTCLIM predicts daytime average and maximum air temperatures with an accuracy of 2.2 °C and minimum temperatures within 3.3 °C. Predictions of solar radiation are only accurate to 100 W m^{-2} and relative humidity to 11 percent. Precipitation predictions are accurate within 3.8 mm when two base stations are used.

Running et al. (1987) combined GIS-integrated topography, soil, vegetation, and climatic data at 1.1-km scale with LAI from AVHRR for a mountainous region 28 × 55 km of mainly coniferous forest in western Montana. The daily microclimate of each cell was estimated from ground and satellite data and interpolated using MT-CLIM.

ET and photosynthesis were calculated for each cell using a forest ecosystem simulation model. Across the landscape, LAI ranged from 4 to 15 m^2 m^{-2}, ET from 25 to 60 cm, and net photosynthesis from 9 to 20 Mg ha^{-1} yr^{-1}.

Over the last two decades there have been several attempts to develop gridded topoclimatic data sets (TCDs) that take account of factors such as elevation, potential for cold air drainage, and coastal influences operating on horizontal scales < 10 km. "Daymet" was developed by Thornton et al. (1997) and the Parameter-elevation Relationships on Independent Slopes Model (PRISM) by Daly et al. (2002, 2008). Both approaches use weather station data and a digital elevation model (DEM) to incorporate the effects of topoclimatic factors and statistically interpolate climate variables to a regular grid. Daymet takes account only of elevation, while PRISM has a sophisticated station-weighting scheme to take account of other topoclimatic factors. Oyler et al (2015) developed a statistical framework termed TopoWx ("Topography Weather") for modeling topoclimatic air temperature with a 30-arcsec (~800 m) resolution for the continental United States (CONUS). The daily minimum and maximum temperatures are derived for 1948–2012. The spatiotemporal interpolation procedures include geostatistical kriging, geographically weighted regression (GWR), and the application of remotely sensed land skin temperature taken from the MODIS 8-day, 1-km product as a spatial predictor of topoclimatic air temperature. Overall, the mean absolute errors for annual normal minimum and maximum temperature are 0.78 °C and 0.56 °C, respectively. There was a general tendency for Daymet and PRISM to show higher valley and lower mountain minimum temperatures than TopoWx.

D. Physically based Models

Shelterbelts

The effects of shelterbelts on agriculture were first recognized in the nineteenth century in Germany and subsequently in Russia and North America. The influence of windbreaks for agriculture was documented by Bates (1911) of the United States Department of Agriculture. Shelterbelts and their effects on wind flow and microclimate were described in detail by Caborn (1957, 1965) of the United Kingdom Forestry Commission, Gloyne (1956) and by Cleugh (1998), among many others. Gloyne showed that eddying flow in the lee of the barrier extends vertically to about twice the barrier height, *h*, and downwind for about 10–15 *h*. Undisturbed flow is restored at between 50 *h* and 100 *h*. The eddying region is present until the density of the barrier falls below 40 percent.

Wang and Takle (1995) developed a model to simulate flow fields in the vicinity of shelters, and Wang et al. (2001) subsequently applied this to shelters of varying porosity, shape, and orientation. The model simulates characteristics of all three zones of airflow passing over and through shelterbelts: (i) the zone of wind speed reduction on the windward side, (ii) the over-speeding zone above the shelterbelt, and

(iii) the leeward zone of wind speed reduction. Typically, the flow upwind below the barrier height begins to decrease at about 5 times the height of the barrier upwind. Some air penetrates the windbreak, but most of the air is displaced upward over the barrier and so its speed increases in a layer extending to 1.5 times the barrier height.

The dynamic pressure that results from the convergence/divergence of the flow field alters the perturbation pressure field. This disturbed pressure controls the formation of the separated flow, but also the location of maximum wind speed reduction, streamline curvature, speed-up over the shelterbelt, and leeward wind speed recovery rate. The flow patterns can be divided into two regimes: unseparated flow with shelter porosities of 0.40–0.99 and separated flow with shelter porosities of 0.06–0.30. The size of the separated recirculating eddy increases as the shelterbelt porosity decreases. For a shelterbelt with 0.50 porosity, the extent of downwind influence is about 20–25 times the barrier height. The shape of shelterbelts was found to have only minor effects on the wind speed reduction. The possible effect of shelterbelts in reducing evaporation of soil moisture and plant transpiration has been identified in some studies, but challenged in other work (Brenner et al., 1995). Wang et al. (2001) show that medium-dense shelterbelts have the maximum evapotranspiration-shelter efficiency. To couple multiple parallel shelterbelts optimally, a spacing of about 20 times the shelterbelt height is recommended.

In addition to field experiments, simulations have been performed in wind tunnels, where the flow conditions can be more readily controlled. Judd et al. (1996), for example, found that the "quiet zones" behind each windbreak are smaller in multiple arrays than single ones, as a result of the enhanced turbulence in the rough-wall internal boundary layer that develops over the multiple arrays. However, the overall shelter effectiveness is higher for multiple arrays than single windbreaks because of the "local shelter" induced by the array as a whole.

Barriers have other effects through shading, which may be beneficial to livestock in summer, but is usually unfavorable in winter and spring. North-south oriented barriers will of course have minimal shading effects. Shaded areas in midlatitudes will also retain snow drifts, which as well as cooling their vicinity may provide soil moisture in spring. Absolute and relative humidities also tend to be higher in the sheltered zone whereas evaporation is reduced, but the quantification of these effects is difficult.

Alterations of the microclimate downwind of a barrier not only include changes in the radiation and water balances, but also the redistribution of snow. After snow has fallen, its redistribution depends on the grain size of the snow and the horizontal wind speed. These two factors together determine the extent of snow drifting downwind of a barrier such as a hill, vegetation, or man-made structures. Analogous to the movement of sand grains by wind, studies of the movement of snow crystals (grains) by wind have shown that there are two primary transport mechanisms; saltation and suspension. Snow grains that are too large for the wind to lift them off the surface roll or creep along the top of the snow surface, becoming rounder as they roll, and form waves or dunelike formations as the snow migrates downwind. Smaller, lighter snow grains momentarily

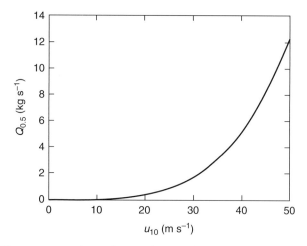

Figure 6.6. The transportation of snow in the first 5 m above the ground (Q_{0-5}; kg s^{-1} m^{-1}) increases exponentially with wind speed measured at a height of 10 m (u_{10}; m s^{-1}); $Q_{0-5} = u_{10}{}^{3.8}/233847$ (redrawn from Tabler 2002).

jump into the air in a parabolic arc in a process known as saltation. The smallest snow grains can remain suspended in the atmosphere and become smaller as they evaporate, thus remaining aloft longer and higher in the process known as suspension.

In all these processes, wind speed is a key determinant of snow transportation close to the ground (Figure 6.6); therefore, barriers, either structural or designed vegetation (living snow fences), have been effectively used to mitigate snow drifting effects. In terms of the standard wind speed measurement height of 10 m above the ground (u_{10}), saltation usually begins at a threshold u_{10} of 5–8 m s^{-1}, and the transition from saltation to suspension at a u_{10} of 7–11 m s^{-1} (variation due to the snow properties). Since blowing and drifting snow pose a significant transportation hazard, controlling snow transportation by decreasing wind speed through structural or living vegetation is an active area of applied research.

A combination of measured snow properties, either in situ or by remote sensing, and modeling can be used to design effective snow drift mitigation structures. The basic concept is to design a structure with enough porosity to allow the wind to pass through, yet reduce the wind speed sufficiently to reduce creep, saltation, and suspension so that the snow is deposited on the leeward side of the barrier. Improper design or placement, however, can result in creating more of a hazard than intended.

Snow properties that must be determined are the snow accumulation season and the snow water content (S_{we}). The snow accumulation season determines the period of drift growth and can be estimated simply from the number of days when the mean monthly air temperature remains below 0 °C (D). The potential for snow transport based on snowfall (Q_{spot}) is then estimated from the total S_{we} over the accumulation season and the maximum transportation distance (T_{max}), typically set equal to 3000 m (Tabler, 2002):

$$Q_{spot} = 0.5 T_{max} S_{we} \qquad\qquad 6.4$$

The S_{we} either is measured in situ using a device known as a snow pillow, where a pressure transducer measures the weight of the accumulating snow as the snow pushes on the pillow pad (these are used in the SNOwpack TELemtery sites operated by the Natural Resources Conservation Service in the western United States), or can be estimated over large spatial areas using remote sensing products such as aircraft-mounted gamma radiation sensors or satellite-mounted visible radiation sensors that can detect snow cover but still require ground truthing of S_{we}. Measurements of S_{we} from manually taken snow cores sampled along transect lines that are routinely revisited, known as a "snow course," often form the basis for verification of remotely sensed estimates of S_{we}.

The potential for snow transportation from wind speed (Q_{upot}) can be estimated from the length of the snow accumulation season, D (in number of days), and u in each wind direction (Tabler, 2002):

$$q_{i,j} = f_{i,j} D 86400 u_{i,j}^{3.8} / 233847 \qquad 6.5$$

where q (kilograms per meter) is the potential snow transport for each u class i for each wind direction class j, and f is the frequency of observations. Equation 6.5 provides Q_{upot} in each wind direction section, and the summation provides Q_{upot} for all wind directions.

Comparing Q_{upot} to Q_{spot} indicates whether snow transport is limited by wind ($Q_{upot} > Q_{spot}$) or snowfall ($Q_{spot} > Q_{upot}$), and then the snow fence can be designed appropriately. An example of the design of a living (vegetated) snow fence for drift control in Rocky Mountain National Park is provided by Blanken (2009). At a location in the park where drifting posed a chronic driving hazard, a living snow fence 1.61 m tall with a porosity of 50 percent, 56.4 m upwind of the road, would trap 79 percent of the blowing snow.

Heat and Water Balance Models

The simultaneous heat and water (SHAW) model was originally developed at the Northwest Watershed Research Center of the USDA Agricultural Research Service in Boise, Idaho, to simulate soil freezing and thawing (Flerchinger and Saxton, 1989). In its current version it simulates heat, water, and solute transfer within a one-dimensional profile, which includes the effects of plant cover, dead plant residue, and snow to a specified depth in the soil (Flerchinger, 2000).

Daily or hourly weather data of air temperature, wind speed, humidity, solar radiation, and precipitation at the upper boundary and soil conditions at the lower boundary are used to define heat and water fluxes into the system. Solar and longwave radiation exchange among the canopy layers, residue layers, and the snow or soil surface are computed by considering direct and upward and downward diffuse radiation being transmitted, reflected, and absorbed. The "leaves" can be oriented randomly or vertically (e.g., stubble). Transmission of direct and diffuse radiation and reflected and scattered radiation within residue layers is calculated, as are longwave transmission and absorption.

Sensible and latent heat fluxes are computed from temperature and vapor gradients between the canopy-residue-soil surface and the atmosphere. The resistance to convective heat transfer is determined. The ground heat flux is calculated from the residual of the energy balance.

Heat and vapor fluxes within the canopy are determined by computing transfers between layers of the canopy and considering the source terms for heat and transpiration from the leaves for each layer within the canopy. Gradient-driven transport (K-theory) is used for transfer within the canopy and from the canopy elements to the air space within canopy layers. Heat flux within the snow, residue, and soil is modeled. The equations for vapor flux are similar to those for heat flux. Precipitation and snowmelt are calculated at the end of each time step. Canopy interception and infiltration into the soil are also determined.

Kearney et al. (2014a) integrate a general microclimate model (Niche Mapper) with gridded continental-scale soil and weather data. A niche is the set of environmental factors that allows a species to exist in a given geographical region. Niche Mapper performs hourly calculations of solar and infrared radiation; aboveground profiles of air temperature, relative humidity, and wind velocity; and soil profiles of temperature, typically to 2 m depth. These values can be made as a function of vegetation shading, soil properties, thermal properties, and terrain factors. Kearney et al. (2014a) extend the microclimate model of the software package Niche Mapper to capture spatial and temporal variation in soil thermal properties and integrate it with gridded soil and weather data for Australia at 0·05° resolution. They demonstrate how hourly microclimates can be modeled mechanistically over decades at the continental scale with biologically suitable accuracy. When tested against historical observations of soil temperature, the microclimate model predicted 85 percent of the variation in hourly soil temperature over 10 years, from the surface to 1 m depth, with an accuracy of 2–3·3 °C (*ca.* 10 percent of the temperature range at a given depth) across an extremely climatically diverse range of sites. Kearney et al. (2014b) report hourly microclimate data for the globe.

ENVI-met is a three–dimensional microclimate model designed to simulate surface-plant-air interactions in an urban environment. The model simulates the flow around and between buildings, turbulence, the exchange properties of heat and vapor at the ground surface and at walls, the exchange at vegetation and vegetation parameters, bioclimatology, and pollutant dispersion.

It has a typical resolution of 0.5–10 m. The parameters treated in four domains – atmosphere, soil system, vegetation, and surface – are the following:

Atmosphere	Soil system	Vegetation	Surface
Wind	Temperature	Foliage temperature	Ground surface fluxes
Temperature	Water flux	Heat exchange	Fluxes at walls and roofs
Vapor	Water bodies	Vapor exchange	Heat transfer through walls
Pollutants		Water interception	
Turbulence		Water transport	

The model calculations include

- Shortwave and longwave radiation fluxes with respect to shading, reflection, and reradiation from building systems and the vegetation;
- Transpiration, evaporation, and sensible heat flux from the vegetation into the air, including simulation of all plant physical parameters (e.g., photosynthesis rate);
- Surface and wall temperature for each grid point and wall;
- Water and heat exchange inside the soil system;
- Biometeorological parameters such as mean radiant temperature;
- Dispersion of inert gases and particles including sedimentation of particles at leaves and surfaces.

An online manual of ENVI 3.1 is available at www.envi-met.com/. Huttner and Bruse (2009) provide a summary of the features of ENVI-met 4.0.

Of all the various components of the Earth's climate system, the cycling of carbon, water, and radiation in vegetation is one of the most difficult to measure and model. This is because as with most living objects, there is tremendous variety in vegetation structure, biology, phenology, and vegetation's responses to both biotic and abiotic variables. Therefore, the remote sensing and modeling of vegetation in climate models remain an active area of research.

Using the Ohm's law analogy (see Box 2.5), the concept of stomatal resistance (and the reciprocal, stomatal conductance) is common in nearly all land surface models to quantify the flux of CO_2 and H_2O through the leaf stomata. Stomata are small pores typically on the underside of leaf surfaces that expose the inside tissue to the atmosphere, allowing carbon to dissolve into water. Cells on each side of the stoma, guard cells, actively regulate the stomatal aperture by enlarging or decreasing their size through controlling their water solute potential in response to the ambient atmospheric and soil moisture conditions. As a consequence of direct exposure of water to the atmosphere in the stomata, water is lost to the atmosphere (transpiration). This represents the "transpiration dilemma" for the plant: how to maximize carbon uptake while minimizing water loss. The ratio of the mass of carbon gained to water lost through transpiration is the plant's water use efficiency (WUE). Most terrestrial plants (including trees) are vascular, meaning they have special tissues (xylem) to move water and the nutrients it contains through the plant. Plant such as mosses, fungi, and lichens are non-vascular and do not have stomata to regulate photosynthesis and transpiration actively.

To model a plant's carbon uptake (photosynthesis) and water loss (transpiration), first the primary class of photosynthesis must be known. There are three main classes of photosynthesis in terrestrial vascular vegetation; C3, C4, and CAM. By far, C3 is the most common photosynthetic pathway, used in all trees and most vascular plants in temperate regions (~85 percent), with the stomata open during the day and closed at night. C4 photosynthesis can be found in fast-growing plants (e.g., maize) and those found in arid environments (e.g., high elevation grasslands). C4 plants have specialized cells that recapture respired carbon before it reaches the atmosphere and, like C3 plants, have open stomata during the day and closed at night. Finally, CAM plants (crassulacean acid metabolism) such as pineapple are capable of closing their stomata during the day and opening them at night, resulting in a high WUE. The acid malate is formed and stored during the day, then used at night as an alternative energy source for photosynthesis.

Once the photosynthesis plant functional type is known (C3, C4, or CAM), then the response of the plant to ambient conditions must be quantified. Empirical functions that attempt to define vegetation's response to key atmospheric variables represent some of the first attempts. Using the ecological concept of limiting factors, Jarvis (1976) directly measured the stomatal conductance (g_s) of several species under a wide range of natural ambient conditions. His idea was that with adequate sampling, a boundary line (the uppermost limit) of plots of stomatal conductance versus important ambient conditions (PAR, T_a, D) would define the stomatal response under non-limiting conditions in the response to each variable. When points fell below the upper boundary line, some other factor was limiting, so if the maximum stomatal conductance for a given species was known, empirical functions could be used to lower the maximum stomatal conductance. This approach has been used successfully for many vegetation types and ecosystems worldwide, including wetlands in the Hudson Bay Lowland region of North America (Blanken and Rouse, 1995). The benefit of the Jarvis-type approach is that natural ambient conditions are the basis for measurements, not artificial laboratory conditions. Conversely, many laborious measurements are required to capture the full spectrum of conditions to define the boundary line properly, and the method is empirical, not offering insight into the true biophysical processes.

Moving forward from the Jarvis approach, other vegetation model efforts take more of a plant physiological approach. For example, the Ball-Berry stomatal conductance model (Collatz et al., 1991) uses the relationships among leaf-level net photosynthesis (A_n), CO_2 partial pressure at the leaf surface (c_s), atmospheric pressure (P), and relative humidity at the leaf surface (h_s):

$$g_s = m \frac{A_n}{c_s / P} h_s + b\beta_t \qquad 6.6$$

The variables m and b are empirically determined constants that vary by plant function group, and β_t represents a soil water stress function that ranges from 0 to 1. Although still semi-empirical, this relationship for C3, C4, and CAM plants is used to define better the combined stomatal response by plant functional type; the variables m and b vary with plant functional type ($m = 9$ and $b = 10,000$ for C3 plants; $m = 4$ and $b = 40,000$ for C4 plants; Sellers et al., 1996). This approach has been tested in many vegetation types (e.g., boreal aspen forest; Blanken and Black, 2004) and is used in many climate models, including the Simple Biosphere Model (Sellers et al., 1996), and in the land surface component of the community climate model (described below).

The models described previously are only a sample of the many available that use remotely sensed and/or in situ data to quantify the microclimate of various surfaces at various spatial and temporal scales. In an effort to provide improved communication and synergy among the various modeling efforts, various Community Climate Models (CCMs) have been developed that promote the open exchange of ideas in the climate modeling community. An example of such a model is the Community Land Model (CLM) (Oleson and Lawrence, 2013). Each grid cell in the CLM can have different land units, and each land unit can have a different number of columns and plant types (if vegetated). This "nested sub-grid hierarchy" approach allows a great deal of flexibility for the user to specify land use and land cover characteristics within each grid cell. In the current version of CLM (version 4.5), land units that can be specified are glacier, lake, urban, vegetated, and crops. Detailed characteristics within each unit can then be further specified: for example, the density in the urban areas, irrigated or non-irrigated crops, 15 possible plant functional types.

Similar to CLM, the Weather Research and Forecasting (WRF) model features a core fine-scale weather forecast model with components that can be coupled on- or off-line with the core WFR model (Janjic et al., 2014). The WRF modeling system is freely available and amenable to user-derived updates, bug fixes, and user support. As the name implies, WRF can be used for real-time weather forecasting on a global scale, or for research purposes. Currently, key components of the core WFR model are the Data Assimilation System (WRFDA), Atmospheric Chemistry Model (WRF-Chem), and Hydrological Modeling System (WRF-Hydro). The WRFDA is designed to combine, or assimilate, observations with a first guess (or background forecast) to provide an improved forecast. The WFR-Chem is designed to simulate chemistry and aerosols within the WRF model, for applications such as testing air pollution abatement strategies and predicting the impacts of aerosols on the radiation balance. The WFR-Hydro is designed to simulate the terrestrial hydrologic cycle, either independently of WRF (stand alone, or off-line version) or coupled to WRF.

E. Downscaling

Downscaling refers to procedures that enable model output available at large scales (50 km or more) to be extended to local scales (1–10 km) (Hewitson and Crane, 1996). There are two basic approaches – dynamic and statistical downscaling. Dynamic downscaling involves the use of high-resolution climate models that are run on regional subdomains as Regional Climate Models (RCMs) using observational data, or lower-resolution climate model output, as boundary conditions. RCMs have the advantage that they can provide a large suite of climate variables that are physically consistent, and they can generate regional-scale feedbacks. The disadvantage of dynamic downscaling is that it is computationally intensive. Statistical downscaling requires; (i) the development of statistical relationships between local climate variables and large-scale predictors such as pressure fields and (ii) the application of these relationships to the output of global climate model experiments to simulate local climate characteristics. For example, surface air temperature can be interpolated using a mean atmospheric temperature lapse rate based on surface and 700 hPa fields and the height difference between the model grid point and a high-resolution topographic grid.

An example of statistical downscaling to derive high-resolution precipitation data over the Tibetan Plateau is provided by Shi and Song (2015). They first used a downscaling method involving a machine-learning algorithm to estimate annual precipitation with six predictors: enhanced vegetation index, elevation, slope, aspect, latitude, and longitude. Second, they generated maps of annual precipitation over the Tibetan Plateau at 1-km spatial resolution by applying the downscaling procedure to annual TRMM $0.25 \times 0.25°$ data for 2001 to 2012, and then produced maps of monthly precipitation by disaggregating annual data using a simple fraction method. Finally, differences between the downscaled values and gauge data at 91 stations were computed and interpolated to apply adjustments.

Winkler et al. (2011) introduced a threefold classification of downscaling: dynamic downscaling, empirical-dynamic downscaling, and disaggregation approaches. "Empirical-dynamic" downscaling refers to approaches that empirically relate local or regional surface climate variables to large-scale airflow and other atmospheric state variables that represent relevant dynamic and physical atmospheric processes. Typically, GCMs best simulate airflow and atmospheric variables in the free air that represent large spatial scales. Downscaling here is performed for each climate variable and can result in inconsistencies among the variables. Disaggregation methods focus on the interpolation of a climate variable from a coarse-resolution field to either a fine-resolution grid or a specific location, or the inference of a finer time resolution from monthly or seasonal averages of a climate variable. Spatial disaggregation can be accomplished by developing linear regression models between the large-scale field and values at a particular location. This is usually done for monthly or seasonal averages. Another approach is to interpolate values in the large-scale field to smaller domains ($0.5° \times 0.5°$, $0.1° \times 0.1°$, etc.). The temporal disaggregation of monthly temperature or precipitation fields can be accomplished by using stochastic weather generators. These employ Markov processes to simulate wet/dry days and then estimate other variables conditioned on precipitation occurrence. Considerations for evaluating different downscaling options are detailed by Winkler et al. (2011).

References

Asner, G. P., Scurlock, J. M., and Hicke, J. E. 2003. Global synthesis of leaf area index observations: implications for ecological and remote sensing studies. *Global Ecol. Biogeog.* 12, 191–205.

Bateni, S. M. et al. 2014. Coupled estimation of surface heat fluxes and vegetation dynamics from remotely sensed land surface temperature and fraction of photosynthetically active radiation. *Water Resour. Res.* 50(11), 8420–40.

Bates, C. G. 1911. *Windbreaks, their influence and value.* Forest Service Bulletin, 86. Washington, DC: U.,S. Dept. of Agriculture.

Best, M. J, et al. 2011. The Joint UK Land Environment Simulator (JULES), model description. Part 1. Energy and water fluxes. *Geosci. Model Dev*. 4, 677–99.

Blanken, P. D. 2009. Designing a living snow fence for snow drift control. *Arctic, Antarct., Alp. Res*. 41(4), 418–25.

Blanken, P. D., and Black, T. A. 2004. The canopy conductance of a boreal aspen forest, Prince Albert National Park, Canada. *Hydrol. Process*. 18(9), 1561–78.

Blanken, P. D., and Rouse, W. R. 1995. Modelling evaporation from a high subarctic willow-birch forest. *Int. J. Climatol*. 15(1), 99–106.

Brenner, A. J., Jarvis, P. G., and van den Beldt, R. J. 1995. Windbreak-crop interactions in the Sahel. 2. Growth response of millet in shelter. *Agric. For. Met*. 75, 235–62.

Budyko, M. 1956. *The heat balance of the earth's surface*. Washington, DC: U.S. Weather Bureau.

Caborn, J. M. 1957. Shelterbelts and microclimate. Forestry Commission, Bulletin No. 29. Edinburgh: HMSO.

Caborn, J. M. 1965. *Shelterbelts and windbreaks*. London: Faber & Faber. 288 pp.

Cai, D-L. et al. 2014: Climate and vegetation: An ERA-Interim and GIMMS NDVI analysis. *J. Climate* 27, 5111–18.

Cleugh, H. A. 1998. Effects of windbreaks on airflow, microclimates and crop yields. *Agrofor. Systems* 41, 55–84.

Collatz, G. J., Ball, J. T., Grivet, C., and Berry, J. A. 1991. Physiological and environmental regulation of stomatal conductance, photosynthesis, and transpiration: A model that includes a laminar boundary layer. *Agric. For. Meteor*. 54, 107–36.

Courault, D., Seguin, B., and Olioso, A. 2005. Review on estimation of evapotranspiration from remote sensing data: from empirical to numerical modeling approaches. *Irrigation Drainage Systems* 19, 223–49.

Curran, P. J. 1983. Multispectral remote sensing for the estimation of green leaf area index. *Phil. Trans. Roy. Soc. London, A* 309, 257–70.

Daly, C. et al. 2002. A knowledge-based approach to the statistical mapping of climate. *Clim. Res*. 22, 99–113.

Daly, C. et al. 2008. Physiographically sensitive mapping of climatological temperature and precipitation across the conterminous United States. *Int. J. Climatol*. 28, 2031–64.

ENVI-met: www.envi-met.com/.

Fisher, J.B., Tu, K.P. and Baldocchi, D.D. 2008. Global estimates of the land–atmosphere water flux based on monthly AVHRR and ISLSCP-II data, validated at 16 FLUXNET sites. *Rem. Sens. Environ*., 112, 90119.

Flerchinger, G. N. 2000. The Simultaneous Heat and Water (SHAW) model: Technical documentation. Tech. Rep. NWRC 2000–09. Boise, ID: Northwest Watershed Research Center, USDA, Agricultural Research Service.

Flerchinger, G. N., and Saxton, K. E. 1989. Simultaneous heat and water model of a freezing snow-residue-soil system I. Theory and development. *Trans. ASAE* 32(2), 565–71.

Gloyne, R. W. 1956. Some effects of shelterbelts upon local and micro-climate. *Forestry* 27, 85–95.

Hewitson, B. C. and Crane, R. G. 1996. Climate downscaling: Techniques and application. *Clim. Res*. 7, 85–95.

Huang, N., Gu, L., and Niu, Z. 2014. Estimating soil respiration using spatial data products: A case study in a deciduous broadleaf forest in the Midwest USA. *J. Geophys. Res. Atmos*. 119. doi:10.1002/2013JD020515.

Hungerford, R. D. et al. 1989. MTCLIM: A mountain microclimate simulation model. Ogden, UT: Inter Mountain Research Station, Res. Pap. INT 414. U.S. Dept. of Agriculture, Forest Service.

Huttner, S., and Bruse, M. 2009. Numerical modeling of the urban climate – a preview on ENVI-met 4.0. *Seventh International Conference on Urban Climate, 29 June – 3 July 2009, Yokohama, Japan*.

Janjic, Z. et al. 2014. WRF-NMM Version 3 Modeling System User's Guide, April 2014.

Jarvis, P. G. (1976) The interpretation of the variations in leaf water potential and stomatal conductance found in canopies in the field. *Phil. Trans. R. Soc. Lond. B*. 273, 593–610.

Judd, M. J. et al. 1996. A wind tunnel study of turbulent flow around single and multiple windbreaks. Part 1. Velocity fields. *Boundary-layer Met.* 80, 127–65.

Kearney, M. R. et al. 2014a. Microclimate modelling at macro scales: A test of a general microclimate model integrated with gridded continental-scale soil and weather data. *Methods Ecol. Evolution* 5, 273–86.

Kearney, M. R. 2014b, Microclim: Global estimates of microclimate based on long-term monthly climate averages. www.readcube.com/articles/10.1038/sdata.2014.6.

Lagouarde, J-P. 1991. Use of NOAA-AVHRR data combined with an agrometeorological model for evaporation mapping. *Int. J. Rem. Sens.* 12, 1853–64.

Mu, Q-Zh., Zhao, M-Sh., and Running, S. W. 2011. Improvements to a MODIS global terrestrial evapotranspiration algorithm. *Remote Sens. Environ.* 115, 1781–1800.

Mu, Q-Zh. et al. 2009. Satellite assessment of land surface evapotranspiration for the pan-Arctic domain. *Water Resour. Res.* 45, W09420.

Nagler, P. et al. 2005. Evapotranspiration on western U.S. rivers estimated using the Enhanced Vegetation Index from MODIS and data from eddy covariance and Bowen ratio flux towers. *Remote Sens. Environ.* 95, 337–51.

National Research Council 1986. *Remote sensing of the biosphere.* Washington, DC: National Research Council, Committee on Planetary Biology.

Oleson, K. W., and Lawrence, D. M. 2013. Technical Description of version 4.5 of the Community Land Model (CLM). NCAR Technical Note, NCAR/TN-503+STR.

Oyler, J. W. et al. 2015. Creating a topoclimatic daily air temperature dataset for the conterminous United States using homogenized station data and remotely sensed land skin temperature. *Int. J. Climatol.* 35, 2258–79.

Pitman, A. 2003. The evolution of, and revolution in, land surface schemes designed for climate models. *Int. J. Climatol.* 23, 479–510.

Raupach, M. R., and. Finnigan, J. J. 1988. Single-layer models of evaporation from plant canopies are incorrect but useful, whereas multilayer models are correct but useless: Discuss. *Aust. J. Plant Physiol.* 15, 705–16.

Running, S.W., Nemani, R.R. and Hungerford, R.R. 1987. Extrapolation of synoptic meteorological data in mountainous terrain and its use for simulating forest evapotranspiration and photosynthesis. *Can. J. For. Res.*, 17, 472–83.

Sellers, P. J., Randall, D. A., Collatz, G. J., Berry, J. A., Field, C. B., Dazlich, D. A., Zhang, C., Collelo, G. D., and Bounoua, L. 1996. A revised land surface parameterization (SiB2) for atmospheric GCMs. Part I. Model formulation. *J. Climate* 9, 676–705.

Shi Y., and Song, L. 2015. Spatial downscaling of monthly TRMM precipitation based on EVI and other geospatial variables over the Tibetan Plateau from 2001 to 2012. *Mountain Res. Devel.* 35, 180–94.

Tabler, R. D. 2002. Design Guidelines for the Control of Blowing and Drifting Snow. National Cooperative Highway Research Program, Project 20–7(147). Washington, DC: Transportation Research Board of the National Academies. Tesemma, Z. K. et al. 2014. Leaf area index variation for crop, pasture, and tree in response to climatic variation in the Goulburn–Broken catchment, Australia. *J. Hydromet.* 15, 1592–1606. doi: http://dx.doi.org/10.1175/JHM-D-13–0108.1.

Tessema, Z. K. et al. 2014. Effect of year-to-year variability of leaf area index on variable infiltration capacity model performance and simulation of streamflow during drought *Hydrol. Earth Syst. Sci. Discuss.*, 11, 10, 515–552.

Thornton, P. E., Running, S. W., and White, M. A. 1997. Generating surfaces of daily meteorological variables over large regions of complex terrain. *J. Hydrol.* 190, 214–51.

Verger, A. et al. 2015. GEOCLIM: A global climatology of LAI, FAPAR, and FCOVER from VEGETATION observations for 1999–2010. *Remote Sens. Environ.* 177, 126–37.

Viterbo, P., and Beljaars, C. M. 1995. An improved land surface parameterization scheme in the ECMWF model and its validation. *J. Climate* 8, 2716–48.

Wang, H., and Takle, E. S. 1995. A numerical simulation of boundary-layer flows near shelterbelts. *Boundary Layer Met.* 75, 141–73.

Wang, H., Takle, E. S., and Shen, J. 2001: Shelterbelts and windbreaks: Mathematical modeling and computer simulation of turbulent flows. *Ann. Rev. Fluid Mech.* 33, 549–86.

Winkler, J., Guentchev, G.S., Perdinan, Tan, P., Zhong, S., Liszewski, M., Abraham, Z., Niedźwiedź, T. and Ustrnul, Z. 2011. Climate scenario development and applications for local/regional climate change pmpact assessments: An overview for the non-climate scientist. Geog. Compass, 5, 275-300.

7

Microclimates of Different Vegetated Environments

In this chapter we consider the primary characteristics of microclimates in different environmental settings. We examine in turn the landscapes of arctic and alpine tundra, grassland, farmland, wetlands, and forests. Small-scale atmospheric and edaphic processes that operate in the surface boundary layer are determined by these local environments, giving rise to contrasts in the radiation balance, temperature, humidity, and wind among the different surfaces and with height above and below the ground surface. It is these microclimates that provide the environmental controls for plants, insects, and other animals inhabiting that region (Monteith, 1976; Jones, 1992). Where appropriate data are available, we will include ecological information.

A. Tundra

Tundra is a term that originated in northern Finland, where it referred to treeless plateaus. Subsequently it has been extended to designate the land between the northern boundary of the boreal forest and high Arctic ice caps. It is also found on sub-Antarctic islands and in high alpine environments (Wielgolaski, 1997).

Tundra accounts for about 8 percent of the Earth's land surface. Microclimates are responsible for much of the diversity of the tundra environment. Low Sun angles mean that there is a great variety of local exposure to solar radiation. At the regional scale, the presence of a snow cover for nine months of the year has a major impact on the absorbed solar radiation.

Arctic Tundra

Lewis and Callaghan (1976) note that tundra embraces many different vegetation types ranging from dense willow scrub, through scrub heath and wet meadow, to sparse lichen and moss cushions. They selected wet meadow and dry fell field as extremes that were extensively studied during the International Biological Programme (IBP), 1964–1974. Peak LAI reaches unity in the meadow with a maximum canopy height of 20 cm. The full canopy develops within a few weeks of snowmelt. As little as 11 percent of solar insolation and 2–6 percent of PAR penetrate the full canopy. Dead leaves form an increasing portion of the canopy after midseason.

At fell field sites the LAI of higher plants is only about 0.1 m^2 m^{-2}. The canopy is discontinuous with large areas of bare ground. Wind is a major factor in fell field sites where the winter snow cover is discontinuous. The effects of wind exposure are manifest in snow redistribution, the erosion of soil and litter, mechanical effects on individual plants (injury due to ice-crystal abrasion, wind pruning), increased convection and evaporation, and the dispersal of propagules.

On the coasts of the Arctic Ocean in summer the prevalence of low-level stratus cloud offsets the effect of 24-hour sunlight. Single-layer cloud has a mean cloud top albedo of 55 percent, and multilayered cloud has an albedo of 65 percent. This is apparent in the results of Weller and Holmgren (1974) for the energy balance at Barrow, Alaska. For wet tundra with an albedo of 0.15, the absorbed radiation at the surface is 39 percent for single-layer cloud and 30 percent for multilayered cloud of the clear sky value. For 12 cm tall vegetation, the incoming radiation is reduced from 100 percent at the top of the canopy to 60 percent at 5 cm, 35 percent at 2 cm, and 15 percent at the soil surface.

During summer 1971 and 1972 at Barrow, air temperatures at a height of 16 m averaged 0.8 °C lower than at 5 m, and surface temperatures were 2.5 °C higher than at 0.5 m. Typical profiles of air and soil temperature through vegetation and snow are shown in Figure 7.1. Air temperatures within the vegetation canopy are 2–3 °C higher than in the overlying atmosphere. The soil thaws to a maximum depth of 40 cm. The winter snow cover insulates the ground surface, which is 10–15 °C higher than the temperature of the snow surface. In early spring the snow surface is 5–10 °C higher than the underlying ground surface.

The components of the energy balance for six stages of the year are shown in Table 7.1. The stages are pre-melt, snow melting, post-melting, midsummer, freeze-up, and midwinter. The heat input into the atmosphere increased by an order of magnitude from early spring to summer. The post-melt latent heat flux corresponds to an evaporation rate of 4.6 mm d^{-1}.

At 83 °N in Peary Land, northern Greenland, Mølgaard (1982) reported that slope aspect has major impacts on diurnal temperature variation. On a 20° north-facing slope, the diurnal variation is the opposite of that on level ground or south-facing slopes, and soil surface temperatures may be 10 °C higher on the former. However, the diurnal variation at the soil surface was about 10 °C on the north-facing slope and 18 °C on the south-facing one. Diurnal variations of plant and soil temperatures are much greater than those in the air. Under sunny conditions, plant temperatures may be 20–25 °C compared with 5 °C at 2 m in the air.

On a polar semi-desert site at King Christian Island (77.7 °N, 101.2 °W), Addison and Bliss (1980) determined summer conditions in 1973 and 1974. For July and August the mean air temperature was 2.4 °C and the cloud cover was 84 percent. The global radiation (0.28–2.8 μm) for July 15 to August 14, 1973, was

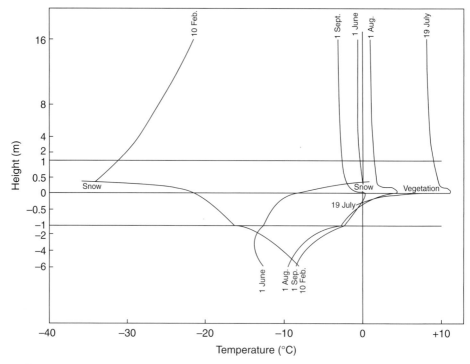

Figure 7.1. Typical temperature profiles at Barrow, Alaska, in air, vegetation, snow, and soil. Note the expanded vertical scale between −1 and +1 m (Weller and Holmgren, 1974).
Source: *Journal of Applied Meteorology*, 13: p. 860, fig. 6. Courtesy American Meteorological Society.

Table 7.1. Energy Balance at Barrow, Alaska, for Six Stages of the Year 1971–72 (Daily Averages in Watts Per Square Meter)

Date	R_n	G	H	λE
3/29–06/3	18	11	3	3
06/4–13	67	15	13	40
06/14–17	184	16	34	134
07/15–28	115	2	37	76
08/25–09/1	27	0	11	37
02/5–10/72	17	−2	−14	−1

Source: After Weller and Holmgren, 1974.

1.25×10^{-7} MJ m^{-2} day^{-1} and the net radiation 0.89×10^{-7} MJ m^{-2} day^{-1}, 71 percent of the global amount. At moss, lichen, and bare soil microsites, representing 18, 40, and 33 percent of the surface, respectively, the energy budget components were as follows:

Surface	λE (percentage of R_n)	H	G
Moss	17	75	8
Lichen	28	58	14
Bare soil	29	57	14

Hence, the differences in energy dissipation between the sites are relatively small. Surface temperatures were 0.8–6.2 °C higher than at 1.5 m with a mean difference of only 1.8 °C. The active layer reached a depth of about 50 cm.

For the subpolar Ural mountains of eastern Russia, Dymov et al. (2013) have characterized the microclimatic regime (temperature of organogenic soil horizons and air temperature) in dominant landscapes of mountain-tundra and mountain forest belts beginning in 2010. Litter temperature largely followed the air temperature, but soil temperatures depended on landscape position and vegetation belt location. Disposition and altitude of plots play major roles in the microclimatic characteristic. The minimum temperatures of soil organogenic layers located at 600–730 m above sea level (a.s.l.) are −18 °C, but minimum soil temperature at similar organogenic horizons at soils located at 400–450 a.s.l. are −7 to −10 °C. Maximum average daily temperatures during the summer season in tundra soils vary from +7 up to 21 °C depending on landscape position. The temperature of organogenic horizons in the mountain-forest belt is somewhat higher than in tundra soils – the minimum value in the winter season is −4 °C. The maximum average daily temperature of litter in forest soils in summer is 24 °C. Forest cover influences the diurnal temperature amplitude. Daily temperature amplitudes during the summer season in forest soils are 10–12 °C, while the daily amplitudes in mountain-tundra soils are 17–20 °C. The average daily temperature below zero observed in mountain-tundra soils averaged 200–240 days per year, and in mountain-forest soils 145–208 days per year.

At an upland tundra site and a black spruce forest site near Churchill, Manitoba (58.7 °N), Rouse (1984a, b) measured the radiation budget and temperature conditions during 1979. The solar and net radiation conditions are shown in Table 7.2. The tundra with a snow free albedo of 0.15 absorbs about half of the incoming solar radiation, whereas the forest with an albedo of 0.12 absorbs about two-thirds.

Tundra soils are somewhat warmer in summer and much cooler in winter than their forest counterparts (Rouse, 1984b). Figure 7.2 illustrates the seasonal course of soil temperatures at the two sites. The maximum depth of the active layer is 2.1 m in the tundra and 1.9 m in the forest. In late winter 1982 (late April–late May) the open forest canopy (6 m tall black spruce) averaged 11.4 °C compared with an air temperature of 4.8 °C and a snow surface temperature of −1.2 °C.

The forest soils are much wetter than those at the tundra site as a result of the large input of snow melt water. The tundra snow cover is much thinner. Heat storage in the

Table 7.2. Solar and Net Radiation (Megajoules Per Square Meter for the Period Indicated) over Tundra and Spruce Forest Near Churchill

Surface Type	Period	Incoming Solar	R_n Tundra	R_n Forest
Snow cover	05/16–23	192.5	47.1	108.6
Snow free tundra	05/14–06/19	526.0	344.3	359.8
Snow free	06/20–09/05	1366.2	751.6	836.7
	05/16–09/15	2085.1	1142.9	1305.1

Source: From Rouse, 1984a.

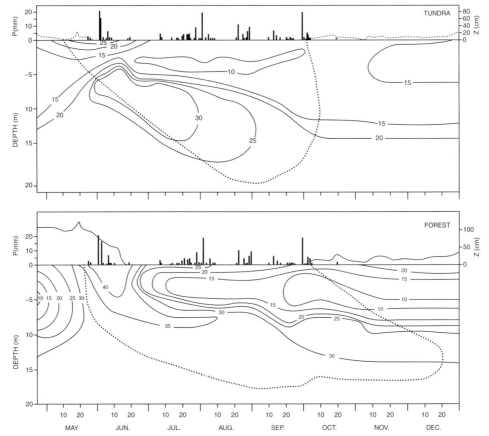

Figure 7.2. Time trends of soil temperature at tundra and forest sites near Churchill, Manitoba. The heavy dotted line denotes the position of the frost table. The soil moisture isopleths are in percentage volume. The vertical bars show the precipitation. The upper graph is a plot of the air temperature (W. R. Rouse 1984).
Source: *Water Resources Research*, 20, p. 70, fig. 3. Courtesy of the American Geophysical Union.

Table 7.3. Heat Storage (Megajoules Per Squre Meter for the Given Periods) in Tundra and Forest Soils at Churchill

Surface	Exchange	Days	H	λE	G	R_n
Tundra	Warming	123	26.1	174.4	200.5	1077.1
	Cooling	101	−31.3	−145.8	−177.2	82.4
	Both	224	−5.2	28.6	23.3	994.7
Forest	Warming	131	19.6	183.1	202.7	1228.8
	Cooling	109	−22.0	−128.9	−150.9	−
	Both	240	−2.4	54.2	51.8	−

Source: From Rouse, 1984b.

soil as sensible, and latent heat is shown in Table 7.3. Some 80–90 percent of the heat storage is in the form of latent heat.

Alpine Tundra

Tundra in alpine regions shows similar characteristics to arctic tundra, except that its characteristics are derived from the high elevation; thus alpine tundra exists on every continent. Alpine tundra is typically more isolated and exists in patches compared to its arctic counterpart; the complex high-elevation terrain in which it is situated results in a more complex microclimate with large spatial gradients in soils and vegetation.

The tree line denotes the lower boundary of alpine tundra, and this is more of a gradual transition between the subalpine forest and the tundra, so is often referred to as the forest-tundra ecotone. This boundary, marking the edge of tolerance for tree species, exhibits an abrupt transition between the shelter of subalpine forests and alpine tundra (Figure 7.3). This transition from trees to shrubs and short-statured vegetation is primarily a result of the high wind speeds and low temperatures in the alpine tundra.

High and persistent wind speeds are characteristic of alpine tundra as a result of the high elevation and lack of tall vegetation to reduce wind speed by absorbing momentum. Until recently, the highest measured wind speeds on Earth have been recorded in mountain tundra environments, 372 km hr^{-1}, on Mount Washington, New Hampshire. Near Niwot Ridge in Colorado, 24-hr average wind speeds in alpine tundra (13 m s^{-1}) were roughly double those observed above a subalpine forest (7 m s^{-1}) located only 4 km away during the winter months (Blanken et al., 2009). During the summer, wind speeds were similar at both sites, but still relatively high (4 and 3 m s^{-1} 24-hr average, alpine tundra and subalpine forest, respectively). The steep topography in which alpine tundra is located strongly influences not only wind speeds, but wind direction. Both thermal and katabatic flows are common. For example, downslope katabatic winds occurred throughout the winter months at both the alpine tundra and subalpine forest sites on Niwot Ridge, but in summer, thermally driven upslope flows during

Figure 7.3. The alpine tundra ecotone is the region above the dark green subalpine forest vegetation (background) and the barren summits (foreground), as viewed from Torreys Peak, Colorado, August 10, 2014.
Source: Photograph provided by P.D. Blanken.

the daytime were common at the forest site but did not reach the alpine tundra site (Blanken et al., 2009).

These persistent high wind speeds and wind directions have a profound microclimatological effect on alpine vegetation. As shown previously, the momentum force, or shear stress (τ), created by wind is given by

$$\tau = K_\mathrm{m} \rho_\mathrm{a} du / dz \qquad\qquad 7.1$$

where K_m is the eddy viscosity, ρ_a is air density, and du/dz is the change in the horizontal wind speed (u) with height (z). The drag force (F_d) can be calculated as

$$F_\mathrm{d} = \frac{1}{2} \rho_\mathrm{a} u^2 A C_d \qquad\qquad 7.2$$

where A is the cross-sectional area and C_d is the drag coefficient. Unlike solid surfaces, vegetation is aeroelastic and able to bend and become more streamlined as wind increases; thus C_d is dependent on u (Blanken et al., 2003). In addition to this dependence, differences in species type, age, stand density, soil type, depth, and moisture, all make it difficult to determine the threshold u for F_d to exceed the vegetation's limits and break or blow over. For example, a 50-year-old Sitka spruce stand is predicted to break at sustained wind speeds of 32 m s^{-1} and overturn at wind speeds of 28 m s^{-1} (Gardiner and Quine, 2000).

In alpine tundra, the high sustained wind speeds with strong gusts and thin soils limit the capacity for tall vegetation growth. Trees that do exist at the upper elevation

Figure 7.4. A krummholz, or "tree island," located at the upper limit of the forest-tundra ecotone on Quandary Peak, Colorado. The dense, green vegetation is located beneath the winter snow cover, and the desiccated "flag trees" protrude above with branches predominant on the leeward side.
Source: Photograph provided by P.D. Blanken.

limit of the forest-tundra ecotone display a stunted, island-like morphology, known as krummholz, from the German word *krum* meaning twisted or bent. Most of the branches are located beneath the snow where they are protected from air temperature extremes and sheltered from the winds and abrasive ice pellets that can damage tissue, resulting in a "flag tree" with above-snow branches prevalent on the leeward side (Figure 7.4).

Alpine vegetation is short to avoid not only the forces created by high wind speeds, but the temperature and humidity effects associated with high wind speeds. First, air temperatures are already low at high elevations because of the decrease in temperature with height following the dry adiabatic lapse rate of 9.7 °C km^{-1}. High wind speeds generate enough turbulence to mix the air near the surface, thus eliminating any temperature gradients that may exist. As a result, the atmosphere often has a neutral stability profile, and therefore there is negligible sensible heat exchange between the surface and atmosphere during very windy conditions.

Low air temperatures in alpine regions also have a direct impact on humidity through the saturation vapor pressure–temperature relationship. As shown in Figure 2.8, there is an exponential relationship between air temperature and the maximum, or saturation, vapor pressure (e_s). At low temperatures, the saturation vapor pressure is low, and alpine regions tend not to have very much available liquid water because of sub-zero temperatures, which act to keep the ambient vapor pressure (e_a) low. Overall, the difference between e_s and e_a, the vapor pressure deficit ($D = e_s - e_a$) tends to be high in alpine regions.

The dry and cold atmosphere has a direct effect on alpine vegetation by reducing the stomatal conductance, since stoma close when D gets too large. The reduced stomatal conductance limits both the photosynthetic uptake of carbon and water release through transpiration. The low stomatal conductance together with the small leaf area index of alpine vegetation (short statured plants with small, succulent leaves) means that alpine ecosystems tend to have low biological productivity.

The surface energy balance is also directly impacted by the dry, cold, windy alpine conditions. The 24-hr average evaporative fraction, $E_F = \lambda E/(\lambda E + H)$ for alpine tundra near Niwot Ridge, Colorado, was only 0.39, a value more typical of dry grasslands and rangelands (Knowles et al., 2012). The sensible heat flux tended to be small because of the lack of air temperature gradients in the windy atmosphere, and the latent heat flux was also small as a result of the dry conditions. The soil heat flux, however, was large compared to other ecosystems because of the sparse vegetation and bare, rocky surface.

Most of the precipitation that falls in alpine regions is in the form of snow, and the characteristic high winds, low temperatures, and rugged terrain create additional unique microclimate environments due to drifting snow. With the moisture-limited, cold alpine environment, trapped snow has the benefits of insulating plants and soils from the extreme fluctuations in air temperatures, providing shelter from the wind, and contributing a source of water upon melting. Any object, such as vegetation, that can effectively reduce the wind speed below the snow's deposition velocity may result in a snowdrift. There is probably a positive feedback between this snow trapment and vegetation growth (Figure 7.5), which likely plays an important role in the long-term movement of the alpine tree line.

B. Grassland

About 40 percent of the Earth's land surface is under grassland. There are several types, mainly differentiated by latitude. Tropical and subtropical grasslands include numerous savannas in Africa, and the llanos of northern South America. Temperate grasslands include the prairies of North America, the steppes of Eurasia, and the pampas of Argentina, Brazil, and Uruguay.

Grasslands span a wide latitudinal range from about 25° to 55° latitude in the Northern Hemisphere with mean annual air temperatures ranging from 0 °C in the north to 20 °C in the south (Ripley and Redmarsh, 1976). Mean annual precipitation increases from west to east in North America from about 300 mm near the Rocky Mountains to 1000 mm near the Great Lakes. The contribution of snowfall increases northward. The snow cover insulates the ground and prevents winter soil temperatures from dropping below about −10 °C (Ripley, 1972). Much of the snowmelt enters the soil in spring.

Figure 7.5. Snow drifts created behind emerging vegetation on Niwot Ridge, Colorado, provides important insulation from extreme temperature fluctuations and wind and provides moisture.
Source: Photograph provided by P.D. Blanken.

Typical profiles of net radiation, temperature, wind speed, and vapor content in a grass canopy during the day are shown in Figure 7.6. Net radiation is highest at the top of the canopy, while wind speed continues to increase in the free air. Moisture is highest at the surface as a result of evaporation from the soil. The temperature is highest within the canopy (Ripley et al., 1996). Over a grassland surface in western North Dakota, the diurnal range on a hot, clear summer day increased from 25 °C above the canopy to 43 °C just above the soil surface (Whitman, 1969). Nighttime profiles for radiation, temperature, and in most cases humidity are the mirror image of those by day,

Studies were undertaken in native prairie at Matador, Saskatchewan, as part of the International Biological Programme during 1968–1972 (Ripley and Redmarsh, 1976). Solar radiation levels in summer were high because of the generally clear skies and low specific humidity. Mean incoming solar radiation in July averaged about 25 MJ m^{-2} day^{-1} and net all-wave radiation about 13 MJ m^{-2} day^{-1}. Within the 25-cm high canopy the extinction coefficient for solar radiation had a minimum value of ~ 0.5 around noon. The leaves of many grasses at Matador have the ability to roll up in response to changes in turgor and plant water status. This has the effect of reducing the leaf area exposed to radiation. In late July 1970, R_n averaged ~ 150 W m^{-2}, λE accounted for ~100 W m^{-2}, and H ~ 40 W m^{-2}.

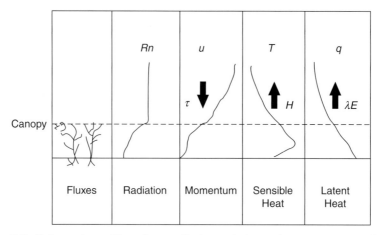

Figure 7.6. Schematic profiles of net radiation, wind speed, temperature, and vapor content in a grass canopy (after Ripley et al., 1996).

Canopy temperature profiles for June 28, 1972, showed a diurnal amplitude at mid-canopy of 25 °C, compared with 1.5 °C at 20 cm in the soil. The warmest section of the canopy was at 7 cm at 10:00 moving down to 2 cm at 14:00 as a result of the deeper penetration of sunlight. The vapor pressure on the same day was 10–12 hPa at 30 cm but reached a maximum at the surface throughout the day that was up to 9 hPa larger.

In an ungrazed northern grassland near Lethbridge, Alberta, Wever et al. (2002) monitored evapotranspiration (ET) during 1998–2000. Most of the water for ET is supplied by summer rainfall. In the much wetter than average growing season of 1998, peak ET was 4.5 mm day^{-1}, compared with 3 mm day^{-1} in the average year of 1999 and the dry year of 2000. ET in this grassland was strongly controlled by surface conductance.

In the southern Harz Mountains of Germany, dry grasslands at 12 sites at equal altitudes along a steep west-east precipitation gradient were compared (Bruelheide and Jandt, 2007). Macro- and microclimatic conditions with respect to temperature were quite homogeneous, but floristic composition reflected the precipitation gradient.

On chalk grasslands in southern England, Bennie et al. (2008) estimate cumulative air temperatures above 5 °C (as a measure of growing season) and 30 °C (important for thermophilic plants and invertebrate) and potential evaporation across complex terrain using 5-m horizontal resolution digital elevation models (DEMs) to calculate solar radiation. Canopy temperature (T_c) of a 12-cm high *Festuca sward* is determined empirically from observations as a function of air temperature (T_a) and net radiation (R_n) scaled by 2-m wind speed (u_2):

$$T_c = T_a + aR_n \qquad\qquad 7.3$$

where $a = 0.022$ if $u_2 \leq 1$ m s^{-1} and 0.013 if $u_2 > 1$ m s^{-1}. These relationships will change with vegetation canopy structure that will alter LAI, foliar projected cover, and roughness length. Potential evaporation is calculated by a version of the Penman-Monteith equation. Soil moisture was observed at 50-cm depth and modeled from the calculated evaporation and a four-layer soil model. In the shallow chalk soil, rapid vertical drainage predominates. Under direct sunlight, south-facing sward temperatures can vary by 29 °C over 24 hours and can exceed air temperatures by 14 °C.

An intensive study of grassland near Manhattan, Kansas, was undertaken under the First ISLSCP Field Experiment (FIFE), 1987–1989, by Betts and Ball (1998). Surface flux measurements were made at between 10 and 22 sites. Soil moisture was measured by gravimetry for the near-surface layer 0–10 cm and using neutron probes to depths of up to 2 m for volumetric soil moisture. Daily averages were calculated for the FIFE site. Clear sky values of solar radiation ($S\downarrow$) ranged from a daily mean of 350 W m^{-2} in midsummer to a little above 100 W m^{-2} in midwinter. On rainy and cloudy days, $S\downarrow$ falls by about 60 percent. The mean albedo is 0.18. The corresponding values of daily average R_n are 200 W m^{-2} in midsummer to 25 W m^{-2} in midwinter. Peak noontime values of R_n reach 550–580 W m^{-2}. The ground heat flux has a peak daytime ratio in May 1988 of 0.15 to R_n with a minimum in September of 0.08. In 1987 the peak ratio was 0.18 in June. The ratio of the evaporative fraction $E_F = \lambda E/(\lambda E + H)$ shows a clear dependence on volumetric soil moisture (θ). For $\theta < 18.5$ percent, E_F is a little more than 50, whereas for $\theta > 29$ percent E_F is 77.

In the grasslands of the Transdanubian Middle Mountains, Bauer and Kenyeres (2007) sampled 84 sites during June to September, 2000–05. These included 28 hayfields, 12 steppe grassland, 12 semi-dry grassland, 11 rush fen, and 11 rocky grassland. Temperature and humidity were observed on calm, clear days from 10:00 to 16:00 hours. The vegetation was typically < 60 cm tall and the measurements were made at 120 cm height. *Orthopteran* (grasshopper) samples were collected at 54 sites by sweep netting four times during the season. In semi-dry grasslands there were significant positive correlations between the number of *orthopteran* species and surface temperature and humidity, and humidity at 10 and 20 cm. Humidity variations were also important in rush fens and drying fens. There was an absence of correlations for the hayfields.

C. Farmland

The areas of cropland, pasture, and irrigated land are shown in Box 7.1.

Box 7.1. Cropland, Pasture, and Irrigated Land

The global extent of cropland in 2000 amounted to 10.6 percent of the global land surface and of pasture to 24.3 percent (HYDE, 2010). Ramankutty et al. (2008) estimated 12 percent (15.1 million km^2) and 22 percent (28 million km^2), respectively.

Box 7.1 (*cont.*)

A world map of cropland in 2000 is available at
  ftp://ftp.pbl.nl/../hyde/supplementary/land_use/crop2000ad.tif
and of pasture in 2000 at
  ftp://ftp.pbl.nl/../hyde/supplementary/land_use/gras2000ad.tif.
  Fritz et al. (2015) provide a new 1 km global cropland map for 2005 and a map of global field sizes.

A global Historical Irrigation Dataset (HID) has been developed by Siebert et al, (2015). It provides estimates of the Area Equipped for Irrigation (AEI) between 1900 and 2005 at 5 arc-minute resolution (~ 85 km at the equator). Subnational irrigation statistics were collected from various sources and it was found that the global extent of AEI increased from 63 million ha (M ha) in 1900 to 112 M ha in 1950 and 306 M ha in 2005.

Arable land (annual and perennial crops) accounts for nearly 11 percent of the Earth's land surface. Obviously the category of farmland refers to a very wide range of crops. The types of problem that have been investigated are similarly diverse. Crop specific studies are relatively numerous (rice, corn, wheat, soybeans, and so on). There has also been research on edge effects, hedgerows, and shelterbelts (Caborn, 1957), for example.

An early study of the microclimate of a potato crop at Rothamstead was reported by Broadbent (1950). The rows were 70 cm apart and the plants 40 cm apart. The soil was clay with flints. Air temperatures at a height of 15 cm within the crop were compared with screen temperatures during 1947–49. In dry sunny weather, maximum temperatures in the crop were up to 7 °C higher than in the screen and over 11 weeks in 1946 averaged 3.3 °C higher, while crop minima were about 1 °C lower. The average daily temperature range was 4.4 °C greater than in the screen. Profile measurements at 10. 20, 30, and 60 cm were made in 1948 and 1949. On sunny days, it was hottest at 10 cm in an open crop and at 30 cm in a dense crop. In both cases it was usually coolest at 60 cm when the crop was dry, but over wet soil the lowest temperature was at 10 cm. An inversion developed in the crop during clear conditions before sunset. Humidity was generally highest at 10 cm, but in a dense crop over dry soil, transpiration from the leaves caused the air at 30 cm to be more humid.

Sugar beet and potatoes are considered together by Brown (1976) in view of their comparable characteristics of canopy height and row spacing. Typical temperature profiles in 60 cm tall sugar beet on a clear summer day at Scotts bluff, Nebraska, show lapse conditions from 10:00 to 13:00 hours, from 13:00 to 15:00 the temperature profile went through a neutral state, and subsequently it became an inversion. Vapor pressure gradients increased during the day and were steepest (0.064 hPa cm^{-1}) with inversion conditions.

Sugar beet and potatoes have stomata on both surfaces of the leaf. The ratio of the number of stomata on the upper and lower surfaces is about 5:8 for sugar beet and

1:3 for potatoes (Burrows, 1969). Both crops are sensitive to decreasing soil water supply. A potato crop can exhibit water stress when soil water potential drops below −0.25 bar. Burrows (1969) showed that the decrease of the rate of evaporation to potential evaporation with soil water deficit for potatoes occurred at smaller deficits than for sugar beets. For a relative evaporation of 0.7, the soil water deficit was 14 cm for potatoes compared with 20 cm for sugar beet.

For mid-August 1966 the average daytime (06:00–18:00) energy balance components above an irrigated sugar beet crop in the Great Plains were approximately as follows: solar radiation 27, net radiation 18, λE 15, H 2, and G 0.5 MJ m^{-2} (Brown and Rosenberg, 1971).

Paddy rice occupies about 163 million ha (1.4 percent) of the global land surface. It grows to 1.0–1.8 m tall. For rice fields, the roughness length (z_0) and zero-plane displacement (d) can be approximated by the canopy height h (Uchijima, 1976):

$$z_0 = 0.062 \, h^{1.08}$$

$$d = 1.04 \, h^{0.88}$$

However, the variation of z_0 and d with wind velocity is irregular because of the plants' waving in the wind.

The heat balance of a rice field in Japan for July through September was as follows (Uchijima, 1976): R_n = 921.8; λE = 725.7; H = 196.1 MJ m^{-2}. The Bowen ratio was 0.27.

Canopy profiles show that upward λE flux is mainly between $0.5h$ and $0.8h$. The flux of H is upward above about $0.4h$, but is downward below this level and is used for evaporation from the water surface.

Denmead (1976) provides an overview of temperate cereal crops, paying particular attention to water relationships. He notes that there are large resistances to water transport in the roots, stems, and leaves of wheat crops. Irradiance exerted strong control on stomatal opening in leaves until turgor water potential dropped to ~ 8 bars, when stomata began to close. The critical water potentials at which stomata began to close were −19, −13, and −7 bars for top, middle, and low leaves on the stem, respectively. Closure began first in the lower leaves. Denmead reports different modes of transpiration control in wheat depending on the source of the water stress. When there was excessive evaporation demand, stomata adjusted to keep leaf water potential at their critical levels. The fastest rates of soil evaporation plus transpiration in wheat are about 8.4 kg m^{-2} day^{-1} (Fritschen, 1966). When there is insufficient soil water supply, the feedback control is ineffective. Leaf water conductance then depends on irradiance rather then leaf water status (Denmead, 1976).

The water and energy cycle of winter wheat on the Loess Plateau of Northwest China was studied by Zhang et al. (2014) during the period April 8–July 15, 2006. The upper 40 cm of the soil is 78 percent silt. Latent heat flux was the main consumer of available energy during the period. The daily energy partitioning of λE, H, and G averaged 45, 34, and 10 percent of R_n, respectively, and the Bowen ratio is estimated at 0.90. The latent heat flux was primarily controlled by precipitation, with several high latent heat flux days occurring after intensive rain events, whereas the sensible heat flux was mainly controlled by net radiation.

The case of λE from sparse crops has received much attention from agronomists as well as theoreticians. Shuttleworth and Wallace (1985) developed one-dimensional equations for the transition between evaporation from a bare soil and a complete vegetation canopy, which incorporates a bulk stomatal resistance as well as a soil surface resistance. The equation for total crop evaporation is

$$\lambda E = C_c \, PM_c + C_s \, PM_s \qquad\qquad 7.4$$

where PM_c and PM_s are expressions like the Penman-Monteith combination equation for evaporation from a closed canopy and bare substrate, respectively. The form of PM_c is

$$PM_c = \frac{\Delta A + (\rho_a c_p D - \Delta r_a^c A_s)/\left(r_a^a + r_a^c\right)}{\Delta + \gamma[1 + r_s^s /\left(r_a^a + r_a^c\right)]} \qquad\qquad 7.5$$

The coefficient $C_c = \{1 + R_c \, R_a / R_s \, (R_c + R_a)\}^{-1}$ \qquad 7.6

where

A_s = total energy flux (H and λE) leaving the crop and substrate,
D = vapor pressure deficit at the reference height,
r_a^a = aerodynamic resistance between canopy source height and reference level,
r_a^c = bulk boundary layer resistance of the canopy,
r_s^s = surface resistance of the substrate,
$R_a = (\Delta + \gamma) \, r_a^a$,
$R_s = (\Delta + \gamma) \, r_a^s + \gamma r_s^s$,
$R_c = (\Delta + \gamma) \, r_a^c + \gamma r_s^c$,

where r_a^s = aerodynamic resistance between the substrate and canopy source height, and

r_s^c = bulk stomatal resistance of the canopy.

The equations for PM_s and C_s are analogous.

They note that the mean boundary layer resistance of the canopy and its bulk stomatal resistance are both influenced by the surface area of the vegetation and vary inversely with the total leaf area of the vegetation (the LAI). The model shows limited sensitivity to the aerodynamic resistance values chosen. Also, there is very low sensitivity to the extinction coefficient for net radiation entering the canopy.

Study of a cotton row crop was conducted in 1989 at the Texas Agricultural Experiment Station (Ham et al., 1991). The rows had a north-south orientation and 1-m spacing. The crop was flood irrigated throughout the growing season. The energy balance was determined by the Bowen ratio method. λE_c from the canopy was obtained by sap flow measurements of transpiration and the soil λE_s flux was obtained from the difference between total λE and λE_c. Daily energy balance was strongly

Table 7.4. Energy Components of a Sorghum Crop in
Kansas during Water Stress (MJ m^{-2})

Day	Rn	λE	H	G
1	17.2	13.8	2.5	0.9
10	15.3	18.5	−4.4	1.1

influenced by the transport of H. When the soil surface was dry, the canopy absorbed H from the soil and above-canopy air, accounting respectively for 21 and 12 percent of λE_c. After irrigation, λE_c accounted for more than 50 percent of λE, even when the LAI was >2 m^2 m^{-2}. Soil evaporation was the primary mode of latent heat flux when the soil was wet. When the soil was dry, canopy evaporation increased and the total λE, therefore, changed little between wet and dry soil conditions. Between 12 and 21 percent of λE occurred at night.

Study of a sorghum crop (LAI = 3 m^2 m^{-2}) in Kansas was undertaken by Kanemasu and Arkin (1974). The canopy underwent water stress for 10 days in 1972, during which time the leaf-water potential decrease from −12 bar to −18 bars and the plants wilted. Table 7.4 shows the energy components on day 1 and day 10.

Evaporation is increased over the period and the flux of sensible heat is reversed. The sorghum had the ability to extract soil water and keep pace with the energy supply.

Crop residue or mulch effects on surface energy and water balance and soil climate have received considerable attention. A review is provided by Horton et al. (1996). Daytime soil temperatures in summer are much lower under mulch (7–10 °C) than bare soil and diurnal amplitudes are reduced. However, soil water dominates the surface energy exchanges and it is only when soils dry that soil temperature effects become apparent. In temperate latitudes in spring mulch can have negative effects on soil conditions by keeping soils cold and wet. Under wet conditions most of the available energy is used in evaporation.

D. Wetlands

Wetlands are found on all continents, especially in flat areas with poorly drained soils (e.g., permafrost areas). They cover about 5.3 million km^2 or nearly 4 percent of the total land surface, occurring primarily in the cool temperate, boreal, and subarctic zones. Their principal characteristic is the presence of water at or near the ground surface (Roulet et al., 1997). There are five wetland types: shallow open water, marsh, swamp, fen, and bog (see Table 7.5).

Wetlands are often "biological hotspots," containing rare plant, invertebrate, and vertebrate species. Additionally, wetlands play important roles in flood mitigation and biogeochemical cycling. As such, drainage of wetlands that was common in the

Table 7.5. *Features of Wetland Classes from the Canadian Wetland Classification System*

Class	Vegetation	Peat Depth (m)	Acidity
Shallow open water	Aquatic	0	Alkaline to acidic
Marsh	Graminoids, sedges, reeds, grasses, rushes, trees, and shrubs	0–0.4 mineral to peatland	Alkaline to neutral
Swamp	Conifers and deciduous trees, tall shrubs	0–2 Mineral to peatland	Neutral to acidic
Fen	Graminoids, sedges, shrubs, mosses, conifers	>0.4 peatland	Alkaline to neutral
Bog	Mosses, shrubs, graminoids, sedges, conifers	>0.4 peatland	Acidic

Source: After Roulet et al., 1997.

early to mid-twentieth century has now been replaced with conservation and even wetland reconstruction efforts. Microclimate studies of wetlands have largely focused on evaporation, since that aspect of the water balance is important in the maintenance of the water table position. Other studies have focused on carbon cycling in wetlands, since carbon accumulation (and/or methane release) is also an important key feature in the microclimate of wetlands.

Wetlands are unique in their microclimate because they display properties of both terrestrial and open-water systems. Although a wetland is generally defined as an area with predominantly saturated organic soils, some wetlands have annual wet and dry cycles, and complicated hydrology with poorly defined inflows and outflows of water. Seasonal vegetation growth can dramatically change the microclimate of wetlands. For example, open-water evaporation (i.e., no surface resistance) can dominate pre-vegetation emergence, then transpiration can dominate post-vegetation emergence with negligible open-water evaporation due to shading and sheltering of the water surface beneath (Figure 7.7).

Low-level (~45 m) aircraft measurements over the Hudson Bay lowlands show Bowen ratio values ranging from about 0.50 over open fen and open bog, to 0.70 over black spruce peatlands and 0.85 over tamarack fen (Desjardins et al., 1994).

A five-year study of a high-altitude wetland in semi-arid South Park, Colorado, United States, was recently made by Blanken (2014). The 338 ha High Creek Fen is located at 2850 m a.s.l. The vegetation consists of dwarf willow, dwarf birch, sedges, and grasses with an average height of 30 cm. The wetland is maintained by ground-water derived from snowmelt on the Mosquito Range to the west since evaporation far exceeds summer precipitation. During 2000–04, 46 percent of the precipitation fell as snow, except in 2003, when it declined to only 27 percent. The accumulated precipitation measured at a SNOTEL site in the Mosquito Range, 25 km to the west

Figure 7.7. A mix of annual sedges, mosses, and perennial shrubs, together with standing water, represents a complex microclimate surface. This wetland is High Creek Fen, located in central Colorado.
Photo credit: Maria Baden.

at 3399 m elevation, for the water year October 1–September 30, ranged between 765 mm in 2001 and 427 mm in 2002. The energy balance components (watts per square meter) for the period June 15–December 31 between 08.00 and 16.00 hours were as follows:

Components	2000	2001	2002	2003	2004
Total P (mm)	147	132	127	73	181
R_n	281	284	276	272	296
λE	131	105	40	181	159
H	126	150	214	72	113
G	20	16	13	16	19

The shortened snow cover duration and the markedly low SWE during the 2001–02 winter had a large impact on the summer–autumn energy balance in 2003, whereas evaporation was unaffected by low summer rainfall in 2003. The maximum LAI (which occurred between July 1 and 22), varied between 0.62 and 0.76 except in the "drought" year of 2002, when it fell to 0.47 m^2 m^{-2}. Correspondingly, the maximum fractional PAR ranged between 0.30 and 0.36 except in 2002, when it was 0.22.

A 19-year-long energy budget study of a peatland in northern England is reported by Worrall et al. (2015). It was possible to calculate 1662 daily evaporation rates (26 percent of the period). The estimated median evaporation on rain-free days was 1.6 mm day^{-1},

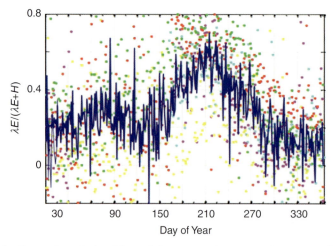

Figure 7.8. Example of the bi-modal evaporation annual pattern from High Creek Fen, Colorado, with the peak around Day Of Year (DOY) 90 from open water evaporation, and the second around DOY 210 from transpiration (Blanken, unpublished data).

and the median energy flux was 44.8 W m^{-2}. Evaporation increased slightly over the study period while sensible heat flux significantly declined, reflecting an increased use of sensible heat energy to meet evaporative demand. The relatively small change in evaporative flux compared to other energy fluxes suggests that this system is "near-equilibrium." On a seasonal basis the sensible heat flux was about 20 W m^{-2} from May to August and was negative (into the ground) from September to February.

In many wetlands, there is a strong seasonal development in the water table's position that determines whether there is standing water present or not, and the seasonal development of vegetation, with either the growth of annuals or the development of leaves on overstory deciduous vegetation. In such cases, the microclimate of the wetland undergoes a large shift from evaporation from the open water surface, to some combination of evaporation and transpiration (Figure 7.8). In some cases, the wetland evaporation efficiency (evaporation/potential evaporation) decreased with vegetation development, since the added transpiration has not offset the reduction in the open water evaporation by vegetation sheltering (Lafleur, 1990). This two-stream evaporation feature has led several researchers to develop means to partition these two streams. As discussed in Chapter 5, the Shuttleworth-Wallace approach of segregating two streams, soil evaporation and transpiration, has been applied to wetlands. For example, Wessel and Rouse (1994) used this approach to model evaporation in a subarctic wetland but found that a simple modified Penman-Monteith approach performed better since evaporation from standing water, not soil evaporation, was important.

In the same subarctic wetland region near Churchill, Manitoba, Blanken and Rouse (1995) used independent stomatal conductance and leaf area index measurement to partition standing water evaporation from transpiration in a subarctic willow-birch shrub wetland. They found that at the peak leaf area index, 80 percent of the total evaporation

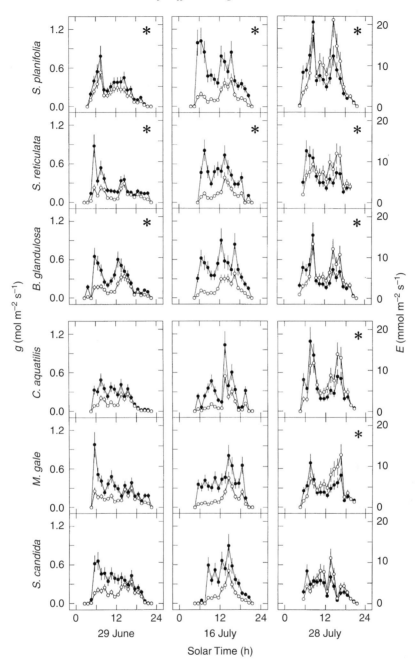

Figure 7.9. Directly measured diurnal course of stomatal conductance (dark circles) and transpiration (open circles) from several subarctic wetland species at Churchill, Manitoba. Asterisks denote midday stomatal closure. From Blanken and Rouse (1996).

was from transpiration. Even when the vegetation has an unlimited water supply (i.e., saturated soils with standing water present), the transpiration stream, thus evaporation, can be influenced by a warm, dry atmosphere. Blanken and Rouse (1996) found that mid-day stomatal closure under a period of high D effectively reduced transpiration in several subarctic wetland species (Figure 7.9). Stomatal closure occurred to reduce the tension (negative xylem pressure potential: see Box 2.2) to prevent cavitation in the xylem tissue.

The anaerobic conditions that exist in many wetland soils often result in a large buildup of organic matter, and wetlands serve as an important sink for atmospheric CO_2 (McGuire et al., 2009). This is especially true for some of the world's largest wetlands such as the Hudson Bay Lowlands and other wetlands spanning the subarctic and Arctic regions. In these regions underlain by permafrost, warming resulting in the recent degradation of permafrost has been found to increase the release of carbon in the form of methane (CH_4) into the atmosphere. As a powerful greenhouse gas (see Chapter 12), CH_4 released from both naturally occurring wetlands and agricultural wetlands (rice paddies) accounts for an estimated 40–50 percent of the total CH_4 emitted annually to the atmosphere (Whiting and Chanton, 1993).

E. Forests

Forests occupy about 31 percent of the Earth's land surface, where "forests" are classified as natural or planted stands at least 5 m tall, excluding agricultural productive systems (The World Bank, 2015). Of the terrestrial surface, 17 percent is classified as coniferous in the boreal forest zone and 14 percent as deciduous. Boreal forests, also known as taiga (a Russian term), occupy the subarctic areas of Canada and Eurasia and are generally evergreen and coniferous. Temperate zones support both broadleaf deciduous forests and evergreen coniferous forests. Warm temperate zones support broadleaf evergreen forests.

Tropical and subtropical forests include tropical and subtropical moist forests, tropical and subtropical dry forests, and tropical and subtropical coniferous forests.

Crowther et al. (2015) estimate that there are 3 trillion trees in the world, of which 1.39 trillion are in tropical and subtropical forests, 0.61 trillion in temperate forests, and 0.74 trillion in boreal forests.

Lee (1978) published a text titled *Forest microclimatology*, but only one chapter specifically treats forest microclimates, although forest examples are incorporated in the seven chapters on physical principles.

Coniferous Forest

The unique features of a coniferous forest climate, compared with other vegetated surfaces, are mostly attributable to their lower albedo, lower stomatal conductance, and higher aerodynamic roughness (McCaughey et al., 1997).

Generally low albedo values are typical of forests because most of the solar radiation is absorbed within the tree canopy. The albedo varies considerably with tree species. Jarvis et al. (1976) cite 0.04–0.06 for *Picea abies*, and 0.08–0.14 for pines and Douglas fir (*Pseudotsuga menziesii*). The extinction coefficient for shortwave radiation around midday during May–September ranges from 0.28 to 0.57 for *Pinus resinosa/strobus* with an LAI of 3.1 (Mukammal, 1971), to 0.28 for *Picea abies* in summer with an LAI of 8.4 (Jarvis et al., 1976). For visible light in *Pseudotsuga menziesii* with an LAI of 5.5, Kinerson (1973) cites 0.79 and for net radiation 0.42. For *Picea sichensis* in June–July with an LAI of 9.8, Norman and Jarvis (1974) report 0.49–0.56 for visible light.

Coniferous canopies affect the spectral distribution of radiation. For direct sunlight, coniferous canopies exhibit a marked increase in the proportion of red light. For diffuse radiation there is a sharp decline in blue light, related to the low reflectivity of needles in these wavelengths, and an increase in near-infrared light compared with the composition of sky light above the canopy. The canopy floor typically received little solar radiation, but Canham et al. (1990) point out that sun flecks, lasting an average of 6–12 minutes, can provide from 37 to 68 percent of the total seasonal photosynthetically active radiation in temperate coniferous and hardwood forests. (The energy used in photosynthesis by a coniferous forest during the day is about 3 percent of the net radiation and is therefore usually ignored in energy balance assessments (Denmead, 1969)).

Net radiation is partially determined by the albedo and the temperature of the radiating surface. R_n decreases with depth within the canopy with most of the decrease taking place in the upper third of the canopy. The importance of the canopy for net radiation is illustrated by the strong effect following clear cutting or burning. In Quebec, McCaughey (1981) reported a 10–22 percent reduction in R_n following clear cutting of a mature 15 m high balsam fir forest. The changes were mainly due to the increased albedo following removal of the trees. Lower reductions were found when the sites were wet and higher ones when they were dry.

The difference between air and needle temperatures is generally small (Jarvis et al., 1976), because of the large aerodynamic roughness of coniferous forests. This results in ample mixing of the air in and around the needles. Rutter (1967) found that under normal transpiration conditions needles in a Scots pine canopy were only 0–0.2 °C above those in adjacent air, while when wet they were 0.3–1.0 °C cooler than the air.

Needle stomatal conductance decreases greatly with age. In *Picea sitchensis* (Sitka spruce) total conductance rates were about 3 mm s^{-1} for the current year, declining to ~ 1.8 for one-year old to 0.7 mm s^{-1} for two-year old needles (Jarvis et al., 1976). Because of waxes on the needles, which merge into the cuticle, the cuticle is impervious to water vapor and CO_2.

The wind profile within forest canopies has received considerable attention. For Thetford Forest (pine trees) in eastern England, Oliver (1971) showed that below the canopy there is a weak secondary maximum that has been attributed to flow through the trunk space (see Figure 7.10). The presence of an understory, however, may cause this maximum to disappear or be considerably reduced. Lowest wind velocities are observed in association with maximum foliage density. Shaw (1977) demonstrates theoretically, however, that the secondary maximum arises through the downward transfer of momentum, whose rate decreases rapidly with decreasing height from its

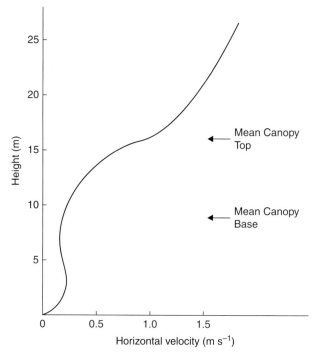

Figure 7.10. Profile of wind velocity with height above the surface at Thetford Forest in eastern England (adapted from H. R. Oliver, 1971).
Source: *Quarterly Journal of the Royal Meteorological Society* 97, p. 549, fig. 1, Courtesy Royal Meteorological Society.

maximum value at the canopy top. He shows that the Reynolds stress $\overline{u'w'}$ (where the prime denotes a deviation from the mean, u is horizontal and w is vertical velocity, and the overbar denotes a time-average value) is considerably larger in the overlying air layer, and this stress is transported downward by turbulence from the upper parts of the vegetation. The turbulent transfer of stress $\overline{u'w'}$ is larger inside the canopy than above and this results in a net transport of Reynold's stress from the upper canopy to lower regions. Hence, the mean wind gradient can reverse direction, giving rise to a secondary wind maximum.

The energy balance of a Scots pine (*Pinus sylvestris*) plantation at Hartheim, Germany, was investigated by Gay et al. (1996) for May to October 1992. The 157-day means for May 11–October 14 of the energy budget components are as follows: in watts per square meter, $S\!\downarrow$ 208, R_n 129, H 60, λE 65, G −4. Monthly Bowen ratios ranged from 0.67 in June to 1.84 in August. The total evaporation was 358 mm and the precipitation 308 mm, about 75 percent of normal.

Ludlow and Jarvis (1971) cite minimum stomatal resistances to water vapor transfer with needle age in Sitka spruce. Values increase from 120 s m^{-1} in the current year to 480 in year 1, 650 in year 2, and 1400 s m^{-1} in year 3 of the study period. In general, these are somewhat higher than comparable values for deciduous leaves, farm crops, and herbaceous plants. Stomatal resistance in conifers decreases in response to increased solar irradiance. Over a wide temperature range the resistance is small, but it increases below 5 °C and above 20 °C according to Nielson and Jarvis (1975). Stomatal resistance increases in response to vapor pressure deficits above about 10 hPa (Ludlow and Jarvis, 1971). Stomatal closure in pines occurred at between −14 and −17 bars; little closure occurred in *Pseudotsuga menziesii* at −19 bars or in *Abies grandis* at −15 bars (Lopushinsky, 1969).

Latent heat flux (λE) measurements were made by Kasurinen et al. (2014) at 65 boreal and arctic eddy-covariance (FLUXNET) sites. The data were analyzed using the Penman–Monteith equation. They stratified the sites into nine different ecosystem types: harvested and burned forest areas, pine forests, spruce or fir forests, Douglas fir forests, broadleaf deciduous forests, larch forests, wetlands, tundra, and natural grasslands. Surface resistance to water vapor transfer(r_s) tightly controlled λE in most mature forests, while it had less importance in ecosystems having shorter vegetation. The parameters of the Penman–Monteith equation were clearly different for winter and summer conditions, indicating that phenological effects on r_s are important. Kasurinen et al. (2014) also compared the simulated λE of different ecosystem types with mean ecosystem parameters under meteorological conditions at one site. Values of λE varied between 15 and 38 percent of the simulated R_n. The simulations suggest that λE is higher from forested ecosystems than from grasslands, wetlands, or tundra-type ecosystems. The highest values of λE were found for deciduous forests, followed by larch forests and fir or spruce forests. Forests usually showed a tighter stomatal control of λE as indicated by a pronounced sensitivity of r_s to atmospheric vapor pressure deficit (*D*). Nevertheless, the surface resistance of forests was lower than for open vegetation types including wetlands. Tundra and wetlands had higher r_s values, which were less sensitive to *D*. The results indicate that the variation in r_s might play a significant role in energy exchange between terrestrial ecosystems and atmosphere.

Forests of black spruce (*Picea mariana*) cover 44 percent of interior Alaska, where they are closely associated with permafrost (van Cleve et al., 1983). They are widespread on gentle north-facing slopes. They are highly fire prone with fires recurring as often as every 30–50 years. Ground layer mosses are abundant and their insulation and low thermal conductivity keep soil temperatures low. The annual sum of soil heat with respect to 0 °C at 10 cm depth is about 500 degree-days. Soil moisture contents are also high (120–240 percent of the average seasonal content in the surface organic layer). Tree productivity is low; aboveground tree biomass is only 5 kg m^{-2}, a fifth of that for white spruce (*Picea glauca*).

Over May–September, 1992, ET was 358 mm (a daily average of 2.28 mm d^{-1}) in a uniform Scots pine forest in the southern Rhine valley, Germany. For the five months, the mean R_n was 129, *H* was 60, and λE was 65 W m^{-2}. Note that precipitation was about 75 percent of normal, resulting in less soil moisture.

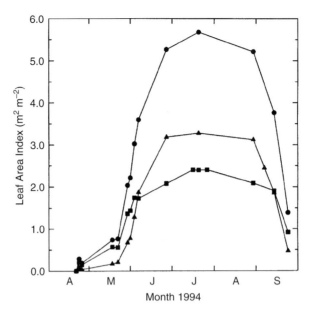

Figure 7.11. Seasonal variation in LAI of the forest (circles), hazelnut understory (triangles), and aspen overstory (squares) of a boreal aspen forest in central Canada. *Source*: Blanken et al., 2001, *Journal of Hydrology,* 245, p. 125, fig. 3.

Deciduous Forest

"Deciduous" refers to the fact that the plant's leaves fall off during part of the year. The seasonal shedding of leaves is a drought-defense mechanism to deal with the lack of available water in winter, although species can shed leaves at any time of year in response to drought. As a consequence, however, photosynthesis, hence growth, is negligible during the leafless periods.

Deciduous biomes are located mainly in the eastern half of the United States, Canada, Europe, southwestern Russia, eastern China, Japan, New Zealand, and southern Chile. Deciduous forests span a wide variety of tree types. They include maple, many oaks, elm, aspen, and birch, as well as a number of coniferous genera, such as larch. Temperate deciduous forests developed in response to seasonal temperature variations, whereas tropical and subtropical forests are a response to seasonal variations in rainfall. Leaf fall in these regions is not seasonal and may occur at any time of year in response to drought stress. In temperate forests the mean annual temperature is around 10 °C and the mean annual precipitation between 750 and 1500 mm.

Although physiologically considered an advanced adaptation to tolerate drought, the annual loss of leaves can have a profound impact on the microclimate both for the forest as a whole, and especially for the understory beneath. The senescence of leaves in the autumn is generally triggered by a decrease in air temperature coupled with a decrease in day length. As a result, the leaf area index of deciduous forests undergoes a dramatic seasonal pattern (Figure 7.11).

Table 7.6. Forest Types in Russia

Type of Tree	Oak (*Quercus robur*)	Maple (*Acer platanoides*)	Aspen[a] (*Populus tremula*)	Linden (*Tilia cordata*)[a]	Birch (*Betula verrucosa*)
Top of crown height (h_1)(m)	6.5	7.0	10.5	11.5	7.6
Bottom of crown height (h_2)	3.0	4.0	5.5	5.5	3.5

[a] These include undergrowth.
Source: After Rauner, 1976.

The range in crown height in various 15- to 20-year-old deciduous forest stands in the Kursk region of Russia is shown by Rauner (1976) (see Table 7.6).

The maximum LAI occurs at values of $(1 - z/h_1)$ of 0.2 for oak, 0.15 for maple, 0.4 for aspen, 0.1 for linden, and 0.2 for birch. The total LAI values were, respectively, 4.6, 5.0, 7.1, 4.8, and 5.3 $m^2\ m^{-2}$. In an oak stand, the LAI reaches a maximum of 5 at 30–50 years and then remains virtually unchanged to 70–80 years. From 90 to 220 years it slowly declines from 4.6 to 3.6 (Rauner, 1976).

The reflection coefficient around midday for oak without foliage in spring is about 0.12, which increases to 0.15 with full summer foliage. Grulois (1968) shows that for 9 to 10 m high oak forest in Belgium the summer albedo is around 0.16 and for 15 to 17 m birch-aspen in the Moscow region it is around 0.13. The downward fluxes of both solar and net radiation in forests have been shown to be well approximated by an exponential function of the cumulative LAI (Rauner, 1976). As a function of LAI, the relative absorption of solar radiation by aspen increases from 0.2 for LAI = 1, to 0.52 for LAI = 3, to 0.82 for LAI = 6. The effect of aging on the relative values of solar and net radiation under oak forest with full foliage can be seen in the following (after Rauner, 1976, table VI, p. 251).

Stand Age (years)	10	40	100	180
Solar Radiation	0.70	0.12	0.21	0.27
Net Radiation	0.58	0.08	0.15	0.20

The aerodynamic characteristics of forest stands depend on their density and height. Rauner (1976) shows that the normalized values of the roughness parameter (z_o/h) and the zero plane displacement (d/h) depend on the wind velocity above the canopy; d/h decreases exponentially to about 0.5 whereas z_o/h increases linearly to about 0.25 as the wind speed increases to 8 m s^{-1}. As the wind increases, the d/h gradient drops to the layer with maximum leaf area density, whereas the roughness increases with the wind speed.

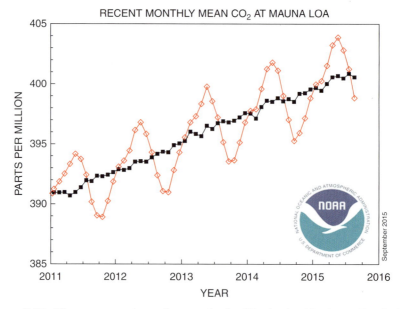

Figure 7.12. The concentration of atmospheric CO_2 in the Northern Hemisphere shows a pronounced annual cycle superimposed on the longer-term steady increase. *Source*: NOAA.

Components of the energy balance in July for a 40- to 60-year-old oak forest were as follows (Rauner, 1976): $R_n = 419$ MJ m^{-2}, $\lambda E = 410$, $H = 0$, $G = 9$, $E = 163$ mm. In May, H was 88 MJ m^{-2}, out of 369 MJ m^{-2} for R_n. The zero value of H in July represents the "oasis" effect of a small diameter forest area amid agricultural land. For a 20- to 25-year-old linden-oak forest in mid-July to mid-August, the components were $R_n = 432$ MJ m^{-2}, $\lambda E = 324$, $H = 104$, $G = 4$, $E = 128$ mm. The relative daytime total evaporation rate as a function of LAI for oak (and aspen plus undergrowth) declined from 0.5 (0.92) for LAI of 0.5, to 0.32 (0.88) for LAI of 1.0, to 0.17 (0.64) for LAI of 3.0, and to 0.08 (0.15) for LAI of 5.0 (Rauner, 1976).

Since most deciduous forest and most of the terrestrial global land area are located in the Northern Hemisphere, the seasonal LAI cycles as shown in Figure 7.12 are so strong that they are apparent in the seasonal concentration of CO_2 on the entire Northern Hemisphere. The dynamics in the microclimate of a deciduous forest that can result in such a strong global signal are detailed in the following.

One of the largest impacts of leaf development is the radiation profile through the canopy. Without leaves, most of the incident solar radiation reaches the forest floor, keeping in mind that solar incident zenith angles are large and day lengths are short in the North American winter season. During the summer season, solar zenith angles are smaller and day lengths longer. The portion of the incident solar radiation used in photosynthesis, PAR (photosynthetic active radiation, 400–700 nm), can also be expressed in the equivalent photosynthetic photon

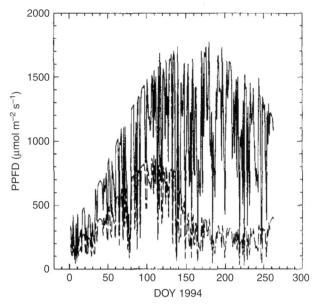

Figure 7.13. Midday PPFD measured above (solid line) and beneath (dashed line) the overstory of a boreal aspen forest.
Source: Chen et al., 1997. *Agricultural and Forest Meteorology,* 86, p 112, fig. 1.

flux density, or PPFD (μmol m^{-2} s^{-1}). Figure 7.13 shows an example of how the PPFD measured both above and below a deciduous forest changes annually as a result of latitude (zenith angles and day length), and how the leafing of the overstory absorbing PAR for photosynthesis drastically decreases the PAR reaching the understory beneath.

The absorption and forward scattering of PAR through the leaves also affect the profile of PAR radiation through the canopy. If we assume a homogeneous canopy (seldom if ever true), the fraction of radiation (*f*) that reaches any height within the canopy (*z*) can be predicted using Beer-Bouguer 's law (see Chapter 4, Section A):

$$f_z = e^{-K L_T} \qquad\qquad 7.7$$

where K is the extinction coefficient and L_T is the cumulative leaf area index from the top of the canopy to *z*. The extinction coefficient specifies the rate of attenuation through the medium (extinction increases as K increases) and varies with wavelength, so usually measurements are required within canopies to determine K for the short- and longwave radiation portions of the spectrum (Figure 7.14).

The timing, or phenology, of both leaf development and leaf shedding affect not only radiation, but also the carbon cycle (e.g. see Figure 7.12). Several studies have found that the net ecosystem exchange (NEE) of carbon for forests is greatest immediately after the spring bud burst. This is because there are optimum conditions at this time: air temperatures are not too high, *D* is low, and there is ample soil moisture derived from snowmelt. Also, ecosystem respiration tends to be small at this time.

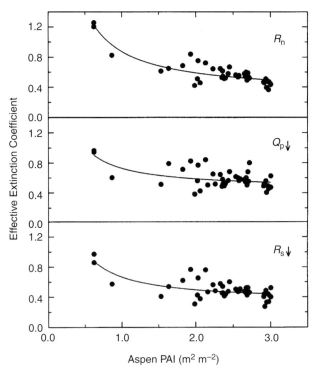

Figure 7.14. The effective extinction coefficient for net radiation (R_n), the photosynthetic photon flux density (PPFD, Q_p), and incident solar radiation (R_s) as a function of the plant area index (PAI; sum of LAI and stem area index) for a boreal aspen forest.
Source: Blanken et al., 2001, *Journal of Hydrology*, 245 p. 126, fig. 7.

Therefore, any shifts in the timing of leaf-out can have dramatic effects for the forests' carbon balance, with deciduous forests more sensitive to the growing season length than evergreen coniferous forests (Richardson et al., 2010).

In addition to radiation and carbon cycling changes induced by deciduous forests, the turbulence (wind) and surface energy balance at the forest floor are influenced by the leaf properties of the canopy above (Figure 7.15).

The development of leaves permits transpiration to occur, and as a result, the amount of net radiation available for sensible heat decreases. Forests tend to become more aerodynamic with leaves than without, resulting in less coupling (or mixing) with air above and within the canopy. Also, the fraction of the soil heat flux relative to net radiation decreases with leaf development since shading of the forest floor tends to minimize any soil temperature gradients (Table 7.7).

The preceding discussion has shown that the seasonal development and shedding of leaves in deciduous forests have pronounced microclimatological effects. Because of this, and the fact that remote sensing estimates of leaf area index are widely available worldwide, several attempts have been made to correlate LAI with various microclimatologic variables (Figure 7.16).

Table 7.7. Daytime Energy Balance Components as a Percentage of Net Radiation Pre- and Post-leaf Emergence for a Boreal Aspen Forest and Understory. Net Radiation Values Used in the Denominator Were 368 (Forest) and 159 (Understory) Watts Per Square Meter

	Pre-leaf		Post-leaf	
	Forest	Understory	Forest	Understory
λE	10	16	61	66
H	73	61	25	10
G	9	20	3	12
S_T	8	3	11	12

Source: Data from Blanken et al., 1997.

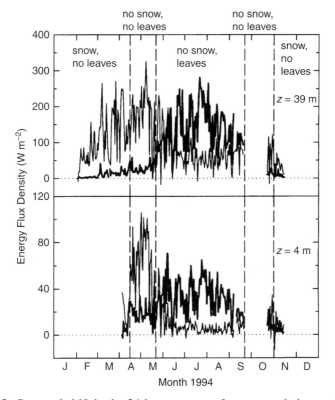

Figure 7.15. Seasonal shift in the 24-hour mean surface energy balance of a boreal aspen forest from dominance by sensible heat in spring (thin line) then latent heat with the leaf development in summer (thick line).
Source: Blanken et al., 2001, *Journal of Hydrology,* 245, p. 130, fig. 10.

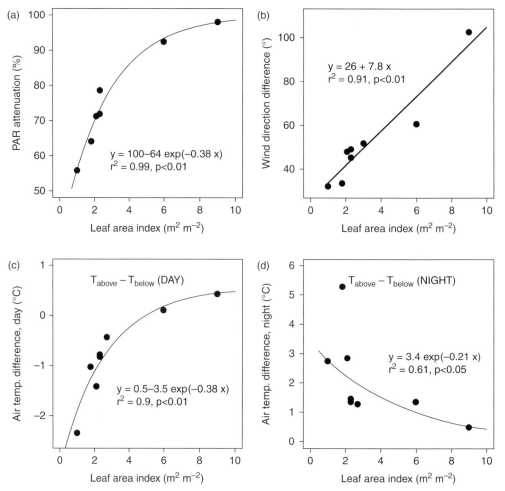

Figure 7.16. Summertime relationships from measured LAI and several micrometeo-rological variables (from Misson et al., 2007, p.19, fig. 1).
Source: *Agricultural and Forest Meteorology*. 144, p. 19, fig. 1 Courtesy of Elsevier.

Overall, most studies report strong relationships, as Figure 7.16 shows. Therefore LAI is a key variable in most micrometeorological models as discussed in Chapter 6.

Tropical Forest

There is a wide variety of tropical forest biomes. They cover 13 percent of the terrestrial land surface with about half the total area in Amazonia.

A critical summary of knowledge of the hydrology of moist tropical was provided by Bruijnzeel (1990) at a time when there were rather sparse data. The average annual ET for 11 sites in Latin America, Africa, and Southeast Asia is 1430 mm. The evaporation from wet canopies is typically around 0.2 mm hr^{-1} but rates that exceed the available net radiation are sometimes observed, pointing to advection of sensible heat to the site. Estimates of annual transpiration range from 885 to 1285 mm yr^{-1} with a mean (9 cases) of 1045 mm. The intercepted rainfall is shown to average 13 percent in lowland tropical forests and 18 percent in montane forests (excluding cloud forests).

Mahli et al. (2002) provide the results of a year-long study in the central Amazon rain forest near Manaus, Brazil. The canopy height was 30 m and the LAI ranged between 5 in the dry season (mid-June to mid-October) and 6 m^2 m^{-2} in the wet season. The mean daily maximum air temperature ranged between 31 °C and 33 °C and the mean daily minimum between 23 °C and 24 °C. Of the 5.56 GJ m^{-2} yr^{-1} of solar radiation supplied over the year, 11 percent was reflected, 15 percent was lost as net thermal emission, 27 percent was transported as sensible heat, and 46 percent was used in evapotranspiration. In relation to net radiation (4.1 GJ m^{-2} yr^{-1}), λE accounted for 63 percent and H for 36 percent. The evapotranspiration amounted to 1123 mm (3.1 mm d^{-1}), accounting for 54 percent of total precipitation.

For the period September 1983 to September 1985, Shuttleworth (1988) determined that the forest canopy in the central Amazon intercepted approximately 10 percent of rainfall, and this accounted for 20–25 percent of the evaporation. The remainder occurred as transpiration from the trees. About half of the incoming precipitation is returned to the atmosphere as evaporation, a process that requires 90 percent of the radiant energy input. These proportions exhibit some seasonal behavior in response to the large seasonal variation in rainfall. The average evaporation over two years was within 5 percent of potential evaporation. Monthly-average evaporation exceeds potential estimates by about 10 percent during wet months and falls below such estimates by at least this proportion in dry months.

A recent study for a semi-deciduous tropical forest at 11.4 °S near Sinop in the Matto Grosso, Brazil, was carried out during January 2000 to January 2007 (Vourlitis et al., 2015). Evapotranspiration was substantially affected by the 2002 El Niño. Prior to this event annual evapotranspiration averaged 1011 mm yr^{-1} and accounted for 52 percent of annual rainfall, while afterward it was 931 mm yr^{-1} and accounted for only 42 percent of the annual rainfall.

Canopy trees in tropical forests are subject to an intense light regime whereas within the canopy there is a particularly low light environment. This is a key factor for ground flora and fauna including birds (Patten and Smith-Patten, 2012).

The Large-Scale Biosphere–Atmosphere Experiment in Amazonia was operated over 3–4 years between 1999 and 2006 at seven flux sites – four in evergreen broadleaf forests, a deciduous broadleaf forest, a savanna site, and two pastures spanning latitudes 2.6–21.6 °S (Gonçalves de Gonçalves et al., 2013). Model simulations can be divided into three different categories based on their feature sets. Nine models contained non-dynamic representations of vegetation of which five were based on the Simple Biosphere (SiB) model. Five models included dynamic vegetation and carbon fluxes and six simulated dynamic vegetation, carbon fluxes, and nitrogen cycling.

De Rocha et al. (2009) showed that ET in the wettest areas of Brazil (central Amazonia) is closely linked to radiation levels, while water availability regulates

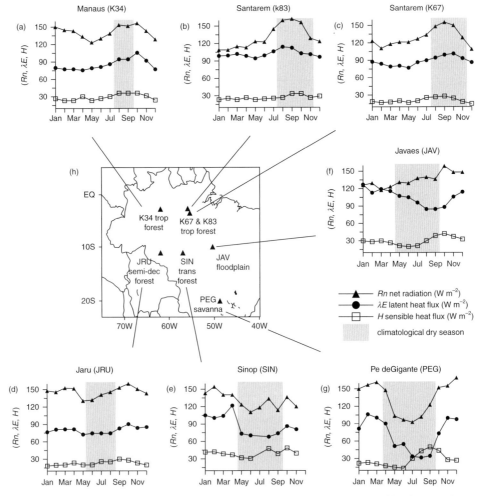

Figure 7.17. Measured above-canopy mean monthly precipitation (white bar, mm), air temperature (heavy line with filled circles, °C), and incoming solar radiation (heavy line) (solar, W m^{-2}, or PPFD, mmol^{-2} s^{-1}) at the flux tower sites displayed in (h) map of Amazonia. The climatological dry season is shaded; climatological mean annual precipitation (mm) and surface air temperature (°C) are displayed at each top panel (from de Rocha et al., 2009).

Source: *Journal of Geophysical Research, Biogeosci.*, 114: p. 4, fig. 2. Courtesy American Geophysical Union.

ET in the drier regions to the south and east. At sites with an annual precipitation >1900 mm and a dry season length less than four months, evaporation rates increased in the dry season, coincident with increased radiation (see Figure 7.17). Evaporation rates were as high as 4.0 mm d^{-1} in the evergreen or semideciduous forests. In contrast, ecosystems with precipitation <1700 mm and a longer dry season showed reduced evaporation in the dry season; rates were as low as 2.5 mm d^{-1} in the transitional forests and 1 mm d^{-1} in the cerrado. The mean annual sensible heat flux at all sites ranged from 20 to 38 W m⁻² and was generally reduced in the wet season and increased in the late dry season, coincident with seasonal variations of net radiation and soil moisture. For the same transect, maximum annual values for both H and λE occur during the dry season, according to Baker et al. (2013). There is little seasonal variation in the wet Northwest.

Poleward of the tropical rain forests there are belts of tropical dry forest between 10 °N and °S and 20 °N and °S. They are predominantly deciduous forests as they occur in warm climates with a long dry season, but some in parts of southern Asia are evergreen. Pinker (1980) studied the microclimate of dry tropical hardwood (dipterocarp) forest on the Korat Plateau of Thailand (14.5 °N). The southwest monsoon begins in May and ends in September with an annual total precipitation of 1500 mm. The forest has a surface layer, 0–5 m; a subcanopy layer 5–25 m; and a canopy layer, 25–35 m in height. The temperature difference around noon between the ground and canopy top is 2 °C in July and September and 4 °C in January and April. The surface temperature difference between a 500 m diameter clearing and below the canopy can be up to 4 °C by day in the dry season. The forest has a significant effect on wind speed reduction. At 32 m height the reduction in the canopy is ~80 percent in December–January and 60–64 percent in February, April, July, and September.

Forest Edges and Clearings

In many forested areas logging has created a landscape mosaic with forest patches and numerous edges that are often a dominant feature. Edge effects in a mixed hardwood forest in North Carolina were studied by Fraver (1994). He examined changes in the percentage cover of individual species, the relative cover of exotic species, and species richness. All indicated that edge effects from adjacent agricultural land penetrate deeper on south-facing edges (50–60 m) than on north-facing edges (10–30 m).

These distances are comparable with those measured by Raynor (1971) in a pine forest about 10.5 m tall in Brookhaven, New York. The forest has an E-W border with a 120 m field to the south. For winds into the forest, speeds at 1.75–3.50 m height tend to approach 15–20 percent of those in the open at distances > 60 m from the edge.

Seasonal and diurnal changes in air temperature, relative humidity, and vapor pressure deficit, soil temperature and volumetric soil moisture, shortwave radiation, and wind velocity are quantified by Chen et al. (1993) for 10- to 15-year-old clear-cut (replanted with conifers), edges, and adjacent interior (> 230 m from an edge) of old-growth Douglas fir (*Pseudotsuga menziesii)* forest in southern Washington State, over two growing seasons. They show that over the growing season, daily averages of air and soil temperatures, wind velocity, and shortwave radiation are consistently lower, while soil and air moisture are higher, inside the forest than in the clear-cut or at the forest edge. Average air and soil temperatures for 35 days in June–September were as follows:

	T_a	T_a max – T_a min	10 cm T_s	10 cm T_s max – T_s min
Forest Clearcut	16.3	14.8	18.3	10.4
Forest Edge	16.7	14.2	17.1	11.1
Forest	15.7	10/1	13.9	2.4

Edge orientation is shown to have important effects on solar radiation and consequently on air and soil temperature and relative humidity. Solar radiation on a south-facing edge displays a bell-shaped curve under cloud-free conditions, whereas for east- and west-facing edges the bell shape is halved and north-facing edges receive only diffuse and reflected light.

The trees surrounding a clearing cast a shadow into the clearing with a length

$$L = H/\tan a \qquad\qquad 7.8$$

where H is tree height and a is the solar altitude angle. The fraction of sky seen from a sloping surface, s – the sky view – is $(1 + \cos s)/2$, meaning that a portion of the sky is blocked.

For a circular clearing, the portion of sky visible from the center of the clearing is $\cos^2 \beta$, where β is the angle between the center of the clearing and the treetops. The remainder, $1 - \cos^2 \beta$, is the terrain view factor.

In tropical rain forest in Costa Rica, Fetcher et al. (1985) studied the microclimate of a 400 m² treefall gap and a 0.5 ha clearing compared with two locations in the primary forest over two years. The mean canopy height was 33 m. Air temperatures were highest in the clearing; intermediate in the canopy and gap, which were similar; and lowest in the understory. Vapor pressure deficits were highest in the clearing, followed by the canopy, the gap, and the understory.

The spatial distribution of net radiation in snow-covered forest gaps near Moscow, Idaho, is shown by Seyednasrollah and Kumar (2014) to be heterogeneous with southern and northern areas of the gap receiving minimum and maximum energy amounts, respectively. For clear sky conditions in the snow season, results suggest that net radiation in the forest gap is a minimum for gaps of size equal to half of the surrounding tree height (Table 7.8). In contrast, when sky cloudiness in the snow season is considered, net radiation shows a monotonically increasing trend with gap size. Slope and aspect of the forest gap floor also impact the net radiation and its variation with gap size. Net radiation is largest and smallest on steep south-facing and north-facing slopes, respectively. Variation of net radiation with slope and aspect is largest for larger gaps. Results also suggest that net radiation in north-facing forest gaps is larger than in open areas for a longer duration in the snow season than in forest gaps on flat and south-facing slopes.

Table 7.8. Statistics of Minimum Net Radiation on Hillslopes of Different Aspects under Cloud-free Skies for Gap Diameter/Width = 0.5

Orientation	Proportion of Radiation in Relation to Very Small Gaps	Proportion of Radiation in Relation to Very Large Gaps
South	91	34
East/west	89	41
North	87	53

Source: From Seyednasrollah and Kumar, 2014, table 3.

References

Addison, P. A., and Bliss, L. C. 1980. Summer climate, microclimate, and energy budget of a polar semidesert on Kimg Christian Island, N.W.T., Canada. *Arct. Alp. Res.* 12, 161–70.

Baker, I. T. et al. 2013. Surface ecophysiological behavior across vegetation and moisture gradients in tropical South America. *Agric. Forest Met.* 182–3, 177–88.

Bauer, N., and Kenyeres, Z. 2007. Seasonal changes of microclimatic conditions in grasslands and its influence on orthopteran assemblages. *Biologia* 62, 742–8.

Bennie, J., Huntley, B., Wiltshire, A., Hill, M.O. and Baxter, R. 2008. Slope, aspect and climate: Spatially explicit and implicit models of topographic microclimate in chalk grassland. *Ecol. Modelling* 216, 47–59.

Betts A. K., and Ball, J. H. 1998. FIFE surface climate and site-average dataset: 1987–1989. *J. Atmos Sci* 55, 1091–1108.

Blanken, P. D. 2014. The effect of winter drought on evaporation from a high-elevation wetland. *J. Geophys. Res. Biogeosci.* 119. doi:10.1002/2014JG002648.

Blanken, P.D., Black, T.A., Yang, P.C., Neumann, H.H., Nesic, Z., Staebler, R. & den Hartog, G. (2001) The seasonal energy and water exchange above and within a boreal aspen forest. *Journal of Hydrology*, 245 (1-4), 118-136.

Blanken, P. D et al. 2000. Eddy covariance measurements of evaporation from Great Slave Lake, Northwest Territories, Canada. *Water Resour. Res.* 35, 1069–77.

Blanken, P.D. and Rouse, W.R. 1995. Modelling evaporation from a high subarctic willow-birch forest. *Int. J. Climatol.,* 15, 97-106.

Blanken, P.D. and Rouese, W.R. 1996, Evidence of water conservation mechanisms in several subarctic wetland species. *J. Appl. Ecol,* 34, 842-50.

Blanken, P.D., Rouse, W.R., and Schertzer, W.M. (2003) Enhancement of evaporation from a large northern lake by the entrainment of warm, dry air. *Journal of Hydrometeorology,* 4(4), 680–93.

Blanken, P.D., Williams, M.W., Burns, S.P., Monson, R.K., Knowles, J., Chowanski, K, and Ackerman, T. 2009 A comparison of water and carbon dioxide exchange at a windy alpine tundra and subalpine forest site near Niwot Ridge, Colorado. *Biogeochemistry*, 95(1), 61–76.

Broadbent, L. 1950. The microclimate of the potato crop. *Quart. J. Roy. Met. Soc.* 76, 439–54.

Brown, K-W. 1976. *Vegetation and the atmosphere.* New York, Academic Press.

Brown, K. W., and Rosenberg, N,J. 1971. Energy and CO_2 balance of an irrigated sugar beet (Beta vulgaris) field in the Great Plains. *Agron. J.* 63, 207–13.

Bruelheide, H., and Jandt, U. 2007. The relationship between dry grassland vegetation and microclimate along a west-east gradient in central Germany. *Hercynia N.F.* 40, 153–76.

Bruijnzeel, L. A. 1990. Hydrology of moist tropical forests and effects of conversion: A state of knowledge review. International Hydrological Programme. Paris: UNESCO.

Burrows, F. J. 1969. The diffusive conductivity of sugar beet and potato leaves. *Agric. Met.* 6, 211–26.

Caborn, J. M. 1957. Shelterbelts and microclimate. Forestry Commission, Bulletin No. 29. Edinburgh: HMSO.

Canham, C. D. et al. 1990. Light regimes beneath closed canopies and tree-fall gaps in temperate and tropical rainforest. *Canad. J. Forest Res.* 20, 620–31.

Chen, J.M., Blanken, P.D., Black, T.A., Guilbeault, M. & Chen, S. (1997) Radiation regime and canopy architecture in a boreal aspen forest. *Agricultural and Forest Meteorology*, 86 (1-2), 107-25.

Chen, J., Franklin, J. F., and Spies. T. A. 1993. Contrasting microclimates among clearcut, edge, and interior of old-growth Douglas-fir forest. *Agric. Forest Met.* 63, 219–37.

Crowther, T. W. et al. 2015. Mapping tree density at a global scale. *Nature.* doi:10.1038/nature14967.

de Rocha, H. R. et al. 2009. Patterns of water and heat flux across a biome gradient from tropical forest to savanna in Brazil. *J. Geophys. Res.* 114, G00B12. doi:10.1029/2007JG000640.

Denmead, O.,T. 1969. Comparative micrometeorology of a wheat field and a forest of *Pinus radiata. Agric. Met.* 6, 357–72,

Denmead, O.T. 1976. Temperate cereals', in Monteitg, J.L. (ed.), *Vegetation and the atmosphere*, Vol. 2. *Case studies.* London;Academic Press. pp. 1- 31.

Desjardins, R. L. et al. 1994. Airborne flux measurements of CO_2 and H_2O over the Hudson Bay lowland. *J. Geophys. Res., Atmos.* 99(D1), 1551–62.

Dymov, A. A., Zhangurov E. V., and Starcev V. V. 2013. Microclimatic characteristics of sub-polar Ural Soils (National Park Yugyd va). International conference "Earth Cryology: XXI Century" (September 29–October 3, 2013, Pushchino, Moscow region, Russia). The Program and Conference Materials, pp. 78–79

Fetcher, N., Oberbauer, S. F., and Strain, B. R. 1985. Vegetation effects on microclimate in lowland tropical forest in Costa Rica. *Int. J. Biomet.* 29, 145–55.

Fritschen, L. J. 1966. Evapotranspiration rates of field crops determined by the Bowen ratio method. *Agron. J.* 58, 339–42.

Fraver, S. 1994. Vegetation responses along edge-to-interior gradients in the mixed hardwood forests of the Roanoke river basin, North Carolina. *Conserv. Biol.* 8, 822–32.

Fritz, S. et al. 2015. Mapping global cropland and field size. *Global Change Biol.* 21(1), 1980–92.

Gardiner, B.A., and Quine, C.P. 2000. Management of forests to reduce the risk of abiotic damage – a review with particular reference to the effects of strog winds. Forest Ecology and Management, 135, (1-3), 261-277.

Gay, L. W., Vogt, R., and Kessler, A. 1996. The May-October energy budget of a Scots Pine plantation at Hartheim, Germany. *Theoret. Appl. Clim.* 53, 79–94.

Gonçalves de Gonçalves, L. G. et al., 2013. Overview of the Large-Scale Biosphere–Atmosphere Experiment In Amazonia Data Model Intercomparison Project (LBA-DMIP). *Agric. Forest Met.* 182–3, 111–27.

Grulois, J. 1968. La variation annuelle du coefficient d'albedo des surfaces superieures du peuplement. *Bull. Soc. Roy. Bot. Belgique* 101, 141–53.

Ham, J. M., Heilman, J. L., and Lascano, R. J. 1991. Soil and canopy energy balances of a row crop at partial cover. *Agron. J.* 83, 744–53.

Horton, R. et al. 1996. Crop residue effects on surface radiation and energy balance – review. *Theoret. Appl. Clim.* 54, 27–37.

HYDE. 2010. History Database of the Global Environment. PBL Netherlands Environmental Assessment Agency. http://themasites.pbl.nl/tridion/en/themasites/hyde/index.html

Jarvis, P. G., James, G. B., and Landsberg, J. J. 1976. Coniferous forest. In J. L. Monteith (ed.), *Vegetation and the atmosphere.* Vol. 2. *Case studies.* London: Academic Press: pp. 171–240.

Jones, H. G. 1992. *Plants and microclimate. A quantitative approach to environmental plant physiology.* Cambridge: Cambridge University Press.

Kanemasu, E. T., and Arkin, G. F. 1974. Radiant energy and light environment of crops. *Agric. Met.* 14, 211–25.

Kasurinen, V., et al. 2014. Latent heat exchange in the boreal and arctic biomes. *Global Change Biol.* 20(11), 3439–56. doi: 10.1111/gcb.12640.

Kinerson, R. S. 1973. Fluxes of visible and net radiation within a forest canopy. *J. appl. Ecol.* 10, 657–60.

Knowles, J.F., Blanken, P.D., Williams, M.W., and Chowanski, K.M. (2012) Energy and surface moisture seasonally limit evaporation and sublimation from snow-free alpine tundra. *Agricultural and Forest Meteorology*, 157, 106-115, doi:10.1016/j.agformet.2012.01.017.

Lafleur, P. 1990. Evaporation from wetlands. Canad. Geogr., 34, 79-82.

Lee, R. 1978. *Forest microclimatology.* New York: Columbia University Press.

Lewis, M.C. and Callaghan, T.V. 1976. Tundra *vegetation and the atmosphere*, 2, London Academic Press. pp. 399-433.

Lopushinsky, W. 1969. Stomatal closure in conifer seedlings in response to leaf moisture stress. *Bot.Gaz.* 130, 258–63.

Ludlow, M.M. and Jarvis, P.G. 1971. Photosynthesis in Sitka Spruce (Picea sitchensis (Bong.) Carr.). I. General characteristics. *J. Appl. Ecol.,* 8, 925-53.

Malhi, Y. et al. 2002. Energy and water dynamics of a central Amazonian rain forest, *J. Geophys. Res.* 107(D20), 8061.

McCaughey, J. H. 1981. Impact of clearcutting of coniferous forest on the surface radiation balance. *J. Appl.Ecol.* 18, 815–26.

McCaughey, J. H. et al. 1997. Forest environments. In W. G. Bailey, T. R. Oke, and W. R. Rouse (eds.), *The surface climates of Canada.* Montreal: McGill University Press. pp. 247–76.

McGuire, A.D. et al. 2009. Sensivity of the carbon cycle ib the Arctic tundra to climate change. *Col. Monogr.,* 79, 523-55.

Misson, L. et al. 2007. Partitioning forest carbon fluxes with overstory and understory eddy-covariance measurements: A synthesis based on FLUXNET data. *Agric. For. Met.* 144, 14–31.

Mølgaard, P. 1982. Temperature observations in High Arctic plants in relation to microclimate in the vegetation of Peary Land, north Greenland. *Arct. Antarct. Alp. Res.* 14, 105–15.

Monteith, J,.L. (ed.) 1976. *Vegetation and the atmosphere. Vol. 2. Case studies.* London: Academic Press.

Mukammal, E. I. 1971. Some aspects of radiant energy in a pine forest. *Arch. Met. Geophys. Bioklim.* B19, 29–52.

Nielson, R. E., and Jarvis, P. G. 1975. Photosynthesis in Sitka spruce (*Picea sitchensis* (Bong.) Carr.). VI. Response of stomata to temperature. *J. appl. Ecol.* 12, 879–92.

Norman, J. M., and Jarvis, P. G. 1974. Photosyntesis in Sitka spruce [*Picea sitchensis* (Bong.) Carr). III. Measurements of canopy structure and interception of radiation. *J. Appl. Ecol.* 11, 375–98.

Oliver, H. R. 1971. *Wind profiles in and above a forest canopy. Quart. J. Roy. Meteorol. Soc.* 97, 548–53.

Patten, M.A. and Smith-Patten. B.D. 2012. Testing the microclimate hypothesis: Light environment and population trends of Neotropical birds. *Biol. Conserv.*, 155, 85–93.

Pinker, R. 1980. The microclimate of a dry tropical forest. *Agric. Met.* 22, 249–65

Ramankutty, N. et al. 2008, Farming the planet: 1. Geographic distribution of global agricultural lands in the year 2000. *Global Biogeochem. Cycles* 22, GB1003

Rauner, Ju.L. 1976. Deciduous forest. In J. L. Monteith (ed.), *Vegetation and the atmosphere.* Vol.2: *Case studies.* London: Academic Press: pp. 241–64.

Raynor, G. S. 1971. Wind and temperature structure in a coniferous forest and a contiguous field. *Forest Sci.* 17, 351–63.

Ripley, E. A. 1972. Man, Matador and meteorology. *Atmosphere* 10, 113–27.

Ripley, E. A., and Redmarsh, R. E. 1976. Grassland. In J. L. Monteith (ed.), *Vegetation and the atmosphere.* Vol.2. *Case studies.* London: Academic Press: pp. 349–98.

Ripley, E. A., Wang, R-Zh., and Zhu, T-Ch. 1996. The importance of microclimate in grassland ecosystems. *J. Northeast Normal Univ., Natural Sci.* 2, 117–37.

Richardson, A. D. et al. 2010. Influence of spring and autumn phenological transitions on forest ecosystem title. *Phil.Trans. Roy. Soc., B.* 365(1555), 3227–46.

Roulet, N. T., Munro, D. S., and Mortsch, L. 1997. Wetlands. In W. G. Bailey, T. R. Oke, and W. R. Rouse (eds.), *The surface climates of Canada.* Montreal: McGill UniversityPress. pp. 149–71.

Rouse, W. R. 1984a. Microclimate of Arctic tree line. 1. Radiation balance of tundra and forest. *Water Resour. Res.* 20, 57–66.

Rouse, W. R. 1984b. Microclimate of Arctic tree line. 2. Soil microclimate of tundra and forest. *Water Resour. Res.* 20, 67–73.

Rutter, A. J. 1967. Evaporation in forests. *Endeavour,* 26, 39–43.

Seyednasrollah, B., and Kumar, M. 2014. Net radiation in a snow-covered discontinuous forest gap for a range of gap sizes and topographic configuration. *J. Geophys. Res. Atmos.* 119. doi:10.1002/2014JD021809.

Shaw, R. H. 1977. Secondary wind speed maxima inside plant canopies. *J. Appl. Met.* 16, 514–21.

Shuttleworth, J. W. 1988. Evaporaion from Amazonian rainforest. *Proc. Roy. Soc. B. London* 233, 321–46.

Shuttleworth J. W., and Wallace, J. S. 1985. Evaporation from sparse crops – an energy combination theory. *Quart. J. Roy. Met. Soc.* 111, 839–55.

Siebert, S. et al. 2015. A global dataset of the extent of irrigated land from 1900 to 2005. *Hydrol. Earth Syst. Sci.* 19, 1521–45.

The World Bank. 2015. Forest area. Washington, DV. www.data.worldbank.org/indicator/AG.LND.FRST.ZS.

Uchijima, J. 1976. Maize and rice. In J,.L. Monteith (ed.), *Vegetation and the atmosphere. Vol. 2. Case studies.* London: Academic Press, pp. 30–62.

Van Cleve, K. et al. 1983. Taiga ecosystems in interior Alaska. *Biosci.* 33, 39–44.

Vourlitis, G. L. et al. 2015. Variations in evapotranspiration and climate for an Amazonian semi-deciduous forest over seasonal, annual, and El Niño cycle. *Int. J. Biomet.* 59, 217–30.

Weller, G., and Holmgren, B. 1974. The microclimates of the Arctic tundra. *J. Appl. Met.* 13, 854–62.

Wessel, D. A., and Rouse, W. R. 1994. Modelling evaporation from wetland tundra. *Bound.-Layer Met.* 68, 109–30.

Wever, L. A., Flanagan, L. B., and Carlson. P. J. 2002. Seasonal and interannual variation in evapotranspiration, energy balance and surface conductance in a northern temperate grassland. *Agric. Forest Met.* 112, 31–49

Whiting, G.J. and Chanton, I. 1993. Rimary Primary production control of metrhane emission from wetlands. *Nature* 3445 794-95.

Whitman, W.C. 1969. Microclimate and its importance in grassland ecosystems. In R. L. Dix and R. G. Beidleman (eds.), *The grassland ecostem: a preliminary synthesis.* Fort Collins, CO: Colorado State University, Range Science Department. pp. 40–64.

Wielgolaski, F. E. (ed.) 1997, *Polar and alpine tundra.* Amsterdam: Elsevier.

Worrall, F. et al. 2015. A 19-year long energy budget of an upland peat bog, northern England. *J. Hydrol.* 520, 17–29.

Zhang, T-T. et al. 2014. Land-atmospheric water and energy cycle of winter wheat, Loess Plateau, China. *Int. J. Climatol.* 34, 3044–53. doi:10.1002/joc.3891.

8

Microclimates of Physical Systems

This chapter continues the presentation of microclimatic conditions in different environments, here treating the physical systems of lakes, rivers, snow cover, mountains, and cities.

A. Lakes

Lakes account for about 3 percent of the Earth's land surface with an annual maximum of 3.8 percent based on AMSR-E data (J. Kimball, p.c. 2011). Verpoorter et al. (2014) use imagery from the Enhanced Thematic Mapper Plus (ETM+) sensor on board the Landsat 7 satellite collected in year 2000 ± 3 years. For all lakes >0.002 km², they estimate ~117 million lakes with a combined surface area of about 5×10^6 km², which is 3.7 percent of the Earth's non-glaciated land area.

The largest in terms of surface area is Lake Superior at 82,400 km² followed by Lake Victoria in East Africa at 69,500 km². The deepest (>1600 m) is Lake Baikal in southern Siberia, which contains 20 percent of the world's freshwater. In the central Mackenzie basin of western Canada, lakes occupy 37 percent of the land surface. Wetzel (2001) provides a comprehensive treatment of limnology – the study of lakes and river systems and their ecology.

Water is an excellent absorber of shortwave radiation (a low albedo of 0.06–0.10); however, the amount of shortwave radiation reflected and hence not absorbed in the water column is affected by the water clarity (turbidity) and the angle of incidence between the Sun's rays and the water surface (solar zenith angle; wind/waves).

In clear water blue and violet light penetrate farthest and red light is absorbed first. However, the opposite is observed in very turbid lakes, or where there is dense phytoplankton (Ganf, 1974). In Lake George, Uganda, red light (0.5–0.7 μm) representing 10 percent of the total surface light reaches a depth of 50 cm and only 1 percent reaches a depth of 90 cm. In Lake Windermere in the Lake District of northwest England, Pearsall and Ullyott (1934) measured the percentages of full daylight penetrating to 1.0 m and 4.3 m depths over a year. The values at the 1-m depth ranged from almost 47 percent in March to 21 percent in August and correspondingly, 6.7 percent

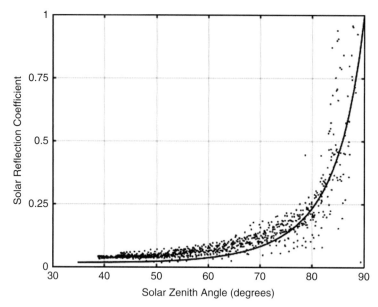

Figure 8.1. Relationship between the reflection coefficient (or albedo) and solar zenith angle measured over Great Slave Lake, Canada, during the ice-free season during cloud-free conditions. Line shows the Fresnel curve for direct-beam reflection from a plane surface (*Source*: P.D. Blanken, unpublished data, 2015).

in March and 1.5 percent in June at a depth of 4.3 m. The penetration is shown to depend on the abundance of phytoplankton in the surface layers. Blue-green algae reduce the light intensity at 4.3 m by more than 50 percent in July-August when they are most active.

For six shallow lakes (1.4–2.6. m deep) in the Danube delta, Cristofor et al. (1994) measured light penetration and factors controlling it over two years. Transparency ranged from 30 to 86 percent. Light extinction was determined 64 percent by phytoplankton, 11 percent by detritus, and 7 percent by zooplankton. Suspended detrital particles were maintained by winds of 3–5 m s^{-1}, which occur with 68–80 percent frequency. The light penetration measured by Secchi disks varied from 0.2 to 2.1 m. In general, 13–15 percent of surface light reached these Secchi depths.

An increase in the reflection of solar radiation from the water surface means that less radiation is available to penetrate through the water column. Water surfaces have the unique property of dramatically increasing the surface albedo when the solar zenith angle exceeds roughly 80° (Figure 8.1). This explains why a lake or ocean surface can have a mirrorlike appearance under certain conditions. These conditions occur when (i) the Sun is positioned with a large zenith angle (e.g., sunrise/sunset, and polar regions); (ii) wind causes waves that continually change the surface from highly absorptive to highly reflective; (iii) the fraction of diffuse versus direct-beam solar radiation becomes large; and/or (iv) the water freezes, and the albedo can again change, depending on whether the ice is covered by snow or not.

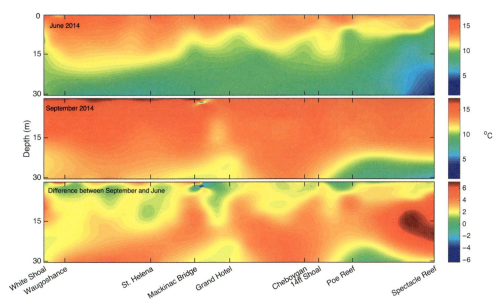

Figure 8.2. Temperatures to 30 m depth in a transect along the Straits of Mackinac between Lakes Michigan and Huron: (a) June 2014, (b) September 2014, and (c) September minus June (*Source*: P.D. Blanken, unpublished data, 2015).

As light energy is absorbed by lake water, it is converted into heat energy. This results in the warming of the surface water. During summer, thermally stratified lakes are warmer at the top and cooler at the bottom. The warm surface layer is called the epilimnion, and the cooler bottom layer the hypolimnion. The zone of rapid temperature decrease in the water column is called the thermocline (see Figure 8.2).

Surface winds mix the surface waters and in stratified lakes the efficiency of this mixing determines the depth of the epilimnion. Lake stratification persists during summer, but as the lake cools in autumn the stratification breaks down and the lake develops a uniform temperature. Freshwater is a unique substance in having its maximum density at 3.98 °C. Hence, as the surface water cools below this temperature, it becomes lighter than the water below. Thus, only a shallow layer needs to cool to the freezing point (0 °C). A water surface skin becomes supercooled at −0.03 to −0.1 °C and ice begins to form.

Ice growth is a function of the number of freezing degree-days (*FDD*) but is also affected by the thickness of overlying snow cover. Michel (1971) presents an equation for ice thickness (h_i in centimeters) using the accumulated freezing degree-days with respect to a base temperature of 0 °C

$$h_i = aFDD^{0.5}$$

where *a* is a coefficient that depends on the characteristics of the water body.

Figure 8.3 illustrates the seasonal temperature regime in deep and shallow lakes. Shallow lakes have no thermocline and in midwinter are nearly isothermal at close to freezing. In the Arctic lakes less than 2 m deep may freeze to the base.

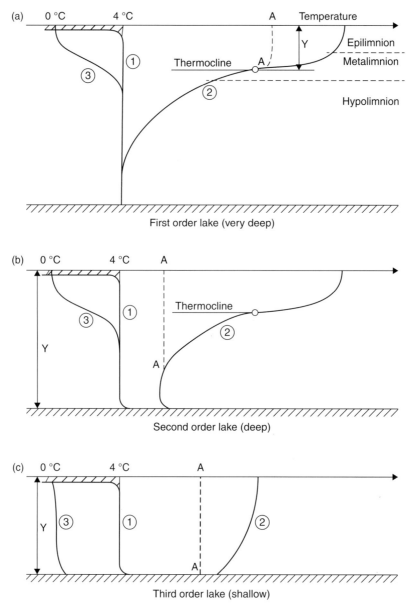

Figure 8.3. The temperature regime in deep and shallow lakes: curve 1 denotes early autumn, 2 midsummer, and 3 late winter (after Michel, 1971).
Source: Barry and Gan, 2011. *The global cryosphere* p. 193, fig. 6.3, Cambridge University Press.

The accumulation of snow cover on the ice depresses the ice surface below the water level, allowing water to flood over it. This saturates the snow and leads to the formation of a slush layer that freezes from the top down. This forms white snow ice in contrast to the black lake-water ice that is below.

Maximum annual ice thickness reaches 2 m in northern North America and Russia. Maximum thickness on lakes in the Russian Far East and eastern Siberia increases from 100 cm at 45 °N to 180 cm at 65 °N and in European Russia from 40 cm at 40 °N to 60 cm at 65 °N, according to Gronskaya (2000), while in western Siberia there is a steeper increase from 100 cm at 65 °N to 200 cm at 75 °N. Compared to open-water areas exposed to winter conditions, water temperatures under the ice cover can be constant and even increase. This is the result of sharp reduction in the turbulent transfer of sensible and latent heat from the lake to the atmosphere, of advection, and of warm water currents from warmer shore areas, and/or warming due to the penetration of solar radiation through the ice when not snow covered.

In spring, when the ice melts, water temperatures again become uniform with depth and the water column undergoes complete mixing. Lakes that experience such spring and autumn overturns are referred to as dimictic. In deep lakes a thermocline develops (see Figure 8.3). Some lakes in the tropics that stratify and mix several times a year are termed polymictic. In high northern latitudes the water temperature may not reach 4 °C and so the water column remains unstable until refreezing occurs. Such lakes are referred to as monomictic.

For the Great Slave Lake in the Northwest Territories of Canada, evaporation measurements by the eddy correlation method were made by Blanken et al. (2000) between July 24 and September 10, 1997, and June 22 and September 26, 1998. The observations were made on a small 10 m high rocky island with a fetch of at least 12 km in all directions. The lake was thermally stratified from mid-July to September with a thermocline depth of 15 m. Daily mean surface water temperatures in summer 1997 ranged between 12 and 16 °C. Extrapolation of evaporation measurements for the entire ice-free periods gave totals of 386 ± 127 and 485 ± 144 mm in 1997 and 1998–99, respectively. Air temperatures in summer 1998 were 4 °C above normal as a result of El Niño conditions, resulting in delayed ice cover and the large evaporative water loss in the fall/winter of 1998–99. Evaporation was mainly a function of the wind speed and the vapor pressure gradient between the water surface and 18 m. Fifty percent of the water loss occurred over only 20–25 percent of the days.

Rouse et al. (2005) provide an assessment of the effects of lakes in the central Mackenzie River valley around Yellowknife on regional energy and water balances. Net radiation is substantially greater over all water-dominate surfaces than for the uplands, which occupy 55 percent of the land surface. The seasonal heat storage increases with lake size. Medium and large lakes are slow to warm in summer, but their large cumulative heat storage at the end of summer, gives large convective heat fluxes in autumn and early winter. The evaporation season for upland and small, medium, and large lakes lasts for 19, 22, 24, and 30 weeks, respectively. The average radiative fluxes for summers 2000–02 for large and medium-sized lakes are shown in Table 8.1. The higher values of S_n and R_n over large lakes are a result of less cloudy conditions in the afternoons. Lakes substantially increase the regional net

Table 8.1. Radiative Fluxes (W m^{-2}) in Summer over Large and Medium-sized Lakes

	S_n	T_w	L_n	R_n
Large lakes	220	9.7	−44	177
Medium lakes	192	15.2	−46	146

Notes: S_n, L_n, and R_n are the net shortwave, longwave, and all-wave radiation, respectively. T_w is water surface temperature in °C.
Source: From Rouse et al., 2005, p. 299, table 6.

Table 8.2. Energy Balance Components (MJ m^{-2}) for Uplands and the Entire Region in the Two "Seasons"

Early Season	R_n	G	λE	H
Uplands	933	57	509	367
Entire region	973	286	433	254
Late season				
Uplands	161	−57	51	167
Entire region	205	−278	305	178

Source: From Rouse et al., 2005, p. 303, table 7.

radiation, the maximum regional subsurface heat storage, and evaporation. Table 8.2 compares the energy balances of uplands with those of the entire region. Evaporation decreases slightly in the first half of the season (April 13–September 2), but undergoes a large increase in the second half (September 3–January 14). The sensible heat flux is reduced considerably in the first half of the season, but changes little in the second half.

The influence of a small (65 km^2) lake in interior Alaska was examined by Kodama et al. (1983) in summer 1979. They found that during May 16–September 14 (excluding June 27–July 14) there was a lake breeze on 13 percent of hourly wind observations and a land breeze on 20 percent. These were absent at a station 6 km from the lakeshore. For easterly winds the air is warmed up to 2.4 °C and for westerly winds up to 1.4 °C. The lakeshore stations are 3.4 °C warmer than 6 km from the lake. The lake had 117 frost-free days compared with 104 at Fairbanks. The energy budget terms (megajoule per square meter per day) for July and August are as follows:

	R_n	H	λE
July	8.96	−1.63	−6.66
August	5.82	−2.43	−8.71

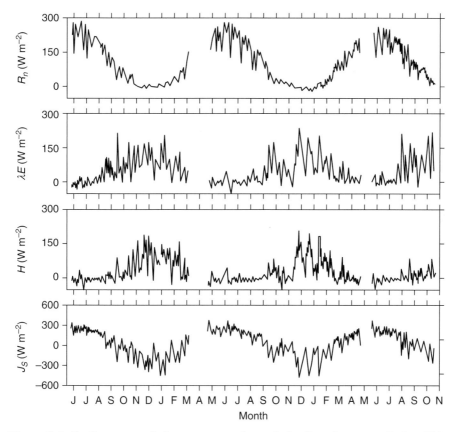

Figure 8.4. Surface energy balance measured over Lake Superior; net radiation (R_n), latent (evaporation; λE) and sensible (H) heat fluxes, and heat storage in the water (J_s). *Source*: From Blanken et al., 2011., *J. Great Lakes Res. 37,* p. 710, fig. 2.

The heat storage in the lake turns from positive in July to negative in August.

Over the North American Laurentian Great Lakes, the direct measurements of the surface energy balance were also made using the eddy covariance method (Blanken et al., 2011; Spence et al., 2011). Similar to the large and deep Great Slave Lake, Lake Superior's evaporative water loss was found to be greatest in the fall and winter seasons. During the summer, nearly all of the net radiation was used to heat the lake water, with virtually no evaporation or sensible heat loss (Figure 8.4). During the fall and winter, the cold, dry, and windy atmospheric conditions coupled with large expanses of ice-free water result in large evaporation spanning periods of 36–72 hours (corresponding to the passage of synoptic-scale cold fronts over the lake).

With large lakes such as Superior and Great Slave, this 6-month delay between summer energy inputs and winter energy outputs has a profound effect on the local climate. Summers tend to be cloud-free (except for summer fog) and cooler, and winters tend to be cloudy and warmer near the lakes. In summer, the stable atmospheric conditions suppress precipitation, whereas in winter, the unstable atmospheric conditions created by the large sensible heat fluxes enhance precipitation. Downwind regions around the Great Lakes experience record snowfalls due to this lake-effect snow. For example, the Upper Peninsula of Michigan receives an average of 635–762 cm of snow per year; the Tug Hill Plateau on the eastern shore of Lake Ontario receives an average of 610 cm of snow annually.

B. Rivers

Ecohydrological processes in river environments are strongly affected by stream flow variability (Naiman and Décamps, 1997). In persistent flow regimes with reduced stream flow variability, vegetation biomass is largely controlled by the pattern of groundwater availability forming a marked transition between aquatic and terrestrial environments (Doulatyan et al., 2014). Conversely, erratic flow regimes exhibit wider aquatic-terrestrial transitions. Stream flow variability plays a major role in riparian vegetation dynamics. Riparian biotic communities are those on the shores of streams and lakes. According to Naiman et al. (1993) the riparian zone comprises "the stream channel between the low and high water marks and that portion of the terrestrial landscape from the high water mark toward the uplands where vegetation may be influenced by elevated water tables or flooding and by the ability of the soils to hold water." For large streams, the riparian zones are characterized by well-developed, but physically complex, floodplains with long periods of seasonal flooding, lateral channel migration, oxbow lakes in old river channels, diverse vegetation, and moist soils.

Flooding can affect riparian vegetation both in a positive manner by providing nutrients, moisture, and seeds and negatively through uprooting, sediment removal, anoxia, and burial. Muneepeerakul et al. (2007) generalized these relationships by assuming that the standard deviation of the flows, σ_Q, and drainage area, A^s, are related as $\sigma_Q \sim A^s$. The scaling exponent s varies between 0.5 and 1. It plays an important role in determining the patterns of the riparian width and vegetation biomass. High values of s, that is, those when the fluctuation grows relatively fast with drainage area, could result in the riparian zones of high-magnitude streams being completely devoid of vegetation, whereas low values of s result in the riparian zones' increasing in width and total biomass with the stream magnitude. For low values of s, the riparian width keeps increasing, exhibiting a power law relationship. For intermediate values of s, the width initially increases for relatively small streams and subsequently decreases for large ones.

The microclimate of riparian zones of small headwater streams in western Oregon has been analyzed by Anderson et al. (2007). Microclimatic gradients were found to be strongest within 10 m of the stream center (average stream width 1.1 m). Forest thinning on the hill slopes resulted in subtle changes in microclimate; mean air temperature maxima were 1 to 4 °C higher than in unthinned stands. With buffer zone $s \geq 15$ m width, daily maximum air temperature above the stream center was less than 1 °C greater, and daily minimum relative humidity was less than 5 percent lower than for unthinned stands. In contrast, air temperatures were significantly higher within patch openings (6 to 9 °C), and within buffers adjacent to patch openings (3 °C) than within unthinned stands. Rambo and North (2008) examined the vertical and horizontal effects of streams in the Sierra Nevada. Stream influence on microclimate was limited to < 5.0 m vertically and < 7.5 m horizontally. In summer and winter, mean daily temperature and vapor pressure deficit (*D*) increased horizontally and vertically from the stream. Maximum absolute differences in temperature and *D* between upland and streamside conditions were greater in summer than winter.

In middle and high latitudes rivers freeze over in winter. The duration of the ice season in Siberia increases eastward from 150–200 days, on average, on rivers in the Ob basin, to 160–220 days in the Yenisei basin, and 180–230 days in the Lena basin (Vuglinsky, 2002). In Canada, the mean ice cover duration (1970–2001) ranges from 65 days on the Thames River at Thamesville, Ontario, to 121 days on the St. John River near East Florenceville, New Brunswick, to 174 days on the Athabaska River below Fort McMurray, Alberta, to 249 days on the Back River, Nunavut (Milburn, 2008).

With a small degree of supercooling (0.01 °C) of the surface water, frazil crystals form in turbulent water, and eventually these amalgamate into pancakes whose shape and rims are due to repeated collisions. These eventually merge into sheets. Near the banks where flow is slower, ice particles develop into a continuous layer of skin ice, which then thickens by thermal cooling. Accumulations of pans, especially at bends in the river, may lead to the formation of ice jams. Freeze up and break up of ice jams may each lead to floods. Break up jams are accumulations of solid ice blocks, whereas freeze up jams are usually unconsolidated ice floes. Large north-flowing rivers such as the Mackenzie River in Canada are prone to flooding induced by ice jams since the southern region of its drainage basin are thawed while the northern regions are still frozen.

Ice thickness on rivers in southern Canada is about 0.3 m, increasing to 1.7 m in the Arctic (Prowse, 1995). In Siberia, corresponding figures are 0.75 and 2.1 m on the Yenisei and 0.65 and 2.5 m on the Lena River (Vuglinsky, 2002). River ice floats with about 90 percept of its thickness submerged, constricting the flow.

C. Snow Cover

The microclimate of snow cover has great importance for the conditions at the underlying surface and in the soil. These in turn impact the behavior of small animals, insects, and microbes living in and beneath the snow pack.

The main control of snow microclimate is its albedo. In the visible wavelengths this is up to 0.95 for fresh snow, declining gradually as the snow ages and forest debris, algae, dust, and soot accumulate on its surface. About eight days after new snow accumulation in the Sierra Nevada, the albedo in spring may decline to about 0.50 (Miller, 1955). Figure 8.5 illustrates the spectral albedo of snow cover and the effects of dust and soot. The albedo drops below 0.4 at wavelengths beyond 1 μm,

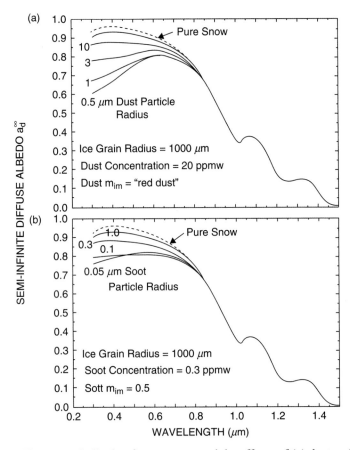

Figure 8.5. The spectral albedo of snow cover and the effects of (a) dust and (b) soot particles (from Warren and Wiscombe, 1980).
Source: *Journal of the Atmospheric Sciences*, 37, p. 2739, fig. 4. Courtesy American Meteorological Society.

where ice particles are moderately absorptive. At short wavelengths, reflectance is not very sensitive to grain size.

Snow cover is notable for its anisotropic angular reflectance pattern, especially in the near-infrared wavelengths (Painter and Dozier, 2004). Most Earth surface materials reflect the Sun's incident radiation back toward the source, but snow, like cloud, scatters light in a forward direction, away from the illumination source. The angular pattern of reflectance is described by a continuous function termed the bidirectional reflectance distribution function (BDRF; see Chapter 4; Schaepman-Strub et al., 2006; Dozier et al., 1988). In practical terms, the BDRF is measured at a discrete and wide range of viewing and illumination angles. The degree to which the BDRF departs from isotropic scattering (where reflectance is the same at all angles) is given by the anisotropic reflectance factor. Dumont et al. (2010) provide detailed angular reflectance measurements over the range 0.5–2.0 μm. They show that the spectral variation of the anisotropic reflectance factor is controlled by the ice absorption coefficient. For wavelengths <1 μm, a noteworthy effect at near-vertical incidence is the darkening at grazing angles. This is a consequence of multiple scattering within the snowpack. In contrast, for wavelengths >1 μm and/or large incident zenith angles, forward scattering is stronger because absorption is high and single scattering prevails.

Some fraction of the incident solar radiation can penetrate into the snowpack. About a quarter of the surface radiation in the 0.4–0.5 μm wavelength range penetrates to 5 cm depth for spherical grains of 200 μm diameter and density of 200 kg m^{-3} and a small percentage to 15 cm depth (Pomeroy and Brun, 2001).

In the central Rocky Mountains of Colorado and Wyoming, fresh snow has a density in the range 10–250 kg m^{-3} with a peak in the range 60–100 kg m^{-3}. Its value varies with crystal size, shape, and the degree of riming (Judson and Doesken, 2000). Densities tend to decrease irregularly with air temperature to about −20 °C. Settled snow is in the range 200–300 kg m^{-3}, and wind-packed snow 350–400 kg m^{-3}.

Snow cover has a major role in insulating the underlying ground because it has a large air content and therefore low thermal conductivity. The typical thermal conductivity for fresh dry snow with a density of 100 kg m^{-3} is 0.045 W m^{-1} K^{-1}, more than six times less than that for soil. Sturm et al. (1997) found that effective thermal conductivity increased to 0.6 W m^{-1} K^{-1} for drifted snow with a density of 500 kg m^{-3}. A 10 cm snowpack in the boreal forest of Saskatchewan was shown to have a temperature of −5 to −10 °C at 1–5 cm above the ground compared with an air temperature just above the snow surface of −8 to −33 °C (Pomeroy and Brun, 2001).

Snow is readily transported by the wind. Estimates for the high plains of the United States indicate that up to 75 percent of annual snowfall is eroded by the wind (Tabler and Schmidt, 1986). Snow transport involves creep (grains rolling on the surface), saltation (snow particles jump up to ~ 50 cm along the surface), and turbulent diffusion (particles are lifted and suspended in the airflow up to tens of meters). However, the greatest mass of snow is typically transported below 1 m height. Dry snow has a threshold wind speed for transport of about 7–8 m s^{-1} (Pomeroy and Brun, 2001). Blowing snow is strongly subject to sublimation and 25–40 percent of annual snowfall may be lost in this manner over the Canadian prairies.

In open environments, terrain irregularities and vegetation can produce large variations in snow accumulation by trapping drifting and blowing snow. Gray et al. (1979) report on the relative snow water retention by different landscapes (see Table 8.3).

Snow falling on trees is partially intercepted by leaves and branches. Snow bridges formed between branches increase the collection area, and cohesion due to ice bonds between snow crystals retains additional snow. Hedstrom and Pomeroy (1998) modeled interception efficiency as a function of leaf area index (LAI) and air temperature. A doubling of LAI from 2 to 4 caused the interception efficiency to increase from approximately 0.4 to 0.5. The large surface area of intercepted snow on trees means that there is considerable potential for sublimation. Pomeroy and Brun (2001)

Table 8.3. Snow Water Retention Relative to Open Level Grassland

Landscape Type	Relative Accumulation
Level grazed pasture	0.60
Gradual slopes, ungrazed pasture	1.25
Steep slopes, ungrazed pasture	2.85
Steep slopes, brush	4,20
Ridges, ungrazed pasture	0.40–0.50
Wide valley bottoms, grazed pasture	1.30
Small, shallow drainage ways, ungrazed pasture, stubble, fallow	2.0–2.15
Level fallow	1.0

Source: From Gray et al.,1979.

indicate that about one-third of the annual snowfall on dense coniferous canopies is lost via such sublimation.

Small clearings that have a diameter of 5 times the height of surrounding forest in Colorado mountains accumulate the maximum amount of snow, whereas with clearing diameter >12 times the forest height the clearing has less snow as a result of wind erosion (Troendle and Leaf, 1981).

Apart from the high albedo of snow cover that greatly limits shortwave absorption, snow cover has high emissivity in the infrared. Warren (1982) shows that it varies from 0.975 to 0.995 with radiation wavelength over the range 3 to 15 μm. This, and the very small transfer of heat from the underlying ground due to the low thermal conductivity of the snow, means that the temperature of the snow surface drops sharply at night.

In spring on the prairies, the energy for snowmelt is provided mainly by net radiation (Granger and Male, 1978). The contribution of turbulent heat fluxes is generally small, but highly variable from year to year. Warm air advection events can occasionally result in large sensible heat contributions leading to early melt. Usually the sensible heat contribution increases as the season advances. The transition from snow cover to bare ground generally occurs rapidly over a 10–15 day period.

For Niwot Ridge (40 °N, 3517 m a.s.l.) in the Colorado Rocky Mountains, Cline (1997) determined energy budgets of snow cover in 1994 and 1995. In 1994, net radiation accounted for 75 percent of the energy available for snowmelt, and sensible and latent heat fluxes accounted for the remaining 25 percent. During the 1995 snowmelt season, which was 1.3 °C higher and slightly more humid than the previous year, the sensible and latent heat fluxes accounted for 54 percent of the energy for snowmelt and net radiation for only 46 percent. The 1994 melt season began on May 5 and lasted 32 days, while the 1995 season began only on June 1 and lasted 45 days.

Snow covers have vertical structure that depends on their history, crystal type, and grain size. A snow surface undergoes some transformation before it is buried by the next snowfall and the pack also settles under its own weight. In dry snow there are vertical temperature gradients as a result of solar radiation absorption and nocturnal cooling at the surface. Temperature gradients of <5 °C m^{-1} produce small rounded grains, whereas gradients of 5–15 °C m^{-1} result in faceted crystals. Gradients >15 °C m^{-1} give rise to the growth of large depth hoar crystals as a result of vapor migration down the temperature gradient. These layers are especially weak and unstable. Wet snow is common when the snow temperature is near the melting point. Snow may have a liquid water content from 0.1 to 8 percent of the snow mass. It percolates into the snow as long as its content exceeds that held by the capillary forces. Water flow through the snow is affected by impermeable layers, zones of preferential flow known as flow fingers (4–5 cm diameter), and large melt water drains.

A seasonal snow cover classification has been proposed by Sturm et al. (1995). It is based on three major variables (air temperature, wind speed, and snowfall) and recognizes seven snow cover classes (tundra, taiga, alpine, prairie, maritime, mountain, and ephemeral). Five representative single value discrimination plots are shown in Figure 8.6 for tundra, taiga, alpine, and maritime snow. A map of the snow classification distributions in North America is presented in Figure 8.7.

Snow depth is highly variable and poorly known. Measurements at weather stations are typically made by inserting a ruler into the snow pack. In Russia, the depth at three fixed stakes is averaged. In mountain environments monthly measurements are collected at fixed snow courses, together with density observations. New instrumentation is, however, becoming available. Deems et al. (2013) report on the measurement of snow depth with lidar. Modern sensors allow mapping of vegetation heights and snow or ground surface elevations below forest canopies. Typical vertical accuracies for airborne data sets are decimeter scale with order 1 m point spacings. Ground-based systems typically provide millimeter-scale range accuracy and submeter point spacing over 1 m to several kilometers.

The snow water equivalent (SWE) is the water content of a snow pack when it is melted. It is expressed as the liquid depth in centimeters. SWE can be estimated from the snow depth (centimeters) multiplied by its density (expressed as a decimal; 100 kg m^{-3} = 0.1). Global SWE has been estimated from SMMR and SSM/I passive microwave data (except where the snow is wet or shallow) for January, 1978 to December, 2007. Armstrong et al. (2005 with updates) provide the Global Monthly EASE-Grid Snow Water Equivalent Climatology. The ESA GlobSnow Snow Water Equivalent (SWE), Version 2 data set combines passive microwave and synoptic weather station data (see Pullainen, 2008) on a daily, weekly, and monthly basis for the Northern Hemisphere from 1979 to the present. It can be accessed at

nsidc.org/data/NSIDC-0595.

In the western United States there is an extensive network of snowpack telemetry (SNOTEL) sites operated by the Natural Resources Conservation Service (NRCS). There are 730 SNOTEL stations in 11 western states including Alaska. Basic SNOTEL

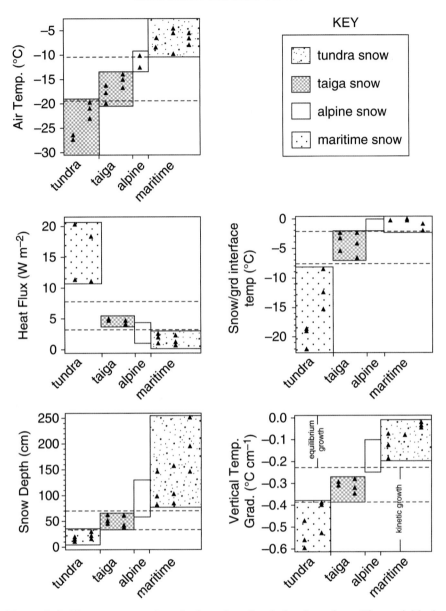

Figure 8.6. Five representative single value discrimination plots. The variable is shown on the ordinate and the snow class on the abscissa. The dashed line on the vertical temperature gradient plot (bottom right) is the critical threshold for depth hoar growth (from M. Sturm et al., 1995).
Source: *Journal of Climate* 8, p. 1270, fig. 5. Courtesy of the American Meteorological Society.

sites have a pressure sensing snow pillow, storage precipitation gauge, and air temperature sensor. Data are recorded every 15 minutes and reported out in a daily poll of all sites by two NRCS master stations operated in Boise, Idaho, and Ogden, Utah. The system uses meteor burst communications technology to collect and communicate

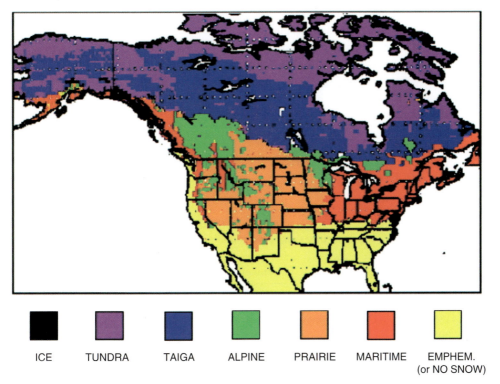

ICE   TUNDRA   TAIGA   ALPINE   PRAIRIE   MARITIME   EMPHEM.
(or NO SNOW)

Figure 8.7. Snow class distribution based on climate variables over North America (from M. Sturm et al., 1995).
Source: Journal of Climate 8, p. 1277, fig. 10; courtesy of the American Meteorological Society.

data in near-real time. VHF radio signals are reflected from the ever-present band of ionized meteorite trails in the ionosphere. Serreze et al. (1999) performed an analysis of SNOTEL data for the 1995/96 season. They showed that the percentage of annual precipitation represented by snowfall is highest for the Sierra Nevada (67 percent), northwestern Wyoming (64 percent), Colorado (63 percent), and Idaho/ western Montana (62 percent). There are much lower percentages for the Pacific Northwest (50 percent) and Arizona/New Mexico (39 percent). For the latter region much of the precipitation falls outside the snow accumulation season.

D. Mountains

Mountain areas account for 20–25 percent of the land surface, according to different definitions, and hence are crucial features in forming topoclimates. They present major obstacles to airflow and set up large contrasts in radiation, temperature, precipitation, and humidity.

Air pressure and density decrease exponentially with altitude, whereas mean air temperature decreases almost linearly though the troposphere. Table 8.4 shows

Table 8.4. Values of Air Pressure (hPa), Temperature (°C), and Density (kg m^{-3}) in the Standard Atmosphere

Height (m a.s.l.)	Air Pressure	Air Temperature	Density
0	1013.25	15.0	1.2250
2000	795.0	2.0	1.0581
3000	701.2	−4.5	0.9093
4000	616.4	−11.0	0.8194
6000	472.2	−24.0	0.6601

Source: COESA, 1962.

Table 8.5. Solar Radiation on a Horizontal Surface in the Alps for Daily Totals (MJ m^{-2}) in December and June (Gradients in Percent km^{-1})

Elevation (m a.s.l.)	December Daily Total	December Gradient	June Daily Total	June Gradient
1500	5.86		28.97	
2000	6.18	9.2	30.42	9.0
2500	6.41	6.8	31.71	7.6
3000	6.62		32.84	
4000	6.94	4.7	34.75	5.6

Source: After Mueller, 1984, tables 3 and 4, pp. 138–41.

average values for a Standard Atmosphere, where mean sea level pressure is 1013.25 hPa, surface temperature is 15 °C, and the lapse rate is −6.5 °C km^{-1}.

Radiation Gradients

The vertical decrease in air density and aerosol content means that atmospheric attenuation of incoming solar radiation decreases with altitude. Hence, solar radiation amounts tend to increase with height. Data assembled by Mueller (1984, 1985) for the Alps illustrate this tendency with gradients ranging from 9 percent km^{-1} at 1500 to 2000 m to 5 percent km^{-1} at 3000 to 4000 m (Table 8.5).

An Alpine Surface Radiation Budget (ASRB) network was established in the Alps between 370 m and 3580 m a.s.l. in 1994 (Marty et al., 2002). For cloud-free conditions, global solar radiation increases by 7 W m^{-2} km^{-1} in winter and 27 W m^{-2} km^{-1} in summer.

The rate of increase with altitude of ultraviolet radiation is significantly greater than for global radiation. For clear-sky conditions, measurements in the Alps showed an altitude effect at Jungfraujoch (3576 m) that for annual totals was 19 percent km^{-1} (UV-B) and 11 percent km^{-1} (UV-A) compared with 9 percent km^{-1} (global radiation) with reference to Innsbruck (577 m) (Blumthaler et al., 1992). In contrast, in the Andes near La Paz, Bolivia, the altitude effect for UV-B irradiance was only about 7 percent km^{-1} between 3420 and 5200 m (Zaratti et al., 2003), because the amounts of aerosols and tropospheric ozone were very small there.

For clear-sky down-welling infrared radiation, Marty et al. (2002) report for the Alps an average 33 W m^{-2} km^{-1} decrease with higher rates in summer than winter. This decrease is in response to the decrease in vapor content with altitude. The outgoing infrared radiation becomes less negative with altitude so that the vertical gradient of net infrared radiation is very small.

Gradients of net radiation are determined mainly by the gradient of solar radiation because of the very small altitudinal variation of net infrared radiation. The annual mean at the ASRB stations is 14 W m^{-2} km^{-1}, although in summer it is almost constant with height (Barry, 2008, pp. 36–51).

Even small altitudinal differences can result in large differences in energy components and soil temperature. At hill and valley sites 5 km apart and 200 m different in elevation near Fairbanks, Alaska, Wendler (1971) measured the energy balance during August, 1966 to July, 1967. R_n at the hill station was only 40 percent of that in the valley and the λE term was only 24 percent. The surface temperature was higher on the hill, which increased the outgoing IR radiation and reduced the net radiation accordingly.

Extreme temperature contrasts due to terrain were demonstrated by Schmidt (1934) at around 1250 m near Lunz in the Austrian Alps. Observations in the famous 150 m deep Gstettneralm doline showed record January minima of −51 °C, compared with around −20 °C on the rim. The vegetation is inverted with grasses in the basin and forest on the higher slopes. A similar phenomenon has been reported at Peter Sinks at 2500 m at the crest of the Bear River Range, Utah. The basin is a 1 km oval-shaped limestone sinkhole, surrounded by higher terrain of about 150 m relief. On February 1, 1985, an extreme reading of −56 °C was recorded at the basin floor. Frost hollows have a similar origin although downslope flows play only a minor role. Clements et al. (2003) examined the Peter Sinks during three clear, dry September nights in 1999. The evolution of cold-pool characteristics depends on the strength of prevailing flows above the basin. On a calm day, a 30 °C diurnal temperature range and a strong nocturnal potential temperature inversion (22 K in 100 m) were observed in the basin. Initially, downslope flows formed on the basin sidewalls. As a very strong potential temperature jump (17 K) developed at the top of the cold pool, however, the winds died within the basin and over the sidewalls. A persistent turbulent sublayer formed below the jump. Turbulent sensible heat flux on the basin floor became negligible shortly after sunset, while the basin atmosphere continued to cool. The cold pool developed fully within 2–3 hours after sunset to a height of around 75 m above the basin floor. Cooling rates for the entire basin near sunset were comparable to the 90 W m^{-2} rate of loss of net longwave radiation at the basin floor, but these rates decreased to around 10 W m^{-2} by sunrise. Temperatures over the slopes, except for a 1–2 m-deep layer, became warmer than over the basin center at the same altitude.

Temperature Gradients

Temperature gradients in mountain environments have been widely investigated. Mean day and nighttime lapse rates for 1968–95 in northern England are examined by Pepin (2001) and shown to average −8.6 °C km^{-1} and −5.5 °C km^{-1}, respectively. Lapse rates of maximum temperature are greatest in March (−10.4 °C km^{-1}) and least in May (−7.3 °C km^{-1}) and December (−7.5 °C km^{-1}). Airflow direction has a large influence with day and night values ranging, respectively, from −10.4 °C km^{-1} and −5.7 °C km^{-1} for westerly flow to −6.7 and −5.0 °C km^{-1} for southeasterly flow. For the European Alps (Tirol, northern Italy, and Trentino-Adige), Rolland (2003) determined maximum and minimum temperature lapse rates for winter and summer for 269 stations over periods of 30 or 55 years. Results, shown in Figure 8.8, all indicate a regular pattern with summer maxima and winter minima. Geographical differences reflect the regional topography.

The seasonal patterns of lapse rates vary considerably with latitude. Kattel et al. (2013, 2015) analyze the temperature lapse rates on the southern slopes of the central Himalaya (70–3920 m) and the northern slope of the eastern Himalaya (3800–4500 m). In both regions the lapse rate is least in summer as a result of the effect of the monsoon in releasing latent heat at higher elevations. On the northern slopes, the rate is greatest in winter when cold air surges give strong cooling at high elevations. The mean lapse rate is −5.8 °C km^{-1} in August and −10.5 °C km^{-1} in January–February.

Seasonal patterns on the southern slopes of the central Himalaya are more complex. There are maxima in the pre- and post-monsoon seasons and minima in winter and summer. The two maxima are associated with clear skies and strong sensible heating. The mean rates are −6.8 °C km^{-1} in April and −5.4 °C km^{-1} in October–November. The lowest mean rate in winter (−4.5 °C km^{-1}) is a result of high level cooling, as on the northern slopes, and the summer minimum (−4.7 °C km^{-1}) is due to latent heat release at higher elevations.

On a smaller space and time scale, mountain and hill slopes form nocturnal patterns due to slope flows of air. Cold air draining down slopes leads to a shallow cold air "lake" in the valley bottom, while the slope above it has higher temperatures due to the mixing by turbulence in the airflow. This forms a temperature inversion.

This zone is known as the "thermal belt" and is located at about one-fourth to one-third the height of the relative relief. These features were first reported from Appalachia, where they were preferred zones to plant fruit trees because of reduced frost frequency. Above the thermal belt there is a normal temperature lapse rate.

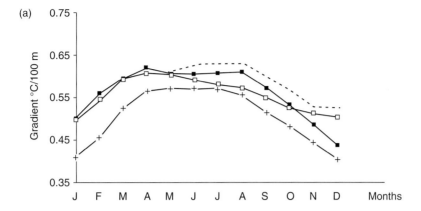

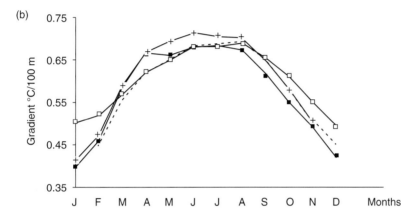

Figure 8.8. Seasonal variations of lapse rates (°C /100 m) for (top) minimum and (bottom) maximum temperatures. For the Tyrol (solid line), northern Italy (dashed line), and in the Trentino–Upper Adige region divided into VB (line with black dots) and SL (line with open circle) (from C. A. Rolland, 2003).
Source: *Journal of Climate* 16: p. 139, fig. 3: 2 of 6 panels; courtesy American Meteorological Society.

Tabony (1985) showed that, for inland stations in Great Britain, minimum temperature and the daily range depend on two scales of topography – the "local" and "large-scale" shelter. The former is determined by the height above the valley within 3 km and the latter by the average terrain height above the valley within a 10 km radius. The height above the valley is the most important variable. For 147 stations it averaged 45.6 m with an average gradient over a 2 × 2 km square of 23.6 m km^{-1}. Its maximum effect on diurnal range occurred in early autumn, when the difference in soil moisture deficit between valleys and summits was a maximum. Large-scale drainage is more important in winter and primarily represents the effects of nocturnal drainage.

Altitudinal gradients of soil temperature in the Santa Catalina Mountains, Arizona, were measured on north- and south-facing slopes at 7.6 cm depth between 2133 m and 2743 m by Shreve (1924) in summer 1922. The maximum soil temperature gradients were the following:

2133 m S–2743 m S	1.3 °C /100 m
2133 m N–2743 m N	1.8 °C /100 m

The gradients decreased upward, particularly on the north-facing slope, as shown in the following:

2133 m S–2627 m S	1.8 °C/100 m
2133 m N–2627 m N	3.0 °C/100 m
2627 m S–2743 m S	0.9 °C/ 100 m
2627 N–2743 m N	0.6 °C/100 m

These gradients average a third less than that of the air temperature. Differences in minimum soil temperatures with altitude were inconsistent.

Soil temperatures at 15 cm depth under forest cover in the Great Smoky Mountains of Tennessee are discussed by Shanks (1956). The data were collected from January to November, 1950, between 420 and 1800 m elevation under hardwood and pine cover at the lowest station and spruce and fir at the highest. The altitudinal gradient for April to September averaged 0.54 °C/ 100 m, close to that of the air temperature. The maximum gradient was 0.58 °C/100 m in July and the minimum 0.4 °C/100 m in November. In spring, the rise in soil temperatures lagged air temperatures by about a month, while in autumn there was a corresponding lag in the soil cooling. The soil-air temperature difference is greatest in summer and at the lower elevations.

Scherrer and Körner (2010) examine alpine surface temperatures with infrared thermometers and microloggers at 3 cm in the soil at six alpine sites in the Alps and arctic-alpine sites in Norway, Sweden, and Svalbard. There are persistent root zone temperatures of 2–4 °C above the air temperature during summer. Surface temperatures show strong positive (2–9 °C) and negative (3–8 °C) deviations from air temperature on bright days and clear nights, respectively, with substantial slope differences. Microtopography can mimic temperature differences of large elevational (or latitudinal) gradients over very short horizontal distances, enabling plants to escape warming climatic conditions.

Vapor Pressure

Unlike temperature, which decreases nearly linearly with altitude, vapor pressure decreases exponentially. Values on Mt. Fuji, Japan, for example, average 14.5 hPa at sea level and 3.3 hPa at 3776 m. Kuz'min (1972) gives an empirical expression for vapor pressure at elevation z

$$e_z = e_0 \, 10^{-az} \qquad\qquad 8.1$$

where e *is* the vapor pressure at sea level, z is the height (kilometers), and the empirical coefficient $a = 0.159$ for central Asian mountains.

Vertical gradients of vapor pressure deficit (D) are an important indicator of drought stress on plants. For Mt. Haleakala, Maui, annual values of D increase from 0.6 hPa at 1600 m on the windward slope to 3.3 hPa at 2500 m; corresponding values on the leeward slope increase from 2.8 hPa between 1000 and 1600 m to 6.5 hPa between 250 and 3000 m (Loope and Giambelluca, 1998).

Precipitation

The variation of precipitation amounts with altitude varies widely around the world. There are broad latitudinal tendencies, as well as seasonal differences, and contrasts between windward and leeward slopes (Barry, 2008, chapters 4.4.2 and 5). In equatorial latitudes annual totals are greatest near the surface as the air is close to saturation and little uplift is needed for condensation and precipitation. In the tropics maxima occur between about 800 and 1500 m near the mean cloud base. In midlatitudes precipitation profiles are complex. On windward slopes amounts may increase to elevations of about 2500 m or, if the mountains are lower, the maximum may occur just east of the crest. In the eastern Alps (11–13.5 °E) maxima occur around 1100 m on the southern slopes and around 700 m on the northern slopes (Schwarb et al., 2001). However, in the western Alps the highest amounts are located over the mountains that reach 2500 m (Spreafico and Weingartner, 2005).

Snow Depth Gradients

Gradients of snow depth with altitude have been examined for six subareas in the eastern Alps and one in the eastern Pyrenees by Grünewald et al. (2014) on the basis of airborne laser scanning (see Deems et al., 2013) and airborne digital photogrammetry. They find that one altitudinal pattern fits 79 percent of subcatchments.

Snow depths increase with elevation up to a certain level, where they have a distinct peak followed by a decrease at the highest elevations. They suggest that there is a generally positive elevation gradient of snowfall that is modified by the interaction of snow cover and topography. These processes involve preferential deposition of precipitation and redistribution of snow by wind, sloughing, and avalanches.

Wind Speed

The variation of wind speed with elevation is at the largest scale a function of the mountain's location with respect to the global wind belts. In the midlatitude westerlies, wind speeds increase steadily with altitude up to the highest mountain summits. Mean monthly wind speeds on Mt. Washington, New Hampshire, in February are 22 m s^{-1} and on Mt. Fuji, Japan, are 21 m s^{-1} in January, for example. On the summit of Long's Peak, Colorado (4345 m), Glidden (1982) reports an average daily maximum on a 3.6 m tower for August 30, 1980, to May 31, 1981, of 29 m s^{-1}, with a maximum gust of 90 m s^{-1}.

In the tropical easterlies wind speeds remain more or less constant or may decrease with altitude. On the tropical margins they may switch to westerlies with altitude.

At the scale of a mountain range, winds are influenced by local topography as well as by synoptic pressure gradients. The air flowing over a mountain range is deflected upward and compressed between the topography and the tropopause inversion layer. This causes the air to accelerate as a result of what is termed the Bernoulli effect.

On lee slopes dynamically induced flows may descend as a result of the presence of a stable layer just above the level of the mountain summits. The descending air warms as a result of adiabatic compression at a rate of 9.8 °C km^{-1} and the relative humidity drops to low values. These winds are known as föhns in the European Alps and chinooks on the eastern slopes of the Rocky Mountains. On the western (lee) side of the Dinaric Alps, cold air may flow from the east in winter crossing the mountains. Although there is some temperature rise due to descent, the air arriving at the coast is colder than that it displaces and this wind is known as a bora.

Another dynamical wind in the lee of steep terrain is the rotor. This occurs when the upper airflow is separated from the ground, allowing a circulation cell to develop beneath it perpendicular to the terrain. Hence, the low-level flow is reversed (toward the mountains). This type of circulation is common in glacial cirques in alpine terrain when the upper flow is strong across the mountains.

Thermally induced winds involve slope flows and valley flows (Barry, 2008). At night, cold slope air drains downslope in a shallow layer (~100 m deep); this is termed a katabatic wind. It has speeds of about 2–5 m s^{-1}. The cold air accumulates in the valley bottom forming a cold air "lake" that can be 100–200 m deep. By day, slope heating creates a horizontal density gradient with the ambient air that generates motion toward the slope. Hence, the air is forced to rise up the slope in a shallow anabatic wind with speeds of 1–3 m s^{-1}. The cold air lake is gradually eroded by day as radiative heating warms the ground and convective mixing leads to the erosion of the upper inversion and its descent. This may take several hours.

The valley air is heated during the day about 2–3 times more than the volume of a box with an equivalent surface area over the adjacent plains. Hence, there is an up-valley airflow toward the mountains, known as a valley wind. At night, cold air from the mountains flows down the valley in a mountain wind. These winds form part of a circulation with reverse flows aloft just above the ridge-top level. They are best developed when the gradient winds are light. The up- and downslope winds feed into these along-valley flows. The circulations reverse just after sunrise and sunset.

E. Cities

Urban climate is treated in detail in Chapter 10 as a major category of topoclimate; here we are concerned with microclimatic processes and phenomena within cities. Arnfield (2003) addresses the scale issue in urban climatology, noting the basic distinction between the Urban Canopy Layer (UCL) and the Urban Boundary Layer (UBL). Within the UCL, roughly from the ground to roof level, microscale processes that are site specific control energy exchanges and airflow. Vertical shear and turbulence are strong here as a result of varying form drag and local advection. The overlying UBL is part of the planetary boundary layer, whose characteristics are determined by the land cover types of the urban surface and processes operating at larger space and time scales (discussed in Chapter 10). The UCL includes the roughness sublayer, where wakes and plumes (of heat, moisture, and pollutants) generated by tall buildings interact. At some height above the buildings, turbulent mixing erases these differences, giving rise to an inertial sublayer where turbulent fluxes are more or less constant with height.

The microclimate of built-up areas involves the effects of the walls, roofs, and ground as a result of the variability in the composition and material properties of the urban fabric (Oke, 1988). Ramamurthy et al. (2014) show that sensible heat fluxes from asphalt pavements and dark rooftops are twice as high as those from concrete surfaces and light-colored roofs. Also, the shift in the peak time of sensible heat flux in comparison with rural areas is shown to be mainly linked to the high heat storage capacity of concrete as well as to radiative trapping in the urban canyon. The vegetated soil surfaces that dot the urban landscape play a dual role: during wet periods they redistribute much of the available energy into evaporative fluxes, but when moisture stressed they behave more like impervious surfaces.

An important measure for radiation in urban areas is the sky view factor – a measure of the degree to which the sky is obscured by the surroundings for a given point. The effect of the sky view factor in urban canyons on nocturnal cooling was simulated in a software model and observed by Oke (1981). Canyon geometry was shown to regulate the rate of longwave radiative loss at night with the sky view factor leading to a decrease in the loss. The calculation of sky view factor for idealized and realistic canyons is discussed by Johnson and Watson (1984). The wall sky view factor (for a symmetrical, infinitely long canyon) is given by

$$\psi_w = 0.5(1 - cos\,\beta) \qquad\qquad 8.2$$

where β = arctan $(H/2W)$; H = building height and W = street width. The sky view factor for the surface element is $\psi_s = 1 - 2\psi_w$.

Table 8.6. Typical Values of Thermal Admittance, μ (W m^{-2} K^{-1})

Natural Materials		Urban Materials	
Snow	240	Brick	1,070
Peaty soil	600	Concrete	1,300
Clay soil	1,800	Asphalt	1,300
Water (20 °C)	1,580	Stone	2,220
Farmland	1,600	Urban	1,200–2,100

Source: After Oke, 1981.

Following Brunt (1941), the temperature at any time *t* after sunset is given by

$$T(t) = \frac{2L * t^{0.5}}{\pi^{0.5}\, \pi} \qquad\qquad 8.3$$

where $L*$ is the net longwave radiative flux and μ is the thermal admittance – a measure a material's ability to transfer heat in the presence of a temperature difference on opposite sides of the material. It is given by the square root of the product of a body's thermal conductivity and volumetric heat capacity, $(\kappa C)^{0.5}$ in W m$^-$K^{-1}. Since $L*$ and μ are fairly constant at night the surface temperature decreases approximately as $t0.5$. Typical values of μ are shown in Table 8.6.

The question of evapotranspiration from city lawns in summer was investigated by Suckling (1980) for a watered suburban lawn in Brandon, Manitoba. Over the 12 day period June 27 to August 1, 1979, λE (154 W m^{-2}) averaged 65 percent of net radiation (235 W m^{-2}). Figure 8.9 illustrates the mean diurnal course of the energy balance components. Hence, evapotranspiration can be a major component of the energy balance of suburban environments.

Radiation Microclimate as a Function of Exposure

For N, E, S, and W exposures at the surface adjacent to downtown buildings and in a residential area in Victoria, British Columbia, Tuller (1973) measured the radiation components on two clear days in March. The sky view factor for the adjacent building for each downtown orientation was 0.48. The daily totals of $S\downarrow$ and R_n are as follows:

Orientation	Solar radiation (MJ m^{-2})	Net radiation (MJ m^{-2})	R_n/S (percentage)
N downtown	1.97	0.68	34.5
E downtown	4.03	2.00	49.7
W downtown	8.19	5.80	70.8
S downtown	10.62	7.63	71.8
Residential	16.01	10.74	67.1

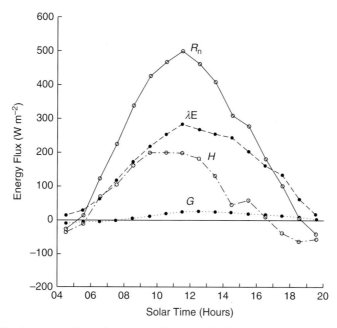

Figure 8.9. Average diurnal course of energy balance components during 12 summer days on a suburban lawn in Brandon, Manitoba (from P. W. Suckling, 1980).
Source: Journal of Applied Meteorology, 29: p. 608, fig. 2. Courtesy of American Meteorological Society.

There is a 5-fold difference in solar radiation between north and south orientations downtown and an 11-fold difference in net radiation. The adjacent and opposite buildings on the east side were about twice the height of those on the west, explaining the E-W contrast in radiation amounts.

A climate model of the urban canopy layer was developed by Mills (1997). The layer that is simulated includes buildings (closed volumes) and canyons (open volumes). Boundary-layer theory is applied to the overlying urban boundary layer and the derived relations are used to parameterize heat and momentum exchanges across the UBL/UCL interface. The exterior energy budgets of the canopy roof, walls, and floor are solved via an equilibrium surface temperature method. A building group is composed of identical structures with height/width (*H/W*) ratio of one regularly spaced on a flat surface. The model has three types of input: (i) latitude, longitude, month, and day; (ii) temperature and wind velocity for a reference level above the UCL; and (iii) UCL characteristics – building orientation, dimensions, and separation distance, thermal substrate, and surface radiative properties. Mills examined the effects of varying *H/W* and building density on energy fluxes and temperatures and undertook an analysis for Los Angeles in June. A few of his results are presented in Table 8.7.

Table 8.7. Energy Budgets (W m^{-2}) for Different Building
Orientations at 16.00 LST at 34 °N in June for $H/W = 1$

Surface	R_n	H	G
South wall	60	41	19
East wall	86	74	13
West wall	367	182	185
North wall	98	47	52
Roof	339	314	26
Canyon floor	228	183	45

Source: From Mills, 1997, table 4b.

Design

Microclimatic landscape design has become an important component of modern landscape architecture. Brown and Gillespie (1995) provide an introduction to the design of comfortable outdoor spaces and reducing the heating and cooling loads on buildings on the basis of microclimatic principles. They emphasize that radiation and wind conditions are most readily modifiable. Planting deciduous trees that provide seasonally varying shade, for example, can modify solar radiation and temperature. Outgoing terrestrial radiation is dependent on surface temperature, which is affected by the surface emissivity of the material. Wind velocity is highly variable in both direction and speed, as well as by time of day and year. Objects (buildings and shelterbelts) locally modify the flow direction and speed, and their relative location and orientation introduce further complications. Wind conditions can be assessed by using scale models in wind tunnels or by making measurements in a variety of generalized situations. Humidity is the most difficult variable to modify since turbulence readily mixes it. However, small enclosed gardens or parks surrounded by tall buildings can modify local humidity and temperature conditions.

Designing to optimize street climate in middle and high latitude cities has four goals, according to Oke (1988a): to maximize shelter, to maximize the dispersal of pollutants, to maximize urban warmth in the cool season, and to maximize access to solar radiation. However, several of the solutions to these objectives are in conflict. Thus, narrow streets and compact layout of building best provide shelter, whereas dispersal of pollutants requires separation and low building density. Similarly, compact buildings promote warmth, but openness is needed to provide radiation access. Oke (1988a) attempts to develop guidelines for resolving these issues.

Airflow around buildings has been extensively researched. There are three main zones of disturbed flow: upstream of the building there is a bolster eddy, while in its rear there is a lee eddy that is drawn into the low pressure cavity caused by the flow separation at the top and sides of the building. Farther downstream is the turbulent wake. With building arrays there are different flow interactions depending on the height and spacing of the buildings.

If the buildings are close to one another, the flow may not descend into the street canyons and there is an upper level "skimming flow" above the buildings. When the buildings are close to one another, the wake flows interact with the bolster eddies, which form on the windward side of buildings, and there are secondary eddies in the canyon spaces. For length/height values typical of canyons, Nakamura and Oke (1988) suggest that a height/width (H/W) ratio of ~ 0.65 should ensure considerable protection, and, therefore, a minimum acceptable value is about 0.4.

Observations in an east-west canyon in Kyoto, Japan, indicate the following conclusions (Nakamura and Oke, 1988): (i) With airflow normal to the long-canyon axis, a simple "reflection" vortex causes the two directions to be 180° apart above the roofs and at the canyon floor. For intermediate angles of attack there is a spiral vortex across the canyon along its length. For flow parallel to the axis there is little effect. (ii) There is a roughly linear relation between wind speeds above and within the canyon. For H/W ~ 1, the diminution factor is about 0.67. (iii) Air temperatures above and within the canyon are similar, but there are large surface-air differences for surfaces receiving direct beam radiation. (iv) The north wall and floor absorb large amounts of solar radiation influencing air temperatures across the canyon. (v) The canyon air remains near neutral stability through the 24 hours.

Shading Effect

Spangenberg et al. (2008) showed that a park in São Paulo, Brazil, in summer was up to 2 °C lower than an adjacent square and street canyon. The effect of adding shading trees to the street canyon was simulated for the same day using the numerical model ENVI-met. The simulations showed that incorporating street trees in the urban canyon had a limited cooling effect on the air temperature (up to 1.1 °C), but led to a significant cooling of the street surface (up to 12 °C) as well as a great reduction of the mean radiant temperature at pedestrian height (up to 24 °C). Although the trees lowered the wind speed by up to 45 percent of the maximum values, the thermal comfort was improved considerably as the physiologically equivalent temperature was reduced by up to 12 °C.

References

Armstrong, R.L. et al. 2005. Updated 2007. *Global monthly EASE-Grid snow water equivalent climatology*. Boulder, CO: National Snow and Ice Data Center. Digital media.

Anderson, P. D., Larson, D. J., and Chan, S. S. 2007. Riparian buffer and density management influences on microclimate of young headwater forests of western Oregon. *Forest Sci.* 53, 254–69.

Arnfield, A. J. 2003. Two decades of urban climate research: A review of turbulence, exchanges of energy and water, and the urban heat island. *Int. J. Clim.* 23, 1–26.

Barry, R. G. 2008. *Mountain weather and climate.* Cambridge: Cambridge University Press.

Bennie, J. et al. 2008. Slope, aspect and climate: Spatially explicit and implicit models of topographic microclimate in chalk grassland. *Ecol. Modelling* 216, 47–59.

Blanken, P. D. et al. 2000. Eddy covariance measurements of evaporation from Great Slave Lake, Northwest Territories, Canada, *Water Resour. Res.* 36(4), 1069–77.

Blanken, P. D. et al. 2011. Evaporation from Lake Superior. 1. Physical controls and processes. *J. Great Lakes Res.* 37(4), 707–16.

Blumthaler M., Ambach W., and Rehwald, W. 1992. Solar UV-A and UV-B radiation fluxes at two alpine stations at different altitudes. *Theor. Appl. Climatol.* 46, 39–44

Brown, R. D., and Gillespie, T. J. 1995. *Microclimatic landscape design: Creating thermal comfort and energy efficiency.* New York: J. Wiley & Sons.

Brunt, D. 1941. *Physical and dynamical meteorology.* London: Cambridge University Press, p. 428.

Clements, C. B., Whiteman, C. D., and Horel, J. D. 2003: Cold -air-pool structure and evolution in a mountain basin: Peter Sinks, Utah. *J. Appl. Met.* 42, 752–68.

Cline, D. W. 1997. Effect of seasonality of snow accumulation and melt on snow surface energy exchanges at a continental alpine site. *J. Appl. Met.* 36, 32–51.

COESA. 1962. U.S. Standard Atmosphere 1962. Washington DC: Comm. Extension Standard Atmosphere.

Cristofor, S. et al. 1994. Factors affecting light penetration in shallow lakes. *Hydrobiologia* 275/276, 493–8.

Deems, J. S., Painter, T. H., and Finnegan, D. C. 2013. Lidar measurement of snow depth: a review. *J. Glaciol.* 59, 467–79.

Doulatyan, B. et al. 2014. River flow regimes and vegetation dynamics along a river transect. *Adv. Water Resour.* doi: 10.1016/j.advwatres.2014.06.015.

Dozier, J. et al. 1988. The spectral bidirectional reflectance of snow. In: Spectral signatures of objects in remote sensing. Proceedings of the conference held 18–22 January, 1988 in Aussois (Modane), France. (T. D. Guyenne and J. J. Hunt, eds.). ESA SP-287. European Space Agency, pp. 87–92.

Dumont, M. et al. 2010. High-accuracy measurements of snow Bidirectional Reflectance Distribution Function at visible and NIR wavelengths – comparison with modelling results. *Atmos. Chem. Phys.* 10, 2507–20.

Ganf, C. G. 1974. Incident solar irradiance and underwater light penetration as factors controlling the chlorophyll A content of a shallow equatorial lake (Lake George, Uganda). *J. Ecol.* 62, 593–609.

Glidden, D. E. 1982. *Winter wind studies in Rocky Mountain National Park.* Estes Park, CO: Rocky Mountain Nature Association.

Granger, R. J., and Male, D. H. 1978. Melting of a prairie snowpack. *J. Appl. Met.* 17, 1833–42.

Gray, D. M. et al. 1979. Snow accumulation and redistribution. In *Proceedings, Modeling of' Snow Cover Runoff* (S. C. Colbeck and M. Ray eds.). Hanover, NH: U.S. Army Cold Regions Research and Engineering Laboratory, pp. 3–33.

Gronskaya, T. P. 2000. Ice thickness in relation to climate forcing in Russia. *Verh. Int. Verein Limnol.*, 27, 2800–2.

Grünewald, T., Bühler, Y., and Lehning, M. 2014. Elevation dependency of mountain snow depth. *The Cryosphere* 8, 2381-94.

Hedstrom, N.R., and Pomeroy, J.W. 1998. Accumulation of intercepted snow in the boreal forest: measurements and modelling. *Hydrol. Processes* 12, 1611–23.

Johnson, G. T., and Watson, I. D. 1984. The determination of view-factors in urban canyons. *J. Climate Appl. Met.* 23, 329–35.

Judson, A., and Doesken, N. J. 2000. Density of freshly fallen snow in the Central Rocky Mountains. *Bull. Amer. Met. Soc.* 81, 1577–87.

Kattel, D. B. et al. 2013. Temperature lapse rate in complex mountain terrain on the southern slope of the central Himalayas. *Theoret. Appl. Clim.* 113, 671–82.

Kattel, D. B. et al. 2015. A comparison of temperature lapse rates from the northern to the southern slopes of the Himalayas. *Int. J. Climatol.* in press.

Kodama, Y., Easton, F., and Wendler, G. 1983. The influence of Lake Minichumina, interior Alaska, on its surroundings. *Arch. Met. Geoph. Biocl. B* 33, 199–218.

Kuz'min, P. P. 1972. *Melting of snow cover.* (Russian version 1961). Jerusalem: Israel Prog. Sci. Transl.

Loope, L. L., and Giambelluca, T. W. 1998. Vulnerability of island tropical montane cloud forests to climate change, with special reference to east Maui, Hawaii. *Clim. Change* 39, 503–17.

Marty, C. et al. 2002. Altitude dependence of surface radiation fluxes and cloud forcing in the Alps: Results from the Alpine Surface Radiation Budget network. *Theoret. Appl. Climatol.* 72, 137–55.

Michel, B. 1971. Winter energy balances of rivers and lakes. *Cold Regions Research and Engineering Laboratory. Monograph III-B1a.* Hanover, NH: U.S. Army Corps of Engineers.

Milburn, D. 2008. The ice cycle on Canadian rivers. In Beltaos, S. (ed.), *River ice breakup.* Highlands Ranch, CO: Water Resources, pp. 21–49.

Miller, D. H. 1955. Snow cover and climate in the Sierra Nevada, California. *Univ. of California Publ. in Geography* 11.

Mills, G. 1997. An urban-canyon layer climate model. *Theoret. Appl. Clim.* 57, 229–44.

Mueller, H. 1984. Zum Strahlungshaushalt im Alpenraum. *Mitteil.Versuchsanstalt Wasserbau. Hydrol. Glaziol. No. 71. Zurich: ETH.*

Mueller, H. 1985. Review paper on the radiation budget in the Alps. *J. Climatol.* 5, 445–62.

Muneepeerakul, R., Rinaldo, A., and Rodriguez-Iturbe, J. 2007. Effects of river flow scaling properties on riparian width and vegetation biomass. *Water Resour. Res.* 43(12). doi: 10.1029/2007WR006100.

Naiman, R. J., and Décamps, H. 1997. The ecology of interfaces: Riparian zones. *Annu. Rev. Ecol. Syst.* 28, 621–58.

Naiman, R.L., Décamps, H. and Pollock, M. 1993. The role of riparian corridors in maintaining regional biodiversity. *Ecol. Appl.,* 3, 209–12.

Nakamura, Y., and Oke, T. R. 1988. Wind, temperature and stability conditions in an E-W oriented urban canyon. *Atmos. Environ.* 22, 2691–2700.

Oke, T. R. 1981. Canyon geometry and the nocturnal urban heat island: comparison of scale model and field observations. *J. Climatol.* 1, 237–54.

Oke, T. R. 1988a. Street design and urban canopy layer climate. *Energy and Buildings* 11, 103–13.

Oke, T. R. 1988b. The urban energy balance. *Prog. Phys. Geogr.* 12, 471–508.

Painter, T. H., and Dozier, J. 2004. Measurements of the hemispherical-directional reflectance of snow at fine spectral and angular resolution, *J. Geophys. Res.* 109, D18115. doi: 10.1029/2003JD004458.

Pearsall, W. H., and Ullyott, P. 1934. Light penetration into freshwater. III. Seasonal variations in the light conditions in Windermere in relation to vegetation. *J. Theoret. Biol.* 11, 89–93.

Pepin, N. 2001. Lapse rate changes in northern England. *Theor. Appl. Climatol.* 68, 1–16.

Pomeroy, J. W., and Brun, E. 2001. Physical properties of snow. In H. G. Jones, J. W. Pomeroy, D. A. Walker, and R. W. Hoham, (eds.), *Snow ecology: an interdisciplinary examination of snow-covered ecosystems.* Cambridge: Cambridge University Press, pp. 45–118.

Prowse, T. D. 1995. River ice processes. In S. Beltaos (ed.), *River ice jams.* Highlands Ranch, CO: Water Resources, pp. 29–70.

Pulliainen, J. 2008. Mapping of snow water equivalent and snow depth in boreal and sub-arctic zones by assimilating space-borne microwave radiometer data and ground-based observations. *Remote Sensing Environ.* 101, 257–69.

Ramamurthy, P. et al. 2014. Influence of subfacet heterogeneity and material properties on the urban surface energy budget. *J. Appl. Met. Climatol.* 53, 2114–29.

Rambo, T. R., and North, M. P. 2008. Spatial and temporal variability of canopy microclimate in a Sierra Nevada riparian forest. *Northwest Sci.* 82, 259–68.D

Rodriguez-Iturbe, I. et al. 2009. River networks as ecological corridors: A complex systems perspective for integrating hydrologic, geomorphologic, and ecologic dynamics. *Water Resour. Res.* 45, W01413. doi:10.1029/2008WR007124.

Rolland, C. 2003. Spatial and seasonal variations of air temperature lapse rates in Alpine regions. *J. Clim.* 16, 1032–46.

Rouse, W. R. et al. 2005. Role of northern lakes in a regional energy balance. *J. Hydromet.* 6, 291–305.

Schaepman-Strub, G. et al., 2006. Reflectance quantities in optical remote sensing definitions and case studies. *Rem. Sens. Environ.,* 1103, 27–42.

Scherrer, D. and Körner, C. 2010. Infra-red thermometry of alpine landscapes challenges climatic warming projections. *Global Change Biol.,* 16, 2602–613.

Schmidt, W. 1934. Observations on local climatology in Austrian mountains. *Quart. J. Roy. Met. Soc.* 60, 345–62.

Schwarb, M. et al. 2001. Mittlere jährliche Niederschlagshöhen im europäischen Alpenraum. In *Hydrologische Atlas der Schweiz.* Bern, Switzerland: Landeshyrdrologie, Bundesamt f. Wasser u. Geologie, Plate 2.6.

Serreze, M. C. et al. 1999. Characteristics of the western United States snowpack from snowpack telemetry (SNOTEL) data. *Water Resour. Res.* 35(7), 2145–60.

Shanks, E. 1956. Altitudinal and microclimatic relationships of soil temperature under natural vegetation. *Ecology* 37, 1–7.

Shreve, F. 1924. Forest soil temperatures as influenced by altitude and slope exposure. *Ecology* 5,128–36.

Spangenberg, J. et al. 2008. Simulation of the influence of vegetation on microclimate and thermal comfort in the city of São Paulo. *Rev. Soc. Brasil. Arbor. Urbana* 3(2), 1–19.

Spreafico, M., and Weingartner, R. 2005. *The hydrology of Switzerland: Selected aspects and results. Water Series no. 7.* Bern, Switzerland: Bundesamt f. Wasser u. Geologie.

Spence, C. et al. 2011. Evaporation from Lake Superior. 2. Spatial distribution and variability. *J. Great Lakes Res.* 37(4), 717–24.

Sturm, M. et al. 1997. The thermal conductivity of seasonal snow. *J. Glaciol.* 43(143), 26–41.

Sturm, M., Holmgren, J., and Liston, G. E. 1995. A seasonal snow cover classification system for local to global applications. *J. Climate* 8, 1261–83.

Suckling, P. W. 1980. The energy balance microclimate of a suburban lawn. *J. Appl. Met.* 29, 606–8.

Tabler, R. D., and Schmidt, R. A. 1986. Snow erosion, transport and deposition. In *Proc. Symposium on Snow Management for Agriculture*, H. Steppuhn and W. Nicholaichuk (eds.). Great Plains Agricultural Council Publication, No. 120. Lincoln, NE: University of Nebraska, pp. 12–58.

Tabony, R. C. 1985. Relations between minimum temperature and topography in Great Britain. *J. Climatol.* 5(5), 503–20.

Troendle, C. A., and Leaf, C. E. 1981. Effects of timber harvest in the snow zone on volume and timing of water yield. In D. M. Baumgartner (ed.), *Interior West Watershed Management Symposium*, Pullman: Washington State University, pp. 231–43.

Tuller S. E. 1973. Microclimatic variations in a downtown urban environment. *Geogr. Annal.* A 54, 123–35.

Verpoorter, C. et al. 2014. A global inventory of lakes based on high-resolution satellite imagery. *Geophys. Res. Lett.* 41, 6396–6402. doi:10.1002/2014GL060641.

Vuglinsky, V. S. 2002. Ice events on the Siberian rivers. Formation and variability. In Squires, V., and Langhorne, P (eds.), *Ice in the environment.* Vol.1. *Proc. 16th IAHR Int. Sympos. on Ice.* Dunedin, New Zealand: Int. Assoc. Hydraulic Engin. Res, pp. 59–66.

Warren, S. G. 1982. Optical properties of snow. *Rev. Geophys. Space Phys.* 20, 67–89.

Warren, S. G., and Wiscombe, W. J. 1980. A model for the spectral albedo of snow. II. Snow containing atmospheric aerosols. *J. Atmos. Sci.* 37, 2734–45.

Wendler, G. 1971. An estimate of the heat balance of a valley and hill station in central Alaska. *J. Appl. Met.* 10, 684–93.

Wetzel, R. G. 2001. *Limnology: Lake and river ecosystems*, 3rd ed. San Diego, CA: Academic Press.

Zaratti, F. et al. 2003. Erythemally weighted UV variations at two high altitude locations. *J. Geophys. Res.* 108(D9), 4263, 1–6.

www.waterencyclopedia.com/Hy-La/Lakes-Physical-Processes.html#ixzz36QaPLOSL

9

Bioclimatology

The majority of life on both terrestrial and aquatic surfaces lives within a few meters of the atmosphere-surface boundary. Therefore, the microclimate at this critical boundary is fundamental for Earth's biosphere, and how life responds and/or adapts to changes in the microclimate is an important topic. In this chapter, we examine how insects, reptiles, and mammals respond to radiation, temperature, wind, and moisture as species constantly react to their environment. Vegetation has been covered in other chapters so is not included here. This topic, broadly known as bioclimatology or biometeorology (depending on the timescale involved), combines our knowledge of microclimates with basic biological principles, to help us understand how species with mobility adapt to their abiotic environment.

Excluding vegetation, there are two basic categories of life: those that have the ability to regulate and maintain a constant internal temperature through metabolism (endotherms or homeotherms) and those that cannot (ectotherms or poikilotherms). This largely coincides with organisms that have an internal skeleton (vertebrates; homeotherms), and those that have an external or no skeleton (invertebrates; poikilotherms). The overwhelming majority of animals are invertebrates (97 percent; May, 1988). Examples of homeotherms include humans and almost all other mammals (the naked mole rat is an exception); examples of poikilotherms include insects, arthropods, reptiles, and all aquatic organisms except marine mammals.

A. The Energy Balance Equation

The energy balance equation for any object is given by

$$\text{Energy Inputs} - \text{Energy Outputs} = \text{Change in Energy Storage} \qquad 9.1$$

For inanimate objects, energy inputs include incident short- and longwave radiation ($S\downarrow$ and $L\downarrow$, respectively) and energy outputs include reflected shortwave and emitted longwave radiation ($S\uparrow$ and $L\uparrow$, respectively), and the non-radiative terms the latent heat flux (λE; the latent heat of vaporization λ times the evaporation rate, E), the sensible heat flux (or convection, H), and the conduction of heat, G. For a living object, however, heat produced by metabolic activity (M) must be included in the basic energy balance equation. The energy balance for an organism in terms of energy inputs and outputs can then be written as

$$\left(S\downarrow +L\downarrow +M\right)-\left(S\uparrow +L\uparrow +\lambda E+H+G\right)=\rho c_p dT/dt \qquad 9.2$$

where the change in energy storage is the rate of temperature change (dT/dt) times the density ρ and specific heat capacity (c_p) of the organism. Note that with all the energy terms in units of watts per square meter (W m^{-2}), the units of $\rho c_p dT/dt$ are W m^{-3}; joules per second per unit volume of the organism. Also note that watts, not watts per area, are used in much of the bioclimatology literature. Conversion between W, W m^{-2}, and W m^{-3} is easy if the organism's surface area or volume is known. In Eq. (9.2), energy gains to the organism are the incident short- and longwave radiation and the heat produced through metabolism. Energy losses are the reflected short-wave radiation and the emitted longwave radiation. The balance between the radiative terms is often referred to as the net absorbed radiation. Non-radiative heat loss terms include the energy lost to evaporate water through breathing or sweating, convective heat loss, and heat loss through conduction.

Since endotherms maintain a constant body temperature that is usually greater than the temperature of objects they are in contact with, conduction (G) is usually a heat loss for the organism. To illustrate this, consider the common experience we have in touching objects around us: a metal surface on a desk or chair and the paper of this book. To us, the metal will feel much colder than the paper book; however, they are almost certainly at equal temperatures. Our brain convincingly tells us that the metal is colder because of the higher thermal conductivity of the metal compared to the book. The temperature difference between our hand and the book or metal is the same, but the conduction that we feel and equate to temperature is far greater for the metal object. Also, consider how it feels to be surrounded by still air with its low thermal conductivity (~0.025 W m^{-1} K^{-1} at 10 °C) compared to submerged in cold water with its large thermal conductivity (~4.18 W m^{-1} K^{-1} at 4 °C, still water); the conductive heat loss while submerged in water is two orders of magnitude larger compared to standing in air.

For many organisms, the challenge that is critical for life is to maintain a near-constant energy balance, meaning that dT/dt is close to 0. This is referred to as "homeostasis" or "thermoregulation." Humans, for example, can tolerate only a 1 °C range in internal temperature (36.5–37.5 °C) before the onset of either hyperthermia or hypothermia. In contrast, lizards can tolerate a 35 °C range (11.0–46.4 °C; Brattstrom, 1965). In the context of the energy balance equation, we will examine how organisms can or cannot adjust the variables in Eq. (9.2) through either behavioral or physiological means to maintain an internal temperature suitable for comfortable life (Table 9.1).

Table 9.1. Capability for Behavioral or Physiological Thermal Regulation for Various Terrestrial Animal Phyla

	Behavioral			Physiological			
	Construct Shelter	Nesting (Incubation)	Basking (Movement / Posture)	Metabolism	Radiation	Panting and/or Sweating	Thermal Conduction
Invertebrates							
Insects	●	●	●	○	◉	○	○
Vertebrates							
Fish	◉	◉	●	○	◉	○	●
Reptiles	◉	◉	●	○	◉	◉	●
Birds	◉	●	●	●	◉	○	◉
Mammals	●	○	●	●	◉	●	○

Notes: Symbols denote the ability or control of the organism to regulate its temperature: ● high capability, ◉ partial capability, ○ poor or low capability.

B. Calculations of the Energy Balance Terms

The combined effects of all the terms in the energy balance equation determine the conditions that the organism must respond to for thermoregulation. For most, expressing these conditions in terms of energy per time (watts), per surface area (watts per square meter), or volume (watts per cubic meter) means little, so various methods have been developed to express the energy balance (or imbalance) in the more traditional and relatable units of temperature. Several of these measures exist, and, as a result, some confusion can arise. Basically, all of these effective or perceived temperatures are the temperatures that the organisms "feel" or respond to; they are based on the ambient microclimate conditions in the organism's habitat. The metal versus paper book example is a clear illustration of this. Most of these perceived temperatures are based on solutions to the energy balance equation, so, first, each term in Eq. (9.2) must be explained.

C. Radiation Terms

Of the energy input terms in Eq. (9.2), the two radiation terms ($S\downarrow$ and $L\downarrow$) are exclusively controlled behavioral actions. Posture and movement can control the incident solar radiation received by the organism. The movement and behavior of most insects and reptiles are directly related to the net solar radiation their bodies receive (e.g., bees: Cena and Clarck, 1972; grasshoppers: Pepper and Hastings, 1952). If too cold, the species could seek sunny areas and adjust posture to maximize solar radiation loads. If too warm, movement to shady areas to remove the direct-beam component of $S\downarrow$, leaving only the diffuse-beam solar radiation, can

Table 9.2. Shortwave Reflection Coefficient for Several Species

Organism	Reflection Coefficient
Mammals	
- Human (female)	0.1920 (blue) 0.2393 (green) 0.4720 (red)
- Human (male)	0.1688 (blue) 0.2278 (green) 0.4509 (red)
- Bison	0.22
- Gray wolf	0.20
- Domestic cat (white)	0.56
- Bobcat	0.30
Birds	
- Stellar's jay	0.12
- Sparrow	0.25
- White swan	0.63
Reptiles	
- Alligator	0.10
- Lizard	0.10

Source: From Gates (1980), except for humans (Jablonski and Chaplin, 2000).

reduce heat stress. For example, Holstein dairy cattle's use of shade structures was positively correlated with $S\downarrow$, and cows under shade had a lower minimum body temperature (Tucker et al., 2007). Exposure to $S\downarrow$, or seeking shelter from $S\downarrow$, can have a major impact on thermal regulation since $S\downarrow$ is often the largest term in Eq. (9.2) and decreases most under shelter. Under a forest canopy, for example, $S\downarrow$ can decrease from nearly 1000 W m^{-2} above the canopy to only 10 W m^{-2} under the canopy (Chen et al., 1997).

Posture and movement also control the incident longwave radiation received by an organism. If too cold, movement close to warm rocks or building a fire can increase $L\downarrow$ especially at night, when longwave radiation is the only form of radiation exchange. During the daytime in the absence of shade, however, $S\downarrow$ overwhelms $L\downarrow$ as an energy input. The incident radiation received by an organism can be fairly easily measured (see Chapter 3, Section E) or estimated from empirically derived equations.

Of the radiation output terms, $S\uparrow$ and $L\uparrow$, $S\uparrow$ is relatively difficult to quantify and $L\uparrow$ is relatively easy. Although the human eye can detect only a portion of the solar radiation spectrum (400–700 nm), our own experiences tell us that there is a large variation in the albedo of organisms, even within one species (Table 9.2). For example, Jablonski and Chaplin (2000) found that the reflection coefficient for humans had a geographic distribution, and that across all latitudes, females were "brighter" than males (i.e., more reflective). The need for protection from harmful UV radiation decreases poleward; thus the decrease in melanized skin poleward, and there is greater need at high latitudes to synthesize vitamin D_3, which requires UVB exposure. The emitted longwave radiation can be calculated using the Stefan-Boltzmann law (see Chapter 4) as a function of surface temperature and emissivity. The organism's surface or internal temperature can be fairly easily measured, and the emissivity (ε) is almost always close to 1 (e.g., 0.98–0.99) because of the high water content.

Overall, the net radiation that an organism absorbs (R_{abs}) is often the largest term in the energy balance and therefore a critical determinant of thermal comfort. Estimates of R_{abs} for a human range from 540–620 W m^{-2} in June under clear skies, to 415–670 W m^{-2} in August under overcast skies, measured in southern Ontario (Kenny et al., 2008). On the basis of R_{abs}, the equivalent temperature (T_e) is the ambient air temperature (T_a) plus the temperature increase due to radiation:

$$T_e = T_a + \frac{\left(R_{abs} - \varepsilon\sigma T_a^4\right)\left(r_R + r_H\right)}{\rho_a c_p} \qquad 9.3$$

where r_R and r_H are the radiative and convective boundary layer resistances, respectively. The radiative resistance is given by

$$r_R = \frac{\rho_a c_p}{4\varepsilon\sigma T_a^3} \qquad 9.4$$

with the emissivity (ε) assumed equal to 1 (i.e., a blackbody). For example, at $\rho_a = 1$ kg m^{-3}, $T_a = 25$ °C (298 K), and $c_p = 1012$ J kg^{-1} K,

$$r_R = \frac{1\frac{kg}{m^3} * 1012\frac{J}{kg\ K}}{4 * 1 * 5.67\times10^{-8}\frac{J}{s\ m^2 K^4} * 298\,K^3} = 0.57\frac{s}{m}\ .$$

The convective boundary layer resistance depends on wind speed (u; m s^{-1}), and d, the characteristic dimension for heat transfer (the downwind length exposed to heat loss), which is equal to 0.17 m for an average person:

$$r_H = 310\sqrt{d/u} \qquad 9.5$$

with r_H in seconds per meter (s m^{-1}) when the constant 310 is used (Robinson et al., 1975). For example, with $u = 1$ m s^{-1} and $d = 0.17$ m,

$$r_H = 310\sqrt{0.17\,\text{m}/1\,\text{m}\,\text{s}^{-1}} = 128\,\text{s}\,\text{m}^{-1}.$$

Using these resistance values and R_{abs} of 500 W m^{-2}, we can calculate the T_e with added temperature increase due to radiation:

$$T_e = 25\,°C + \frac{\left(500\frac{W}{m^2} - 5.67\times10^{-8}\frac{W}{m^2 K^4}\times 298\,K^4\right)\left(0.57\frac{s}{m} + 128\frac{s}{m}\right)}{1\frac{kg}{m^3}\times 1012\frac{J}{kg\,K}}$$

$$= 25\,°C + 6.72\,°C = 31.72\,°C.$$

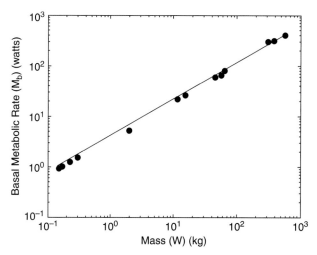

Figure 9.1. Relationship between the mass of various endotherms (*W*; kilograms) and the basal metabolic rate (*M*$_b$; watts). Both axes are shown on a logarithmic scale. The best-fit line is $M_b = 4.24W^{0.72}$, $r^2 = 0.99$ (figure redrawn using the data provided in table 1 from Kleiber (1932), *Hilgardia*, 6(11), p. 317).

Under these conditions, the radiation makes a body operate in a temperature environment that is 6.72 °C above the ambient 25 °C.

D. Metabolic Heat Production

Metabolic heat production (*M*) is controlled by ambient conditions ($S\downarrow$ and T_a) for most insects and reptiles, but for mammals and birds it is controlled by internal physiology to fulfill the requirement of a near-constant body temperature. When food (carbohydrates) are broken down to build tissue, energy in the form of heat is released. The rate of metabolic heat production consists of a basal metabolic rate (*M*$_b$), which is the minimum *M* required for normal body function when the body is at complete rest in a non-stressed thermal environment. Any additional heat created by work performed by (*M*$_w$) is added to *M*$_b$ to give the total *M*:

$$M = M_b + M_w \qquad 9.6$$

The body mass of an organism is the primary determinant of the basal metabolic rate. The basis of modern estimates of *M*$_b$ stems from the work of Kleiber (1932), who plotted the mass of various endotherms against *M*$_b$ (Figure 9.1). The best-fit line to the plot shown in Figure 9.1 is

$$M_b = 4.24W^{0.72} \qquad 9.7$$

where *M*$_b$ is in watts, and *W* is the endotherm's mass in kilograms.

Temperature regulates biochemical reactions through the Boltzmann factor (also known as the Arrhenius term), $e^{-Ei/(kT)}$ where *Ei* is the activation energy (average of

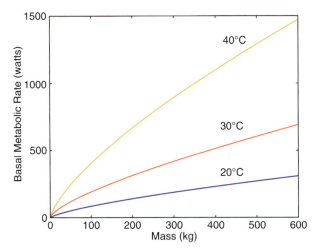

Figure 9.2. The basal metabolic rate for endotherms at temperatures of 20, 30, and 40 °C (293, 303, and 313 K, respectively), calculated using Eq. (9.8) with $b_o = 1.4 \times 10^{11}$ W kg$^{-3/4}$.

0.62 eV), k is the Boltzmann constant (8.62 × 10^{-5} eV K^{-1}), and T is temperature (K), so a temperature effect should also be included in the M_b calculation. Gillooly et al. (2001) showed that the Boltzmann factor can be combined with the Kleiber mass equation to give

$$M_b = b_o W^{0.72} e^{-Ei/(kT)} \qquad\qquad 9.8$$

where b_o is a species-specific normalization constant. Equation (9.8) captures what is known as the Metabolic Theory of Ecology (MTE), which states that M_b regulated by mass and temperature governs most ecological patterns (see Brown et al., 2004, or Price et al., 2010). If we plot M_b from Eq. (9.7) against $W^{0.72}e^{-Ei/(kT)}$ for a given T, b_o can be estimated from the slope of a best-fit line. Using a T of 302 K (the temperature used when M_b was measured in Kleiber's table 1), $b_o = 1.4 \times 10^{11}$ W kg$^{-3/4}$ for a linear best-fit line over a W range of 1 to 600 kg. Using this value for b_o, Eq. (9.8) can then be used to estimate M_b as a function of mass and temperature (see Figure 9.2).

Any physical activity generates additional metabolic heat (M_w) that should be added to the basal metabolic heat (M_b) to provide the total metabolic heat produced (M). As expected, M_w varies greatly for the species, age, sex, conditioning, and type and rate of activity. For humans, many empirical relationships to estimate M_w have been developed, with most including some combination of gender, age, weight, and heart rate as input variables (e.g., Keytel et al., 2005). Recently, wrist-worn accelerometers have become increasingly popular as an alternative to estimate individual caloric output quickly. Or, the MET (metabolic equivalent), defined as the ratio of M_w to M_b (kilocalories per kilogram per hour; 1 kcal = 4182 J; 1 W = 1 J s^{-1}) can simply be multiplied by mass and the duration of activity to give watts (Ainsworth et al., 2011).

Table 9.3. Metabolic Equivalent (MET) for Several Activities

Activity	MET (kcal kg^{-1} hr^{-1})	MET (W kg^{-1})
Sitting-on-job work (light)	1.5	1.74
Standing-on-job work (light)	3.0	3.49
Carpentry work	2.5–7.0	2.90–8.13
Custodial work	2.3–3.8	2.67–4.41
Walking	2.0–4.3	2.32–5.00
Farming (general)	1.3–7.8	1.51–9.06
Golfing (walking carrying clubs)	4.3	5.00
Softball or baseball	5.0	5.81
Construction work (general)	4.0	4.65
Hiking	7.0–7.8	8.13–9.06
Bicycling	4.0–16.0	4.65–18.59
Soccer	7.0–10.0	8.13–11.62
Masonry work	2.5–4.3	2.90–5.00
Hockey	8.0–10.0	9.29–11.62
Basketball	6.0–8.0	6.97–9.29
Forestry work	4.5–8.0	5.23–9.29
SCUBA diving	12.0	13.94

Source: Modified from Ainsworth et al., (2011).

MET values for several activities are shown in Table 9.3. For example, the MET for playing golf is 5.00 W kg^{-1}, so an 80-kg person would generate a M_w of 400 W.

E. Evaporative Cooling

The value of the latent heat of vaporization ($\lambda = 2442$ J kg^{-1} at 25 °C) tells us that a very large amount of energy is required to evaporate water. When that energy is supplied from the skin of an organism, evaporation will then result in a significant decrease in surface temperature. If the organism has ample water, then the total evaporation (sweating, λE_S, plus evaporation of water from the respiratory tract, λE_R) is an effective cooling mechanism:

$$\lambda E_{total} = \lambda E_R + \lambda E_S \qquad 9.9$$

So there is a close connection here between the energy and water balance. Evaporation is not only used by endotherms to help regulate temperature, but also by most ecotherms. For example, foraging honeybees (*Apis mellifera* L.) are able to forage at high ambient desert temperatures by lowering body temperature through the evaporation of regurgitated droplets of honey stomach contents (Cooper et al., 1984).

The evaporation of water through respiration (pulmonary cooling) can be an effective cooling mechanism, especially when the organism's body surface area is small, or ineffective for sweating as a result of fur or feather coverage. The nasal passage is the primary location for pulmonary evaporation. To increase the rate of respiratory cooling at the expense of increased M_w, many organisms employ "panting" or "gaping" to increase airflow across the nasal passages and thus increase evaporative cooling. Dogs, for example, have special nasal glands that secrete fluid to provide 20–40 percent of the respiratory evaporation (Blatt et al., 1972). The magnitude of λE_R is typically small (e.g., 3 W for a 25-kg dog at 40 °C) and proportional to M since breathing requires oxygen consumption. The λE_R also depends on the difference between the vapor pressure of the exhaled versus inhaled air. The inhaled air will be at the ambient vapor pressure (e_a), and the exchanged air at the saturated vapor pressure (e_s) at the nasal temperature (T_n). The metabolic heat production from respiration can be given by the difference in the oxygen partial pressure of the inhaled (O_i) and exhaled air (O_e), and the energy produced per kilogram of oxygen consumed (λ_O; 3.00×10^4 kJ kg^{-1}):

$$\frac{\lambda E_R}{M} = \frac{\lambda \left[e_s \left(T_n \right) - e_a \right]}{\lambda_O \left[O_i - O_e \right]} \qquad 9.10$$

For a human under normal conditions, λE_R is around 10 percent of M (~ 10 W for an adult human). Note that Eq. (9.10) can result in an increased λE_R when T_n is at or above T_a, or act to conserve water loss by reducing λE_R when T_n is less than T_a. The ability to have $T_n < T_a$ is common in many desert species and small birds (notable for their high metabolic rates and body temperatures). In reptiles, for example, Tattersall et al. (2006) found that the nose temperature of tortoises, bearded dragons, and rattlesnakes deceased by as much as 2 °C below their head temperature, indicative of the effectiveness of respiratory cooling.

Evaporation of water on the skin (cutaneous cooling), λE_S, is much more effective for heat dissipation than λE_R, if water is available. For example, for a person at $T_a = 35$ °C and a relative humidity of 75 percent, the measured λE_S required for thermoregulation was 116 W; at $T_a = 49$ °C and a relative humidity of 20 percent, the measured λE_S was 646 W (Shapiro et al., 1982), much more than the λE_R of ~ 10 W. As mentioned previously, most species including honeybees, birds, goats, and most terrestrial mammals use λE_S to supplement λE_R. As with λE_R, λE_S depends on the difference in vapor pressure at the skin surface (the saturation vapor pressure, e_s at skin temperature T_s) and the ambient vapor pressure (e_a):

$$\lambda E_S = \frac{\rho_a \lambda \left[e_s \left(T_s \right) - e_a \right]}{r_{SV} P_a}. \qquad 9.11$$

We also need to know the skin's resistance to vapor transfer (r_{SV}; Table 9.4). This resistance varies with species and ambient conditions.

Note that some sources, especially those concerning human physiology (e.g., Gerrett et al., 2013), report r_{SV} values in units of square meter pascals per watt (m^2 Pa W^{-1}). These units are convenient since the vapor pressure difference, $e_s(T_s)$-e_a in units of pascals, divided by r_{SV} in units of square meter pascals per watt give λE_S in watts per square meter. These r_{SV} units are equivalent to units of seconds per meter since a

Table 9.4. Typical Values for the Total Body Resistance to Vapor (r_{SV}) under Non-heat-stressed Conditions

Species or Material	r_{SV} (s cm^{-1}) and Conditions	Source
Humans		
Modeled	52.1 lightly clothed, mild conditions	Steadman (1979)
Actual (male 75 kg, 23 yr old)	35.9 lightly clothed	Gerrett et al. (2013)
Clothing materials		
Underwear	4.2	Havenith et al. (2007)
100% cotton	3.4	
100% polyester	3.7 - inf	
100% polypropyleneOuterwear	18.6	
Impermeable	5.6	
Semi-permeable		
Permeable		
Birds		
Domestic pigeon (*Columa livia*)	1.4: $T_a = 0\,°C$	Webster et al. (1984)
	3.0: $T_a = 30\,°C$	
	9.0: $T_a = 40\,°C$	
Spotted sandgrouse (*Pterocles senegallus*)	103.0: $T_a = 27\,°C$	Marder et al. (1986)
	20.5: $T_a = 42\,°C$	
	13.5: $T_a = 45\,°C$	
	15.0: $T_a = 51\,°C$	
Amphibians		
Frog (*Rana pipiens*)	0.05: $T_a = 20.7\,°C$	Spotila and Berman (1976)
Salamander (*Desmognathus ochrophaneus*)	0.09: $T_a = 21.7\,°C$	
Reptiles		
Box turtle (*Terrapen carolina*)	77.6: $T_a = 20.8\,°C$	
America chameleon (*Anolis carolinensis*)	196.4: $T_a = 20.6\,°C$	
Spiny soft-shelled turtle (*Trionyx spininferus*)	2.98: $T_a = 15\,°C$	Robertson and Smith (1982)
	4.90: $T_a = 25\,°C$	
	5.38: $T_a = 35\,°C$	
American alligator (*Alligator mississippiensis*)	35: $T_a = 25\,°C$	Robertson and Smith (1982)
Snakes		
Natrix tessellata	117: water snake	Lahav and Dmi'el (1996)
Psammophis scholari	1088: desert biomes	
Lizards		
Agama sinaita	738: $T_a = 35\,°C$; arid habitat	Eynan and Dmi'el (1993)
Agama stellio ssp.	234: $T_a = 35\,°C$; mesic habitat	

pascal is a kilogram per meter per second squared, a watt is a joule per second, and a joule per kilogram is a square meter per second squared.

Using Eq. (9.11) and Table 9.4, we can then calculate λE_S. For example, the r_{SV} for a lightly clothed 75-kg human male is 35.9 s cm^{-1} (3590 s m^{-1}). Under the following conditions, $T_a = 25$ °C, $T_s = 35$ °C, $\rho_a = 1$ kg m^{-3}, $e_a = 1.5$ kPa, and $P_a = 85$ kPa, substitution in Eq. (9.11) gives

$$\lambda E_S = 1.0 \frac{kg}{m^3} \times 2.44 \times 10^6 \frac{J}{kg} \times \left[5.65 kPa - 1.5 kPa\right] \times \frac{1}{3590} \frac{m}{s} \times \frac{1}{85 kPa} = 33.2 \frac{W}{m^2}$$

The relations among surface area (A: square meters), mass (W: kilograms), and height (H: centimeters) is given by the du Bois equation (du Bois and du Bois, 1916):

$$A = 0.007184 W^{0.425} Ht^{0.725} \qquad 9.12$$

Equation (9.12) gives a surface area of 1.96 m^2 with $W = 75$ kg and $Ht = 182$ cm, so λE_S would be 33.2 W m^{-2} × 1.96 m^2 = 65.1 W.

F. Sensible Heat Loss: Convection

The equation for the sensible or convective heat loss from an organism (H) is

$$H = \frac{\rho_a c_P \left(T_a - T_s\right)}{r_H} \qquad 9.13$$

where r_H is the total resistance to convective heat transfer. This r_H is actually equal to the sum of three resistances acting in series: body tissue, the coat, and the external wind environment. The body tissue (e.g., fat) and coat (e.g., fur, feathers, or clothing) can add significant resistance to heat loss, so it should always be included for a proper determination of the total r_H. For now, Eq. (9.5) can be used to approximate r_H unless specific values can be found (Table 9.5).

As with r_{ES}, r_H is sometimes given in units of degrees Celsius (or Kelvin) square meters per watt, convenient when using the equation of the form $H = (T_s - T_a)/r_H$ since H will then be in units of watts per square meter. Conversion to resistance units of seconds per meter can be achieved by multiplying r_H expressed in degrees Celsius square meters per watt by the product of air density (ρ_a; kilograms per cubic meter) times the specific heat of air (c_P; joules per kilogram per degree Celsius). For example, at 0 °C, 1 °C m^{-2} W^{-1} is equal to 1290 s m^{-1} (Clark et al., 1973).

Using Eq. (9.13) and Table 9.5, we can then calculate H. For example, the r_H for a lightly clothed, 182 cm tall 75-kg human male is 200 s m^{-1}. Under the same conditions given for the λE_S example, $T_a = 25$ °C, $T_s = 35$ °C, $P_a = 85$ kPa, $e_a = 1.5$ kPa, $\rho_a = 1$ kg m^{-3}, $c_P = 1012$ J kg^{-1} °C^{-1}, substitution in Eq. (9.13) gives

$$H = 1 \frac{kg}{m^3} \times 1012 \frac{J}{kg\,°C} \times \left[35°C - 25°C\right] \times \frac{1}{200} \frac{m}{s} = 50.6 \frac{W}{m^2}.$$

From Eq. (9.12), this individual has surface area of 1.96 m^2, so H in watts would be 50.6 W m^{-2} × 1.96 m^2 = 99.2 W.

Table 9.5. Total Body Resistance to Convective Heat Transfer

Species	r_H (s m^{-1}) and Condition	Source
Humans		
Actual average of 256 subjects	50: Skin only	Steadman (1979)
Actual (male 75 kg, 23 yr old)	200: lightly clothed; $u = 0$ m s^{-1}	Gerrett et al. 2013
Clothing	215: Per centimeter of cloth thickness	Steadman (1979)
White-crowned sparrow (*Zonotrichia leucophrys gambelii*)	291: $T_a = 1$ °C, $u = 1.68$ m s^{-1} 374: $T_a = 10$ °C, $u = 0.47$ m s^{-1} 405: $T_a = 20$ °C, $u = 0.064$ m s^{-1}	Robinson et al. (1975)
Pigeons (*Columba livia*)		Walsberg et al. (1978)
Depressed plumage	148: $u = 1.0$ m s^{-1}	
Erect plumage	221: $u = 1.0$ m s^{-1}	
Calf	150: $u = 0.4$ m s^{-1}	McArthur (1980)
Steer	160	
Cattle	150	
Cheviot sheep	120	
Blackface sheep	130	
Clun Forest sheep	120	

G. Conduction

Conduction is the transfer of heat achieved through the direct contact of one object with another. As with many of the flux equations, heat conduction (G) can be well represented by Ohm's law, with G proportional to the temperature difference between the two objects (at temperatures T_1 and T_2), divided by the resistance to conductive heat transfer. More often, however, G is given as

$$G = \frac{KA(T_1 - T_2)}{d} \qquad 9.14$$

where G is positive if heat is lost from the object at T_1 to the object at T_2, K is the thermal conductivity (watts per meter Kelvin), A is the contact surface area (square meters), and d is the distance through which heat is transferred (meters). This equation is convenient because K for most materials is given in units of conductance (see Table 9.6), and G is directly proportional to the contact area of the two objects (A).

Recall the discussion at the start of this chapter comparing the sensation of touching a metal chair compared to touching a paper book, and how our minds incorrectly equated G to the temperature of the paper (wood) or steel chair; the steel felt cooler than the book even when they were at the same temperature. This sensation was due to higher K for the steel than the paper, resulting in a much larger G when touching the steel compared to the paper.

Table 9.6. Thermal Conductivity of Human Skin and Various
Materials Typically in Contact with Human Skin

Material	K (W m^{-1} K^{-1})
Air (still, 300 K, 100 kPa)	0.0262
Cotton	0.03
Wool (293–298 K)	0.04
Snow (273 K)	0.05–0.25
Soil (dry, 293 K)	0.15–2.0
Rubber (303 K)	0.16
Human skin (in vivo)	0.3764
Soil (saturated, 293 K)	0.6–4
Water (still, 300 K)	0.609
Wood (oven-dry)	0.12–0.17
Stainless steel (296 K)	18–24

Using Eq. (9.14), consider the heat loss through conduction that occurs when sitting on a steel bench ($K = 20$ W m^{-1} K^{-1}) compared to a wooden bench ($K = 0.12$ W m^{-1} K^{-1}). To use Eq. (9.14), first the total K of the body-bench coupled system must be determined. Thinking back in terms of resistances for a moment, the skin resistance (r_{skin}) and bench resistance (r_{bench}) are acting as resistors in series; therefore, the total resistance $r_{total} = r_{skin} + r_{bench}$. Resistance is equal to 1/conductance, so

$$r_{skin} = \frac{1}{0.4\,\mathrm{W\,m^{-1}K^{-1}}} = 2.5\,\mathrm{W^{-1}\,m\,K}$$

$$r_{steel\,bench} = \frac{1}{20\,\mathrm{W\,m^{-1}K^{-1}}} = 0.05\,\mathrm{W^{-1}\,m\,K}$$

$$r_{wood\,bench} = \frac{1}{0.12\,\mathrm{W\,m^{-1}K^{-1}}} = 8.3\,\mathrm{W^{-1}\,m\,K},$$

and the total resistance for the skin-steel bench is $2.5 + 0.05 = 2.55$ W^{-1} m K, with the reciprocal equal to the total conductance, $1/2.55 = 0.4$ W m^{-1} K^{-1}. For the skin-wood bench, $r_{total} = 2.5 + 8.3 = 10.8$ W^{-1} m K, equal to a total conductance of $1/10.8 = 0.09$ W m^{-1} K^{-1}. With these total skin-bench conductances the conditions $A = 0.15$ m^2, T_1 (body) = 35 °C, T_2 (bench) = 10 °C, and $d = 6$ cm (0.06 m) give

$$G_{skin-steel} = 0.4\frac{\mathrm{W}}{\mathrm{m\,K}} \times 0.15\,\mathrm{m^2} \times (35\mathrm{C} - 10°\mathrm{C}) \times \frac{1}{0.06\,\mathrm{m}} = 25\,\mathrm{W}$$

$$G_{skin-wood} = 0.09\frac{\mathrm{W}}{\mathrm{m\,K}} \times 0.15\,\mathrm{m^2} \times (35\mathrm{C} - 10°\mathrm{C}) \times \frac{1}{0.06\,\mathrm{m}} = 6\,\mathrm{W}$$

showing that the conductive heat loss when sitting on a steel bench is more than four times larger than when sitting on a wood bench.

H. Simplified Energy Balances to Provide Effective Temperatures

As shown, calculating the full energy balance for an organism requires many input variables, most of which are not available or require assumptions (especially the resistances). Since almost none of the required input variables are provided through routine measurements, and few people can relate to them or adjust their behavior when provided with units of watts, many empirical, statistical derivations of the energy balance have been developed.

Perhaps the simplest effective temperature calculation is that developed by Tao and Xin (2003) for domestic chickens:

$$THVI = \left(0.85T_a + 0.15T_w\right)\left(u - 0.058\right) \qquad 9.15$$

where *THVI* is the Temperature-Humidity-Velocity Index (degrees Celsius), with the effect of humidity incorporated simply by weighting the dry- and wet-bulb temperatures, T_a and T_w, respectively, and including a wind speed effect through an empirically determined function.

For humans rather than chickens, other similar empirical indexes have been developed with various levels of detail. One of the first was the discomfort index (*DI*) proposed by Thom (1959):

$$DI = 0.4\left(T_a - T_w\right) + 4.8 \qquad 9.16$$

where the discomfort is classified in Table 9.7. Although simplistic and empirical, such indexes have the advantage that the input data are widely available; thus these equations are still being used (e.g., Tselepidaki et al., 1992). The disadvantage is also that they are empirical and thus do not account for the true processes contained in Eq. (9.2).

The effect of wind has a noticeable and significant influence on human comfort, through the r_H term in Eq. (9.13). Popularly referred to as "wind-chill" or the "wind-chill index" or "factor," this effect is due to the reduction in the thickness of the body's laminar boundary layer as wind speed increases (Eq. 9.5). Through Ohm's law, as the boundary layer thickness decreases, the resistance to heat transfer also decreases, so the sensible heat loss increases for a given temperature gradient.

In a remarkable set of experiments (89 total) conducted in the Antarctic, including using themselves as human subjects, Siple and Passel (1945) developed an equation for the rate of cooling of a small cylinder filled with 250 g of water, as a function of wind speed and temperature, and then related this to human discomfort in the context of wind- chill;

$$K_0 = \left(\sqrt{u \times 100} + 10.45 - u\right)\left(33 - T_a\right) \qquad 9.17$$

where K_0 is the cooling rate (kilogram calories per hour per square meter). In Eq. (9.17), the effect of wind speed on the cooling rate is clear, yet there was uncertainty as to how well a small cylinder represented human morphology.

Table 9.7. Classification of the Discomfort Index

DI (°C)	Classification
DI < 21	No discomfort
21 ≤ DI < 24	10% of the population feels discomfort
24 ≤ DI < 26	50% of the population feels discomfort
DI ≥ 26	Most of the population feels discomfort
DI ≥ 26.7	Discomfort very strong and dangerous
DI ≥ 32	State of medical emergency

Source: From Thom (1959).

Building on the work of Siple and Passel (1945), Steadman (1971) provided the earliest expressions of wind-chill for a clothed person that was based on the energy balance and therefore is much more complex but can still be simplified into a practical form based on readily available measurements. When the Steadman wind-chill (T_{wc}; degrees Celsius) results are plotted in three-dimensional form (with T_a, u, and T_{wc} on each axis), the equation fitted to the three-dimensional surface is

$$T_{wc} = 1.41 - 1.162u + 0.98T_a + 0.0124u^2 + 0.0185(u \times T_a) \qquad 9.18$$

with u in meters per second, and T_a in degrees Celsius (Quayle and Steadman, 1998). Since the human face is the most exposed surface when outside in winter, Osczevski and Bluestein (2005) further refined the T_{wc} calculation by including convective heat loss from the face, approximated by a vertical cylinder:

$$T_{wc} = 13.12 + 0.6215T_a - 11.37u^{0.16} + 0.3965(u \times T_a)^{0.16} \qquad 9.19$$

with T_a in degrees Celsius, and u in kilometers per hour. Both of these T_{wc} equations are expressed in terms of T_a and u as input variables because these are readily available meteorological variables. It is important to remember that Eqs. (9.18) and (9.19) are derived from the full human energy balance equation, hence are statistical "simplifications" of the true T_e. The full T_e calculation requires so many variables that the practical application of those equations would limit the public's acceptance of wind-chill warnings. Imagine, for example, if the wind-chill warning were expressed in units of watts per square meter. The current wind-chill chart used by the National Weather Service in the United States is shown in Figure 9.3.

At the other end of the spectrum, heat indexes have also been developed following the same approach as wind-chill indexes; most are based on the full energy balance of a human, then simplified on the basis of common readily available meteorological variables to enhance practical applicability. Those variables that are primary to human discomfort during hot conditions are T_a and humidity, with humidity expressed as the vapor pressure deficit. The latent heat of vaporization is large, so evaporation from sweat and/or from exhalation is a very effective way to remain cool if not limited by water availability. The vapor pressure gradient between the skin and the ambient air

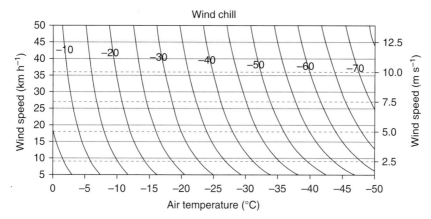

Figure 9.3. Current wind-chill chart adopted by the U.S. National Weather Service. *Source*: Based on Osczevski and Bluestein, 2005 from R. G. Barry and E. Hall-McKim, 2014, p. 87, fig. 4.5, Cambridge University Press.

regulates the evaporation rate via Ohm's law. As T_a increases, e_s increase exponentially through the Clausius-Clapeyron equation (Chapter 2, Section B), so the vapor pressure gradient decreases at high T_a limiting λE, thus the effectiveness of λE as a cooling mechanism. As high wind speeds make us feel cooler as given in the T_{wc} simplified equations, vapor pressure (humidity) makes us feel warmer during hot conditions, and therefore some expression of humidity appears in most heat index equations.

One of the first heat indexes (a sultriness index) was developed by Steadman (1979) as an extension of his T_{wc} derivation. In a similar style and for the same reasons as for T_{wc}, the heat index was simplified to a practical form based on tables of T_a and T_w alone. An example of a humidity index calculation, known as the Humidex (T_H) used in Canada, is

$$T_H = T_a + 0.5555 * \left(e_a - 10.0 \right)$$ 9.20

where e_a is in hectopascals (hPa) (mbar). Note that the units of Eq. (9.20) are senseless, hence the reference to an "index" related to the equivalent dry-air temperature. Conversions to discomfort based on Eq. (9.20) are provided in Table 9.8.

There are several other T_a and humidity-based equations at various levels of complexity used to calculate discomfort; however, Anderson et al. (2013) found that there is little variation among them, and all produce results close to the original Steadman (1979) method.

I. Radiation Impacts on Comfort and Behavior

The exposure to ultraviolet radiation (UVA: 400–315 μm; UVB: 315–280 μm; UVC 280–100 μm) has a significant effect for nearly all organisms. Most obvious is the

Table 9.8. Relationship between the Canadian Humidex Calculation
and the Degree of Discomfort

Humidex	Discomfort
> 46	Dangerous; possible heat stroke
40–45	Great discomfort; avoid exertion
30–39	Some discomfort
20–29	No discomfort

Source: From Environment Canada (2015).

damage to DNA in exposed tissue (skin), resulting in some form of skin cancer. Although the formation of melatonin does offer some natural protection, even minimal exposure to UV is harmful. Location (latitude) and occupation were found to correlate with cancer mortality decades ago (e.g., Applely, 1941). Latitude is often the best predictor of skin cancer, with the incidence rate increasing as latitude decreases (Auerbach, 1961). Increasing altitude also increases UV exposure, coupled with the fact that skies tend to be clearer at higher altitudes. Exposure to UV radiation can result in cataracts in the lens of eyes, and UV exposure can also compromise the human immune system. In other organisms as well, UV exposure can be detrimental. In phytoplankton, UV exposure affects the morphology and biochemistry of phytoplankton, the basis of the aquatic food chain (Hessen et al., 1997).

Alternatively, lack of exposure to radiation in the visible portion of the spectrum (400–700 μm) can be detrimental to human behavior. Recent increases in the occurrence of neonatal jaundice may be caused by a lack of UVB exposure, which limits vitamin D production. Seasonal Affective Disorder (SAD) or "winter depression" affects millions of people annually, especially during winter in high-latitude regions. The cause of SAD is thought to be a lack of vitamin D production due to insufficient UBB exposure.

To inform the public about UV, several indexes have been developed, but few are broadcast to the public. The first UV Index was developed in Canada and broadcast daily starting in 1992; it varies on a scale from 0 (no UV) to 10 (maximum UV) that varies linearly with watts per square meter in the UV portion of the solar spectrum.

J. Examples of Behavioral Modifications: Using the Soil and the Benefits of Burrowing

Historically and today representing the majority of animal life, invertebrates have been and are successful in adaption to the microclimate at the Earth's surface and have taken full advantage of burrowing. Insects in particular compose the majority of the invertebrates; hence they will be the focus of this discussion. A characteristic of all insects is their small body mass, which results in a large relative surface area. Therefore, they are efficient in exchanging energy with their surrounding environment, but inefficient at storing energy (heat). As a consequence, most insects have a narrow range of ambient temperature tolerance suitable for normal functioning. Several insect species have developed clever means of constructing shelter to help maintain a suitable microclimatic environment (Willmer, 1982).

As discussed in Chapter 2, moderated temperature and humidity are observed only a few centimeters beneath the surface in most soils. In tropical and subtropical regions, the annual air temperature variation is muted, further deceasing any diurnal or annual soil temperature fluctuations. Most non-flying insects take full advantage of this fact, using the soil as shelter from excessive temperature fluctuations. Ants, for example, carefully select nest location and depth as a function of soil temperature and humidity. Bollazzi et al. (2008) found that the leaf-cutter ant (*A. lundi*) workers use soil temperature as a clue to decide where to dig for a nest, with a preference for soil surface temperatures of 20–30 °C. They also found that the higher the average soil temperature, the deeper the nest location. The temperature and humidity inside the ant burrows were also found to have a critical impact on the only food source for the developing brood, a fungus cultivated by the leaf-cutting ants that only grows in high humidity and temperatures of 25–30 °C (Bollazzi et al., 2008).

Soil temperature not only affects the nest location, depth, and food supply, but can also affect social behavior; Lu et al. (2012) found that for fire ants in South China forage activity was maximal when the 5 cm deep soil temperature was 27–40 °C. Many species of ants are tolerant of long-term changes in soil temperature such as warming. In Colorado, Menke et al. (2014) found 20 years of continual experimental warming that resulted in a 1.5 °C increase in soil temperature and 10 day earlier snowmelt resulted in no discernible effect on ant communities.

Species such as many toads and frogs spend part of their life belowground, and part aboveground. Where they are depends largely on excessive variations in air temperature and water availability. Since most amphibians lack the ability for physiological thermal regulation, behavioral regulation is the most effective (e.g., nocturnal movement or burrowing during periods of excessive heat or drought). For example, the giant burrowing Australian frog (*H. australiacus*) is active only when the air temperature and relative humidity exceed 10 °C and 60 percent, respectively (Penman et al., 2006). At the other extreme, the wood frog (*Rana sylvatica*) indigenous to interior Alaska and much of the forested region of northern North America has survived temperatures as low as −18 °C (Costanzo et al., 2013). This can be achieved through physiological rather than behavioral means through the accumulation of glycogen (acting as antifreeze) in the frog's body fat and skeletal muscles, in addition to other adaptive responses (Costanzo et al., 2013).

Several other organisms spend part of their time in underground burrows depending on the ambient conditions and their tolerance of changes in their energy balance. Burrowing owls (*Athene cunicularia*) in the desert southwestern United States, for example, maintain a body temperature of 38.0 °C in an air temperature range of 0–38 °C, using pulmocutaneous evaporation and gular flutter to dissipate heat at air

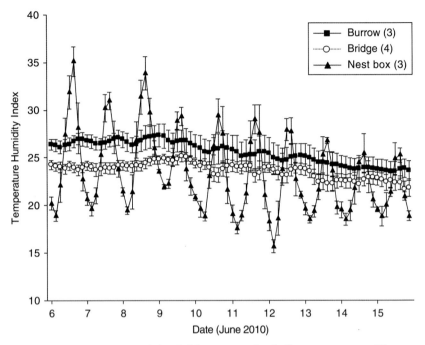

Figure 9.4. Temperature, weighted 85 percent dry-bulb temperature, 15 percent
wet-bulb temperature, for three types of nesting cavities: underground burrows,
bridge cavities, and nest boxes (from M. Amat-Valero et al., 2014).
Source: *International Journal of Biometeorology,* 58, fig. 5., p. 1991. Courtesy of
Springer.

temperatures greater than 38 °C (Coulombe, 1970). Although an effective physiological method of staying cool, evaporation does require ample water, and the existence
of burrows already created by another animal (e.g., black tailed prairie dogs) offers a
metabolically economical opportunity for shelter and nesting for both burrowing owls
and snakes. Temperature and humidity temporal variability in burrows are much lower
than aboveground, an advantage in maintaining thermal regulation. As expected, the
variation in this weighed temperature calculation was least for the burrows and bridge
cavities, and greatest for nest boxes (Figure 9.4).

K. Aboveground Thermal Modification through Shelter

Aboveground structures are also used effectively to buffer against ambient temperature, wind, and radiation fluctuations. Perhaps the most dramatic structures are termite mounds, complex structures thousands of times the size of termites themselves.
Similar to the mechanism for the leaf-cutting ants, the termite mound structure creates a microclimate suitable for the growth of fungi on which the termites feed. The
optimum conditions for fungal growth are constant 30 °C temperatures and low CO_2
concentrations (Korb, 2003). Despite ambient air temperature fluctuations of 35 °C,
the air temperature inside termite mounds is nearly a constant 30 °C with fluctuations

of less than 2 °C (Korb, 2003). Near-constant temperature and low CO_2 concentrations are achieved through ventilation to expel the excess heat created by the high metabolism of the termites and fungi. This ventilation is likely created by solar radiation on the towerlike structures, creating convective currents (Korb, 1997).

Analysis of nesting birds in the Great Plains under a 10 year Landscape Conservation Cooperative shows that lesser prairie chickens (*Tympanuchus pallidicinctus*) on nests were able to maintain relatively consistent average nest temperature of 31 °C and nest relative humidity of 57%, whereas average external temperatures (20–35 °C) and relative humidities (35–75 percent) varied widely throughout the 24 hour cycle (Boal et al., 2013). On native rangeland in Montana, cattle were found to mitigate severe winter conditions (cold and high winds) by remaining in locations with a moderate microclimate (Houseal and Olson, 1995). Six-week intervals in two winters were analyzed. The cattle (*Bostaurus*) selected environments with temperatures above their lower critical temperature of −23 °C when reference temperatures fell below that threshold.

References

Ainsworth, B. E., Haskell, W. L., Herrmann, S. D., Meckes, N., Bassett, D. R. Jr., Tudor-Locke, C., Greer, J. L., Vezina, J., Whitt-Glover, M. C., and Leon, A. S. 2011. 2011 Compendium of physical activities: A second update of codes and MET values. *Med Sci Sports Exerc.* 43(8), 1575–81.

Amat-Valero, M., Calero-Torralbo, M. A., and Vaclav, R. 2014. Cavity types and microclimate: Implications for ecological, evolutionary, and conservation studies. *International Journal of Biometeorology* 58, 1983–94.

Anderson, G. B., Bell, M. L., and Peng, R. G. 2013. Methods to calculate the heat index as an exposure metric in environmental health research. *Environmental Health Perspectives* 121(10), 1111–19.

Applely, FL. 1941. The relation of solar radiation to cancer mortality in North America. *Cancer Research* 1, 194–5.

Auerbach, H. 1961. Geographic variation in incidence of skin cancer in the United States. *Public Health Reports* 76(4), 345–8.

Blatt, C. M., Taylor, C. R., and Habal, M. B. 1972. Thermal panting in dogs: The lateral nasal gland, a source of water for evaporative cooling. *Science* 177, 804–5.

Boal, C. W. et al. 2013. Lesser prairie-chicken nest site selection, microclimate, and nest survival in association with vegetation response to a grassland restoration program. U.S. Geological Survey Open-File, Report 2013–1235, 35 pp.

Bollazzi, M., Kronenbitter, J., and Roces, F. 2008. Soil temperature, digging behaviour, and the adaptive value of nest depth in South America species of *Acromyrmex* leaf-cutting ants. *Oecologia* 158, 165–75.

Brattstrom, B. H. 1965. Body temperature of reptiles. *American Midland Naturalist* 73(2), 376–422.

Brown, J. H., Gillooly, J. F., Allen, A. P., Savage, V. M., and West, G. B. 2004. Toward a metabolic theory of ecology. *Ecology* 85(7), 1771–89.

Cena, K., and Clarck, J. A. 1972. Effect of solar radiation on temperatures of working honey bees. *Nature New Biology* 236, 222–3.

Chen, J. M., Blanken, P. D., Black, T. A., Guilbeault, M., and Chen, S. 1997. Radiation regime and canopy architecture in a boreal aspen forest. *Agricultural and Forest Meteorology* 86(1–2), 107–25.

Clark, J. A., Cena, K., and Monteith, J. L. 1973. Measurements of the local heat balance of animal coats and human clothing. *Journal of Applied Physiology* 35(5), 751–4.

Cooper, P. D., and Schaffer, W. M. 1984. Temperature regulation of honey bees (*Apis mellifera*) foraging in the Sonorian Desert. *Journal of Experimental Biology* 114, 1–15.

Costannzo, J. P., do Amaral, M. C. F., Rosendale, A. J., and Lee, Jr., R. E. 2013. Hibernation physiology, freezing adaptation and extreme freeze tolerance in a northern population of the wood frog. *Journal of Experimental Biology* 216, 3461–73.

Coulombe, H. N. 1970. Physiological and physical aspects of temperature regulation in the burrowing owl *Speotyto cunicularia*. *Comparative Biochemistry and Physiology* 35(2), 307–37.

du Bois, D., and du Bois, E. F. 1916. A formula to estimate the approximate surface area if height and weight be known. *Clinical Calorimetry* 10th Paper, 17, 863–71.

Environment Canada. 2015. www.ec.gc.ca/meteo-weather (accessed March 6 2015).

Eynan, M., and Dmi'el, R. 1993. Skin resistance to water loss in agamid lizards. *Oecologia* 95, 290–4.

Gates, D. M. 1980. *Biophysical Ecology*. New York: Springer-Verlag, 656 pp.

Gerret, N., Redortier, B., Voelcker, T., and Havenith, G. 2013. A comparison of galvanic skin conductance and skin wittedness. *Journal of Thermal Biology* 38, 530–8.

Gillooly, J. F., Brown, J. H., West, G. B., Savage, V. M., and Charnov, E. L. 2001. Effects of size and temperature on metabolic rate. *Science* 293(5538), 2248–51.

Havenith, G., Richards, M. G., Wang, X., Brode, P., Candas, V., den Hartog, E., Holmer, I., Kuklane, K., Meinander, H., and Nocker, W. 2007. Apparent latent heat of evaporation from clothing: Attenuation and "heat pipe" effects. *Journal of Applied Physiology* 104, 142–9.

Hessen, D. O., De Lange, H. J., and Van Donk, E. 1997. UV-induced changes in phytoplankton cells and its effects on grazers. *Freshwater Biology* 38, 513–24.

Houseal, G. A., and Olson, B. E. 1995. Cattle use of microclimates on a northern latitude winter range. *Canadian Journal of Animal Science* 75, 501–7.

Jablonski, N. G., and Chaplin, G. 2000. The evolution of human skin coloration. *Journal of Human Health* 39(1), 57–106.

Kenny, N. A., Warland, J. S., Brown, R. D., and Gillespie, T. G. 2008. Estimating the radiation absorbed by a human. *International Journal of Biometeorology* 52, 491–503.

Keytel, L. R., Goedecke J. H., Noakes T. D.,Hiiloskorpi, H., Laukkanen, R., van der Merwe, L., and Lambert, E. V. 2005. Prediction of energy expenditure from heart rate monitoring during submaximal exercise. *Journal of Sports Sciences* 23(3), 289–97.

Kleiber, M. 1932. Body size and metabolism. *Hilgadria* 6(11), 315–53.

Korb, J. 2003. Thermoregulation and ventilation of termite mounds. *Naturwissenschaften* 90, 212–19.

Lahav, S., and Dmi'el, R. 1996. Skin resistance to water loss in colubrid snakes: Ecological and taxonomical correlations. *Ecosciences* 3(2), 135–9.

Lu, Y. Y., Wang, L., Zeng, L., and Xu, Y. J., Xu, Y. 2012. The effects of temperature on the foraging activity of red imported fire ant workers (*Hymenoptera: Formicidae*) in South China. *Sociobiology* 59(2), 573–83.

Marder, J., Gavrieli-Levin, I., and Raber, P. 1986. Cutaneous evaporation in heat-stressed spotted sandgrouse. *The Condor* 88, 99–100.

May, R. M. 1988. How many species are there on Earth? *Science* 241, 1441–9.

McArthur, A. J. 1980. Thermal resistance and sensible heat loss from animals. *Journal of Thermal Biology* 6, 43–7.

Menke, S. B., Harte, J., and Dunn, R. R. 2014. Changes in ant community composition caused by 20 years of experimental warming vs. 13 years of natural climate shift. *Ecosphere* 5(1), 1–17.

Osczevski, R., and Bluestein, M. 2005. The new wind chill equivalent temperature chart. *Bulletin of the American Meteorological Society* 86, 1453–8.

Penman, T. D., Lemckert, F. L., Mahony, M. J. 2006. Meteorological effects on the activity of the giant burrowing frog (*Heleioporus australiacus*) in south-eastern Australia. *Wildlife Research* 33, 35–40.

Pepper, J. H., and Hastings, E. 1952. The effects of solar radiation on Grasshopper temperatures and activities. *Ecolology* 33(1), 96–103.

Price, C. A., Gilooly, J. F., Allen, A. P., Weitz, J. S., and Niklas, K. J. 2010. The metabolic theory of ecology: prospects and challenges for plant biology. *New Phytologist* 188(3), 696–710.

Quayle, R. G., and Steadman, R. G. 1988. The Steadman wind chill: An improvement over present scales. *Weather Forecasting* 13, 1187–93.

Robertson, S. L., and Smith, E. N. 1982. Evaporative water loss in the spiny soft-shelled turtle *Trionyx spiniferus*. *Physiological Zoology* 55(2), 124–9.

Robinson, D. E., Campbell, G. S., and King, J. R. 1975. An evaluation of heat exchange in small birds. *Journal of Comparative Physiology-B* 105, 153–66.

Shapiro, Y., Pandolf, K. B., and Goldman, R. F. 1982. Predicting sweat loss response to exercise, environment and clothing. *European Journal of Applied Physiology* 48, 83–96.

Siple, P. A., and Passel, C. F. 1945. Measurements of dry atmospheric cooling in subfreezing temperatures. *Proceedings of the American Philosophical Society* 89(1), 177–99.

Spotila, J. R., and Berman, E. N. 1976. Determination of skin resistance and the role of the skin in controlling water loss in amphibians and reptiles. *Comparative Biochemical Physiology* 55A, 407–11.

Steadman, R. G. 1971. Indices of windchill of clothed persons. *Journal of Applied Meteorology* 10, 674–83.

Steadman, R. G. 1979. The assessment of sultriness. Part 1. A temperature-humidity index based on human physiology and clothing science. *Journal of Applied Meteorology* 18(7), 861–73.

Tao, X., and Xin, H. 2003. Acute synergistic effects of air temperature, humidity, and velocity on homeostasis of market-size broilers. *Transactions of the ASAE* 46(2), 491–7.

Tattersall, G. J., Cadena, V., and Skinner, M. C. 2006. Respiratory cooling and thermoregulatory coupling in reptiles. *Respiratory Physiology and Neurobiology* 154, 302–18.

Thom, EC. 1959. The discomfort index. *Weatherwise* 12, 59–60.

Tselepidaki, I., Santamouris, M., Moutris, C., and Poulopoulou, G. 1992. Analysis of the summer discomfort index in Athens, Greece, for cooling purposes. *Energy and Buildings* 18, 51–6.

Tucker, C. B., Rogers, A. R., and Schutz, K. E. 2007. Effect of solar radiation on dairy cattle behavior, use of shade and body temperature in a pasture-based system. *Applied Animal Behavior Science* 109(2–4), 114–54.

Walsberg, G. E., Campbell, G. S., and King, J. R. 1978. Animal coat color and radiative heat gain: A re-evaluation. *Journal of Comparative Physiology B* 126, 211–22.

Webster, M. D., Campbell, G. S., and King, J. R. 1984. Cutaneous resistance to water-vapor diffusion in pigeons and the role of the plumage. *Physiological Zoology* 58(1), 58–70.

Willmer, P. G. 1982. Microclimate and the environmental physiology of insects. *Adv. Insect Physiol.* 16, 1–57.

Part II

Local (Topo-)Climates

This section of the book examines local or topoclimates. As noted in the Introduction (Chapter 1), these have a horizontal scale of hundreds of meters up to about 10 km. They are determined by the influence of topographic elements such as hill slopes, valley bottoms, and ridge tops as well as by water bodies and the built landscapes of towns and cities. They are best developed during calm, clear weather when radiational contrasts are maximized, but also during strong air flow conditions when dynamical effects are most apparent. Their vertical dimension is the planetary boundary layer that varies diurnally between about 500 m at night and 1500 m by day over level terrain.

10

Urban Climates

Urban climatic characteristics are becoming a topic of increased importance, given that more than half the world's population now resides in cities. The building architecture, urban design, and building materials all influence the temperature, humidity, and wind conditions. Heat released by buildings and pollutants from traffic and industries modify the city environment. Urban microclimates are discussed in Chapter 8, Section E. Here we focus on the regional-scale characteristics.

Research into urban climates began with climatic measurements made in the early nineteenth century at a site in the city and three sites around London by Luke Howard. The first edition of his book on the climate of London was published in 1818–20. Mills (2008) describes Howard's major contribution to identifying the urban heat island. An example of a circa 1925 urban weather station is shown in Figure 10.1. More than a century later, Tony Chandler (1965) used station data and nighttime transects made with an instrumented vehicle to analyze the heat island of London. By reversing the route and averaging the two sets of data he was able to determine the climatic conditions across London. Helmut Landsberg (1981) published *The urban climate*, which drew particular attention to pollution aerosols.

There are three components involved in the production of urban climates. These are (i) modification of the atmospheric composition, (ii) modification of the surface energy budget, and (iii) modification of the surface characteristics. The second and third of these modifications are responsible for the development of the urban heat island (UHI).

A. Modification of Atmospheric Composition

The urban atmosphere contains a complex mix of gases and particles in concentrations that differ considerably from those in rural environments. The gases include ozone, sulfur dioxide, carbon monoxide, and nitric oxides. The particles include carbon, complex hydrocarbons, metals, and mineral dust. These are now discussed in turn.

Figure 10.1. An example of an urban meteorological station on the roof of a United States Weather Bureau facility, circa 1925.
Source: National Oceanic and Atmospheric Administration/ Department of Commerce, image ID wea00907.

Gases

Industrial and domestic coal burning generates sulfur dioxide (SO_2), while the combustion of gasoline and oil produces hydrocarbons (Hc), carbon monoxide (CO), nitrogen oxides (NOx), and ozone (O_3). Seinfeld (1989) cites data from 39 U.S. cities during 1984–6 that list concentrations for 48 gaseous compounds: 25 were paraffins, 15 aromatics, and 7 biogenic olefins. The sum of the median concentrations for the paraffins was 266 ppbC, representing 60 percent of the total 48 compounds. Aromatics were 116 ppbC, accounting for 26 percent, and olefins were 47 ppbC representing 11 percent. Under urban conditions, biogenic olefins cause ozone (O_3) to form. Ultraviolet radiation dissociates nitrogen dioxide (NO_2) into NO and monatomic oxygen, O; the latter may then combine with molecular oxygen, O_2, to form ozone. Ozone generation is related to the attack of hydrocarbons by hydroxyl radicals (OH). There are three sources of OH in the urban atmosphere: photochemical dissociation of ozone, of carbonyl compounds, and of nitrous acid (Seinfeld, 1989). Ozone concentrations in unpolluted tropospheric air range from about 20 to 50 ppb but may reach 400 ppb in polluted city air. In the Northern Hemisphere tropospheric ozone levels in Europe increased from about 20 ppb in 1950 to 40 ppb by 2000 (Hartmann et al., 2013, p. 173). In Los Angeles, ozone levels are high because of the large number of vehicles and the high levels of solar radiation. However, ozone concentrations (maximum eight hour averages) decreased from 200 ppb in 1992 to ~125 ppb in 2012, but still well above the federal standard of 70 ppb. In the late 1960s, improved technologies, the phasing out of coal burning, and antipollution regulations led to marked declines in sulfur dioxide pollution in North American and West European cities. This was repeated in central and eastern Europe in the 1990s, when the industrial and domestic use of lignite (brown coal) was greatly reduced. Concentrations of SO_2 in North America and Western Europe seldom exceed 0.04 ppm (125 μg m^{-3}), but in Asia, where coal is widely used, concentrations may be 5–10 times higher. However, as far as air pollution is concerned, European cities have experienced a rise in NO_2 levels as a result of the European Union's push in 1998 to reduce CO_2 emissions by switching to diesel cars. As a result, NO_2 levels now exceed 80 μg m^{-3}, twice the EU limit, in most European cities.

Strengthened regulations in the United States led to reductions between 1970 and 2006 of carbon monoxide emissions from 197 million tons to 89 million tons, of nitrogen oxide emissions from 27 million tons to 19 million tons, and of sulfur dioxide emissions from 31 million tons to 15 million tons.

Aerosols

Suspended particulate matter consists mainly of carbon, lead, and aluminum compounds and silicates. Fine particles, defined as PM_{10} (Particulate Matter <10 μm diameter) or $PM_{2.5}$ (< 2.5 μm), are the primary health risk. Gas-to-particle conversions can occur in several modes. A particle may form by homogeneous nucleation from a single vapor species, or from two or more species, or by heterogeneous nucleation, in which vapor is transferred to preexisting particles. The different types include aerosol sulfate, aerosol nitrate, and carbonaceous aerosol. Sulfates constitute 35 percent of the tropospheric aerosol load (Wolf and Hidy, 1997).

In Europe in the period 2000–9, $PM_{2.5}$ decreased by 3 percent per year, slightly less than in 1990–2009, while in the United States decreases averaged 2–4 percent per year. In Europe, sulfates decreased by 3.1 percent per year from 1990 to 2009 (Hartmenn et al., 2013). There are no comparable data for Asia, but concentrations likely increased as a result of city growth, automobiles, and coal burning.

Concentrations of PM_{10} in polluted urban areas at ground level can exceed 50–100 μg m^{-3}. Highest concentrations occur with low wind speed, low vertical turbulence, temperature inversions, high relative humidity, and air moving from industrial sources within the city. There are strong seasonal and diurnal variations in pollution levels. Highest concentrations are typically observed around 08.00 hours in early winter when pollution that has been trapped by a stable lower troposphere a few hundred meters deep is carried down to the surface as a result of vertical mixing due to thermal convection – a process known as fumigation.

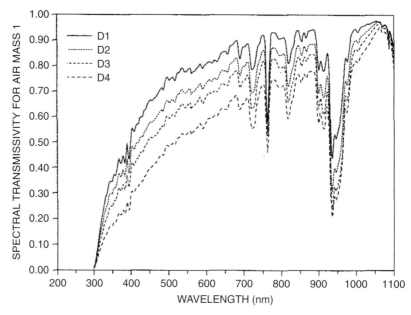

Figure 10.2. Spectral transmissivities, the ratio of surface-to-extraterrestrial solar irradiance, for four pollution categories in Barcelona, Spain, during the period 1989–92. D1 is clean air; D2 is relatively clean air; D3 is moderately turbid air (most frequent); and D4 is extremely turbid air (J. Lorente et al., 1993).
Source: *Journal of Applied Meteorology* 33, p. 410, fig. 5. Courtesy American Meteorological Society.

The primary effect of aerosols is to reduce visibility and decrease incoming solar radiation via reflection, as well as decreasing the duration of bright sunshine. The reductions of solar radiation and UV radiation in St. Louis, Missouri, relative to a rural site in summer 1972 were 2 percent and 7 percent, respectively (Peterson and Flowers, 1977). On cloudless days in Los Angeles in autumn 1973 the average reductions of total solar and UV radiation were 11 percent and 29 percent, respectively. A study by Lorente et al. (1993) in Barcelona, Spain, showed that transmissivity of global solar irradiance (the ratio of the surface to extraterrestrial irradiance) decreased overall for all wavelengths as the aerosol concentrations increased. A decrease in transmissivity of more than 50 percent was observed for very polluted relative to clean air in the UV portion of the spectrum (Figure 10.2). These effects were very evident in European and North American cities in the 1950–60s but had diminished greatly by the 1990s. Now similar problems are apparent in cities in China, India, and Southeast Asia. The average $PM_{2.5}$ level in winter in Beijing since 2008 has been ~100 μg m^{-3}, identified as "hazardous" to health, with some winter days exceeding 500 μg m^{-3}. Levels in winter in New Delhi are even higher, reaching 200–500 μg m^{-3}. The annual average $PM_{2.5}$ level for 2008–13 was 153 μg m^{-3} in New Delhi compared with 59 μg m^{-3} for Beijing, according to the World Health Organization.

A different measure of atmospheric turbidity is given by the aerosol optical depth (AOD), which represents the total amount of aerosol integrated through the atmospheric column. The spectral aerosol optical depth is determined by the depletion of the solar beam at a specified wavelength, often 0.5 μm (that is adopted in Schüepp's turbidity coefficient). Power (2003) gives a detailed review of the different coefficients that are in use. Hsu et al. (2012) analyze SeaWiFS data of AOD for 1997–2010. They show a peak annual average of around 0.22 at 15 °N, reflecting sources in southern and eastern Asia. The corresponding seasonal averages at 50 °N are 0.10–0.15 with a spring maximum. Surface data reported by Gueymard (1994) for Bismark, North Dakota, for 1971–85 and Toledo, Ohio, for 1971–8 show minima of 0.09 and maxima of 0.21 and 0.14 and 0.43, respectively. As reported by Mikalsky et al. (2001), State College, Pennsylvania, for 1992–9 showed AOD values at 0.5 μm, ranging between 0.4 in summer and 0.08 in winter with seasonal averages of about 0.25 and 0.12, respectively.

Analysis of AOD and solar radiation receipts for 20 stations in and around Beijing has been performed by Zhang et al (2015). Regional averages during 1961–2007 show consistent seasonal decreasing trends of −2.29, −3.63, −7.45, and −3.76 W m^{-2} per decade for winter, spring, summer, and autumn, respectively. Solar dimming continued throughout the period despite decreasing cloud cover. This is attributed to increases in AOD, ranging from 0.02 to 0.04 per decade. The largest decrease in solar radiation, ranging from 3.01 W m^{-2} in winter to 9.06 W m^{-2} in summer, were observed at 10 urban stations and were consistent with the largest increases in AOD. The smallest decrease in solar radiation occurred at 5 rural stations, where the trends were 30–54 percent of those at urban stations. Trends at 5 suburban stations were intermediate between urban and rural.

B. Modification of the Energy Budget

In the 1960s and 1970s, observations of the attenuation of solar radiation by atmospheric pollution showed average values of 9–12 percent for industrial cities in Canada (Oke, 1997). The reductions were especially prominent in the UV wavelengths, as also shown in Figure 10.2 for Barcelona. These values have been greatly reduced by the introduction of pollution controls in North America and Europe, as discussed previously. However, even higher reductions are to be expected in the heavily polluted cities of India and China. Apart from this attenuation effect, particles in the air also increase the proportion of diffuse radiation. The infrared radiation balance has been found to respond to the heating effect on the urban atmosphere of the urban heat island rather than to a greenhouse effect of the atmospheric pollutants.

Oke and Fuggle (1972) measured nighttime increases in Montreal of 6–40 W m^{-2} relative to rural locations. In Hamilton, Ontario, Rouse et al. (1973) found average excesses of about 70 W m^{-2}, or 31 percent. However, these excesses in counter radiation were shown to be offset by increases in outgoing IR as a result of the warmer urban surfaces.

Urban surfaces comprise a mixture of concrete, asphalt, brick, grass, and trees. Roofs typically occupy 20–25 percent of the area and pavement about 40 percent. For the most part they are dry as water drains away quickly. The mean albedo of urban areas is about 0.12 and thus much of the incoming solar radiation is absorbed, heating the surfaces, which reradiate longwave radiation. By day, up to 70–80 percent of the net radiation is transferred to the atmosphere via sensible heat. The rough urban surface combined with its heating readily foster turbulent energy transfers; their magnitudes are highly variable in space and time. A two-year study for Łódź, Poland, is presented by Offerle et al. (2006). Building heights typically range from 15 to 20 m and cover 30 percent of the surface. Other impervious surfaces cover 39 percent and vegetation 31 percent. Sensible heat is 52–56 percent of R_n in June–July to September and 1.29–1.48 for October–December and March–April. Corresponding percentages for latent heat are 0.38–0.47 and 1.11–1.34, respectively. Grimmond and Oke (2002) summarize summertime turbulent heat fluxes computed for 12 cities. On average at the residential sites, latent heat constitutes an energy sink of 22–37 percent of daytime and 28–46 percent of daily R_n. Lawn irrigation helps maintain high rates of evapotranspiration. Average daily maximum sensible heat flux is between 200 and 300 W m^{-2}. It is the most important heat sink in the surface energy balance, ranging from 40 to 60 percent of daytime R_n.

Cities also have an additional heat source due to energy produced by combustion. Many midlatitude cities now generate energy through combustion at rates comparable with incoming solar radiation in winter. Both average around 25 W m^{-2}. In the extreme situation of Arctic settlements during polar darkness, the energy balance during calm conditions depends only on net longwave radiation and heat production by anthropogenic activities. In Toulouse, France, combustion gives rise to about 70 W m^{-2} in winter and 15 W m^{-2} in summer. In Montreal, Canada, in the winter of 1961, an average of 153 W m^{-2} was released compared with a net radiation of only 13 W m^{-2} (Oke, 1997).

Energy storage within the urban fabric is a very important factor. It is determined as a residual in the energy balance equation. Grimmond and Oke (1999) determine it for seven North American cities and show that it accounts for 17–58 percent of daytime net radiation (Figure 10.3). Largest values were obtained at downtown and light industrial sites. Grimmond and Oke (2002) state that at dry, built-over locations it accounts for ~50 percent of daytime R_n, while at residential sites it is 20–30 percent. Average daily peaks range from 150 to 280 W m^{-2}. Most of the sites release large amounts of the stored heat during the first third of the night. For the rest of the night there is a balance between the storage source and the radiative loss. There is a phase lag (hysteresis pattern) between the radiative forcing and storage change. However, it is important to recognize that these patterns are also dependent on the particular type of urban surface that is being considered.

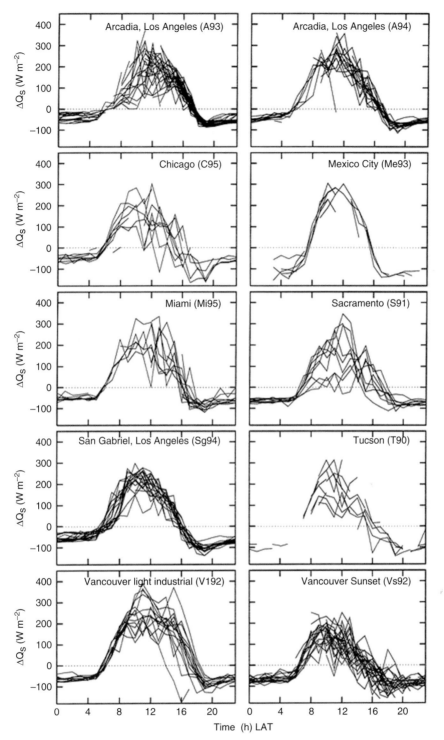

Figure 10.3. Diurnal change in energy storage (ΔQ_s) for several urban centers measured as the residual of the surface energy balance. Positive values indicate the surface is gaining energy from the atmosphere; negative values indicate the surface is losing energy to the atmosphere (from S. Grimmond and T. Oke, 1999).

Source: Journal of Applied Meteorology 38: p. 928, fig. 2.

Courtesy America Meteorological Society.

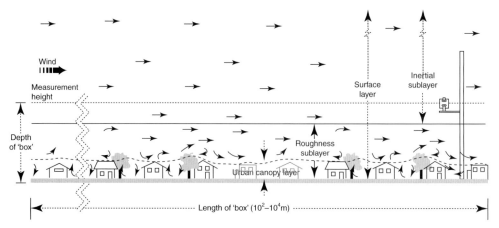

Figure 10.4. Schematic illustration of the boundary layer structure at the local scale over a city (as used in the Local-Scale Urban Meteorological Parameterization Scheme (LUMPS)) (from S. Grimmond and T. Oke, 2002).
Source: *Journal of Applied Metorology*, 41, p. 793, fig. 2. Courtesy of American Meteorological Society.

C. Urban Heat Islands

Luke Howard (1818–20) discovered that the largest heating effects of the city of London were on clear, calm winter nights and pointed out the effect of heat released by buildings and factories. The intensity of the heat island, as well as its extent and temporal characteristics, are determined by city size, morphology, land-use configuration, and the geographic setting (relief, elevation, and regional climate).

The following factors give rise to generally higher temperatures in urban areas:

(1) Changes in the radiation balance due to atmospheric composition.
(2) Changes in the radiation balance due to the albedo and thermal capacity of urban surface materials, and to canyon geometry.
(3) The production of heat by buildings, traffic, and industry.
(4) The reduction of heat diffusion due to changes in airflow patterns caused by surface roughness effects.
(5) The reduction in the energy required for evaporation and transpiration leading to the daytime transfer to the atmosphere of large amounts of sensible heat.

The urban atmosphere has two distinct layers: (1) the Urban Boundary Layer (UBL), which is due to the spatially integrated heat and moisture exchanges between the city and its overlying air (Oke, 1995) and (2) the Urban Canopy Layer (UCL), represented by the surface of the city (see Figure 10.4). Fluxes across this plane comprise those from individual units, such as roofs, canyon tops, trees, lawns, and roads, integrated over larger land-use divisions (city center, suburbs, parks).

The Urban Heat Island [UHI] is the difference between air temperature in the urban and adjacent rural areas. The intensity is mainly a function of the ratio of canyon

Table 10.1. Causes of Urban Heat Island

1. Greater absorption of solar radiation due to multiple reflection and radiation trapping by building walls and vertical surfaces in the city.
2. Greater retention of infrared radiation in street canyons due to restricted view of the radiatively "cold" sky hemisphere.
3. Greater uptake and delayed release of heat by buildings and paved surfaces in the city. This is due not only to the thermal properties of the materials, but also to the solar and infrared radiation "trap" and to reduced convective losses in the canopy layer where airflow is retarded.
4. Differences in the thermal admittance of urban and rural surfaces. The absolute magnitude of the rural admittance sets the maximum heat island and its seasonal variation.
5. A greater fraction of absorbed solar radiation at the surface is converted to sensible rather than latent heat. This effect is due to the replacement of moist soils and plants with paved and waterproofed surfaces, and a resultant decline in surface moisture.
6. Greater release of sensible and latent heat from the combustion of fuels for urban transport, industrial processes, and domestic and commercial space heating/cooling.

Source: After Oke, 1982; Oke et al., 1991.

height to width, though it increases also with an increasing difference in thermal inertia of the urban and rural surfaces, and downward infrared radiation from pollution layers.

The causes of urban heat islands have been detailed by Oke (1982; see Table 10.1).

The heat island effect is especially apparent on calm, cloudless nights during summer and early autumn. For the period 1931–60, the center of London had a mean annual air temperature of 11.0 °C, compared with 10.3 °C for the suburbs and 9.6 °C for the surrounding countryside (Chandler, 1965). Figure 10.5 illustrates the UHI, as shown by minimum air temperatures over London on May 14, 1959, in relation to the built-up area. Note the maximum values near the city center and the relatively steep temperature gradients toward the boundaries of the built-up area. Minimum temperatures in urban areas may be 5–6 °C above those in the surrounding countryside. In London, Kew has an average of 72 more days with frost-free screen temperatures than rural Wisley. Because the heat island is a relative phenomenon, its magnitude also depends on the rate of nocturnal cooling in rural areas. Factors that tend to enhance UHI intensity include fully developed crops, dry soils, and snow cover in the surrounding rural areas, while heavy clay soils, water bodies, wet soils, and bare rock tend to weaken it (Oke, 1997).

An updated analysis for St. James Park, central London, and Wisley by Wilby (2003) found that the mean nocturnal UHI intensity ranged from 1.1 °C in January to 2.2 °C in August, while the daytime UHI ranged from 0.1 °C in April to 0.4 °C in December. Since the urban station was in a park, these differences are probably conservative. The nocturnal UHI is most strongly correlated, negatively, with surface wind strength, indicating the importance of light winds and minimal turbulence. For

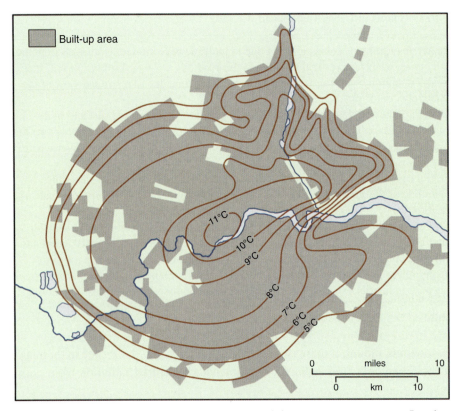

Figure 10.5. The urban heat island as shown by minimum temperatures over London on May 14, 1959 (redrawn after T. J. Chandler, 1965; from R G Barry and R J Chorley, 2010, p. 422, fig. 12.27, London: Routledge/ Taylor and Francis).

1959–98 the nocturnal UHI for London increased by 0.13 °C per decade in spring and 0.12 °C per decade in summer.

A recent study of Melbourne, Australia, by Sachindra et al. (2015) determines the UHI by contrasting temperatures observed in the Central Business District (CBD) with those in less urbanized Laverton, 17 km to the southwest, over the period 1952–2010. Mean nocturnal UHI intensities ranged from 1.4 °C in February to 2.0 °C in September and October. The UHI intensity increased between 1952–71 and 1992–2010 from 1.1 to 1.9 °C in summer, 1.1 to 2.0 °C in autumn, 1.4 to 2.1 °C in winter, and 1.6 to 2.3 °C in spring.

City size has been shown to influence heat island strength. In North America, the maximum urban–rural temperature difference reaches 2.5 °C for towns of 1000, 8 °C for cities of 100,000, and 12 °C for cities of one million people. European cities show a smaller temperature difference for equivalent populations, perhaps as a result of the generally lower building height or differences in urban building density.

The vertical extent of the heat island is poorly known, but it is thought to exceed 100–300 m, especially early in the night. In the case of cities with skyscrapers, the vertical and horizontal patterns of wind and temperature are very complex. Studies of St. Louis, Missouri, showed that the UHI, which had a maximum value of about 1.5 °C, extended vertically to around 1 km and was detectable downwind for 40 km (Auer, 1981). The mixing layer depth is commonly domed over the city center and in the case of St. Louis extended to 150–450 m altitude. Lidar-derived measurements of the mixing layer depth showed much deeper mixing layer depths in central London in the summer compared to rural Chilbolton in the winter (see Figure 10.6).

In some mid-latitude cities there is a reverse "cold island" effect of 1–3 °C in summer. In the United States this effect has been reported in Boston, Massachusetts; Dallas, Texas; Detroit, Michigan; and Seattle, Washington, when corrections are made to temperature for latitude and elevation differences. A suggested reason for this is that micro- and local-scale impacts dominate over the mesoscale urban heat island. A cool island is observed in cities in tropical and subtropical deserts, where it is attributed to the high thermal inertia of the built-up area and sharp diurnal temperature fluctuations.

A study of surface UHIs was performed for 419 large global cities compared with the surrounding suburban areas by Peng et al. (2012) using MODIS surface temperatures. They find that the average annual daytime surface UHI (1.5 ± 1.2 °C) is slightly higher than the annual nighttime value (1.1 ± 0.5 °C). There are different driving mechanisms for day and night UHIs. The distribution of daytime surface UHI is shown to correlate negatively across cities with the difference of vegetation cover between urban and suburban areas, especially during the growing season, whereas the distribution of nighttime surface UHIs correlates positively with the differences in albedo and nighttime city lights between urban area and suburban area.

It should be noted that at night the rural areas tend to develop surface-based inversions as a result of radiative cooling. By early morning this inversion layer may be 100–300 m deep (Oke, 1991).

Over the city meanwhile there is a neutral or slightly stable mixed layer as a result of the surface warming. The rural inversion is eroded as air flows toward the city center. On the downwind side of the city the warmer air rides over the rural cold air layer and forms a downstream "urban plume."

In most tropical cities built land differs from that in higher latitudes; it is commonly composed of high-density, single-story buildings with few open spaces and poor drainage. In such a setting, the composition of roofs is more important than that of walls in terms of thermal energy exchanges. In the dry tropics, buildings have a relatively high thermal mass to delay heat penetration, and this, combined with the low soil moisture in the surrounding areas, makes the ratio of urban to rural thermal admittance greater than in temperate regions. Building construction in the humid tropics is characteristically lightweight to promote essential ventilation. These cities differ greatly from temperate ones in that the thermal admittance is greater in rural than urban areas as a result of high rural soil moisture levels and high urban albedos. The characteristics of tropical heat islands are similar to those of temperate cities, but they are usually weaker, typically 4 °C for the maximum nocturnal difference, and

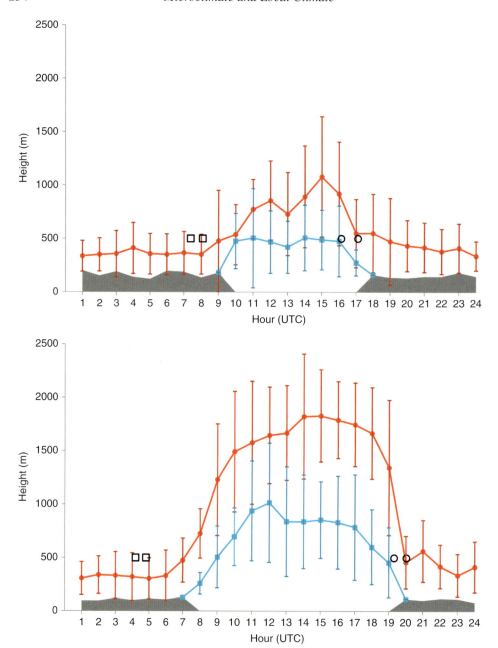

Figure 10.6. Diurnal mixing layer depths derived from lidar measurements over central London (red lines) and rural Chilbolton (blue lines) with standard deviations (bars) during summer (top) and winter (bottom). Open squares and circles note the range of sunrise and sunset times, respectively (from S. J. Bohnenstengel et al., 2015). *Source*: *Bulletin of the American Meteorological* Society 96, p. 706, fig. 3. Courtesy American Meteorological Society.

they are best developed in the dry season. Delhi, India, exhibits a difference of +3 to 5 °C for minimum temperatures and +2 to 4 °C for maximum temperatures.

Stewart (2011) provides a systematic review and scientific critique of methodology in UHI literature published between 1950 and 2007. He finds that in a sample of 190 urban heat island studies the mean quality score of the sample is just 50 percent, and nearly half of all urban heat island magnitudes reported in the sample are judged to be scientifically indefensible. One-half of the sample studies do not sufficiently control the confounding effects of weather, relief, or observation time on reported "urban" heat island magnitudes. Nevertheless, he cites 19 studies that are identified as "top tier" and notes that half of these were carried out between 1998 and 2007.

D. Modification of the Surface Characteristics

The urban surface differs from the surrounding rural environment in several ways. The surface albedo is generally lower (0.12 versus 0.15–0.25), the heat capacity of the building materials is different, the surface roughness is greater, the surface is drier, and rainwater drains off quickly.

Airflow

The generation of turbulence as a result of the surface roughness and the channeling effect of buildings affect air movement. Also, urban heat islands can set up convergent flow patterns over the city. Data on the surface roughness length for momentum (z_0) show a wide range of values. A detailed study of four urban areas in North America by Grimmond et al. (1998) indicates typical summer values in the range 0.5–1.0 m with winter values about 85 percent lower. The corresponding range of zero-plane displacement values (d) is 3.0–4.5 m.

Average urban wind speeds are generally less than in rural areas as a result of the sheltering effect of buildings. In city centers wind speeds average at least 5 percent below those in the suburbs. However, there are considerable diurnal and seasonal variations. By day, city wind speeds are less than in the surrounding rural areas, but at night greater mechanical turbulence over the city leads to downward turbulent mixing that transfers higher wind speeds aloft down to lower levels.

In a study of New York (Bornstein and Johnson, 1977) wind speeds are found to be decreased below (increased above) those at sites outside the city during periods with regional wind speeds above (below) about 4 m s^{-1} at night and 3 m s^{-1} by day. A network of 97 anemometer sites was operated during 1964–7 in a rectangle 220 km E-W and 110 km N-S centered on the west side of Midtown Manhattan. By day, wind speeds decreased nearly linearly from 4.8 m s^{-1} at 32 km upwind of Central Park to 3.9 m s^{-1} at 32 km downwind. At night speeds increased from 2.5 m s^{-1} at 32 km upwind to 3.5 m s^{-1} at 16 km downwind of Central Park and then decreased again. The peak at night is due to the nocturnal heat island.

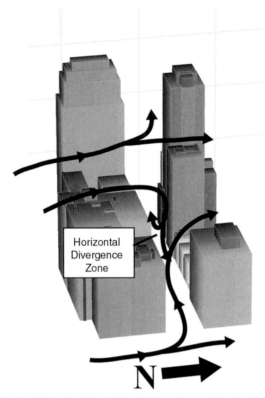

Figure 10.7. Since building height in North American cities is higher and more heterogeneous compared to European cites, increased turbulence and wind speeds are likely, as conceptually shown in the region labeled "Horizontal Divergence Zone" (from M. A. Nelson et al., 2007).
Source: *Journal of Applied Meteorology and Climatology* 46, p. 2048, fig. 7. Courtesy American Meteorological Society.

For Łódź, Poland, Fortuniak et al. (2006) analyzed data for an urban and a rural site for 1997–2002. Wind speed at the urban station was on average lower by about 34 percent at night and 39 percent during the day. Values are 4.0 m s^{-1} at the rural station versus 2.5 m s^{-1} in the city by day and 2.5 m s^{-1} versus 1.7 m s^{-1}, respectively, at night.

The difference in the strength of the UHI between North American and European cities could be related to difference in airflow characteristics. The variability in building heights in North American cities is greater than that in European cities because of building height restrictions in European cities that result in more of a homogeneous urban landscape (Figure 10.7). For example, the mean (standard deviation) building height in downtown Los Angeles is 12.0 m (+/– 22.7 m) (Burian et al., 2002) compared to 8.8 m (+/– 3.0 m) for central London (Wood et al., 2010). As a result, North American cities could be more prone to localized regions of high wind speeds and turbulence, as shown by Nelson et al. (2007) in Oklahoma City.

The deceleration caused by increased drag over urban areas leads to mass convergence and compensatory uplift within the urban boundary layer. The slowdown also reduces the Coriolis parameter, and hence imbalance with pressure gradient force causes the air to turn to the left over the city (in the Northern Hemisphere). Downwind of the urban area, the flow resumes its upstream direction.

In addition to the effects of built-up areas on wind speed, the vertical extent of urban influences on wind velocity is considerable. Under neutral stability it has been shown to reach around 500 m compared with 200 m over rural areas. Typically, the wind speed, u, at a height z is computed using the power law (Oke, 1997):

$$\bar{u} = \overline{u_{ref}}(z / z_{ref})^a$$ 10.1

where $\overline{u_{ref}}$ is the wind speed at a reference station. The exponent a depends on roughness and stability and is about 0.16 in open rural areas and up to 0.35 for downtown high-rise areas.

The airflow in urban areas is often affected by the city's location at a coast, on a lakeshore, or on valley slopes. Each of these situations can give rise to thermally induced breezes (sea and land breezes, mountain-valley winds) that interact with urban circulations that are set up by the thermal gradients created by the city-rural contrasts. These effects can shift the timing and intensity of UHI diurnal patterns.

Moisture Conditions

Urban surfaces have few large water bodies and precipitation that falls on them rapidly runs off into drains. This lowers evaporation, as does the generally limited vegetation cover. The net result is an increase in urban heating. Hence, city air tends to have lower absolute humidity than its surroundings. Relative humidity can be as much as 30 percent less in the city by night, compared with rural surroundings, as a result of the higher urban temperatures.

The microclimates of a suburban landscape near Denver, Colorado, are examined by Bonan (2000) for summer conditions. Infrared thermometer measurements showed that native grass surfaces were considerably higher (36.1 °C) than residential lawns (31.0 °C) and parks (30.1 °C), which were watered. The results demonstrate the importance of evapotranspiration as a cooling agent in the dry, semi-arid Colorado environment. Modeling studies also show the effect of "greening" on the surface energy balance and surface temperatures in urban areas. Using the Regional Boundary Layer Model (RBLM), Yang et al., (2015) found that increasing tree coverage in Suzhou, China, from 0 to 20 to 40 percent decreased the Bowen ratio from 8.78 to 1.20 to 0.43, respectively (Figure 10.8). The maximum air temperatures were reduced by 0.6 °C and 2.0 °C with a 20 percent and 40 percent increase in tree cover compared to the 0 percent case, respectively.

Urban areas are considered to be responsible for local conditions that, in summer especially, can trigger excesses of precipitation under marginal conditions. A recent review by Shepherd (2005) surveys the observational evidence and modeling studies. The triggering involves both thermal effects and the increased frictional convergence

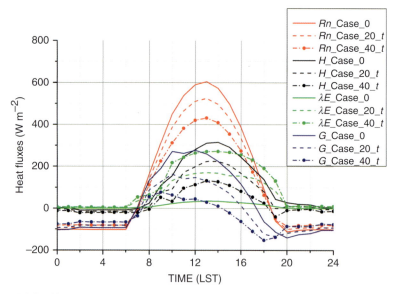

Figure 10.8. Simulated changes in the surface energy balance with tree coverage of 0 percent (Case_0), 20 percent (Case_20), and 40 percent (Case_40) (from J. Yang et al., 2015). LST = Local Solar Time.
Source: *Journal of Applied Meteorology and Climatology* 54, p. 150, fig. 8. Courtesy American Meteorological Society.

caused by built-up areas. European and North American cities tend to record 6–7 percent more days with rain per year than their surrounding regions, giving a 5–10 percent increase in urban relative to rural precipitation. During 1951–60, London's thunderstorm rain (5–15 percent of total precipitation) was 200–250 mm greater than that in rural Southeast England (Atkinson, 1985). Areas in eastern metropolitan Atlanta, Georgia, received 30 percent more rainfall during days of maritme tropical air in June–August 2002–6 than areas west of the city. Both precipitation amount and frequency were enhanced up to 80 km to the east of the urban core of Atlanta. Likewise, more frequent thunderstorms and hail occur for 30–40 km downwind of industrial areas of St Louis, Missouri, compared with rural areas upwind (Huff and Chagnon, 1972).

A detailed study of Beijing's convective precipitation for summers 2008–12 uses precipitation data from 53 Automatic Weather Station (AWS) sites (Dou et al., 2015) covering an area of ~200 km². Temperature, humidity, and wind data were also used. The population of Beijing is 21 million people. The summer UHI was multicentered and averaged 1.7 °C at night and 0.8 °C by day. Correspondingly, specific humidities were lower in the city than surrounding rural areas by 2.4 g kg⁻¹ by day and 1.8 g kg⁻¹ at night.

Low-level flows converged into the UHI in early morning but bifurcated in the afternoon and evening as a result of urban building-barrier-induced divergence. As a consequence, urban thunderstorms bifurcated around the city center. This produced a regional-normalized rainfall minimum in the urban center and directly downwind of up to 35 percent, with maximum values along its downwind lateral edges of greater than 15 percent. However, strong UHIs (>1.25 °C) induced or enhanced day and night thunderstorm formation, which produced a normalized rainfall maximum in the most urbanized area of up to 75 percent.

References

Atkinson, B. W. 1985. *The urban atmosphere.* Cambridge: Cambridge University Press.

Auer, A. H. Jr. 1981. Urban boundary layer. *Met. Monogr.,(Amer. Met. Soc.)* 18, 41–62.

Bohnenstengel, S. I. et al., 2015. Meteorology, air quality, and health in London: The ClearfLo Project. *Bull. Amer. Meteor. Soc.* 96, 779–804. doi:10.1175.

Bonan, G. B. 2000. The microclimates of a suburban Colorado (USA) landscape and implications for planning and design. *Landscape Urban Planning* 49, 97–114.

Bornstein, R. D., and Johnson, D. S. 1977. Urban – rural wind velocity differences. *Atmos. Environ.* 11, 597–604.

Burian, S. J., Brown, M. J., and Velugubantla, S. P. 2002. Building height characteristics in three U.S. cities. Proceedings of the *Fourth Urban Environment Symposium, American Meteorological Society*, Norfolk, VA.

Chandler, T. J. 1965. *The climate of London.* London: Hutchinson.

Dou, J-J. et al. 2015. Observed spatial characteristics of Beijing urban climate impacts on summer thunderstorms. *J. Appl. Met.* 54, 94–105.

Fortuniak, K., Kłysik, K., and Wibig, J. 2006. Urban–rural contrasts of meteorological parameters in Łódź. *Theor. Appl. Climatol.* 84, 91–101.

Grimmond, C. S. B. et al. 1998. Aerodynamic roughness of urban areas derived from wind observations. *Bound.-Layer Met.* 89, 1–24.

Grimmond, C. S. B., and Oke, T. R. 1999. Heat storage in urban areas: Local-scale observations and evaluation of a simple model. *J. Appl. Met.* 38, 922–40.

Grimmond, C. S. B., and Oke, T. R. 2002. Turbulent heat fluxes in urban areas: Observations and a Local-scale Urban Meteorological Parameterization Scheme (LUMPS). *J. Appl. Met.* 41, 792–810.

Gueymard, C. 1994. Analysis of monthly average atmospheric precipitable water and turbidity in Canada and northern United States. *Solar Energy* 53, 57–71.

Hartmann, D.L. et al. 2013: Observations: Atmosphere and surface, in *The physical science basis. Contribution of Working Group I to the Fifth Assessment Report of the Intergovernmental Panel on Climatec change.* Cambridge, United Kingdom; T.F. Stocker, D. Qin, G.-K. Plattner, et al. Editors, pp. 159-254.

Howard, L. 1818–20. *The climate of London deduced from meteorological observations made at different places in the neighbourhood of the metropolis.* 2 vols. London: W. Phillips.

Hsu, N. C. et al. 2012. Global and regional trends of aerosol optical depth over land and ocean using SeaWiFS measurements from 1997 to 2010. *Atmos. Chem. Phys.* 12, 8037–53.

Huff, F., and Changnon S. Jr., 1972. Climatological assessment of urban effects on precipitation at St. Louis. *J. Appl. Met.* 11, 823–42.

Landsberg, H. 1981. *The urban climate.* New York: Academic Press.

Lorente, J., Redaño A., and De Cabo, X. 1993. Influence of urban aerosols on spectral solar irradiance. *J. Appl. Meteor.* 33, 406–15.

Mikalsky, J. et al. 2001. Multiyear measurements of aerosol optical depth in the Atmospheric Radiation Measurement and Quantitative Links programs, *J. Geophys. Res.*, 106, 12099–107.

Mills, G. 2008. Luke Howard and the climate of London. *Weather* 63(6), 153–7.

Nelson, M. A. et al. 2007. Properties of the wind field within the Oklahoma City Park Avenue Street Canyon. Part I. Mean flow and turbulence statistics. *J. Appl. Met. Clim.* 46, 2038–54. doi:http://dx.doi.org/10.1175/2006JAMC1427.1.

Offerle, B. et al. 2006. Temporal variations in heat fluxes over a central European city centre. *Theor. Appl. Climatol.* 84, 103–15.

Oke, T. R. 1982. The energetic basis of the urban heat island. *Quart. J. Roy. Met. Soc.* 108(455), 1–24.

Oke, T. R. 1995. The heat island effect of the urban boundary layer: Characteristics, causes and effects. In J. E. Cernmak et al (eds.), *Wind climate in cities*. New York: Kluwer Academic, pp. 81–108.

Oke, T. R. 1997. Urban environments. In W. G. Bailey et al. (eds.), *The surface climates of Canada*. Montreal: McGill-Queens University Press, pp. 303–27.

Oke, T. R., and Fuggle, R. F. 1972. Comparison of urban/rural counter- and net radiation at night. *Boundary-layer Met.* 2, 290–308.

Oke, T. R. et al. 1991. Simulation of surface heat island under ideal conditions at night. 2. Diagnosis of causation. *Boundary-Layer Met.* 56, 339–58.

Peng, S. et al. 2012. Surface urban heat island across 419 global big cities. *Environ. Sci. Tech.* 46, 696–703.

Peterson, J. T., and Flowers, E. C. 1977. Interaction between air pollution and solar radiation. *Solar Energy* 19, 23–32.

Power, H. C. 2003. The geography and climatology of aerosols. *Progr. Phys. Geog.* 27, 502–47.

Rouse, W. R., Noad, D., and McCutcheon, J. 1973. Radiation, temperature, and atmospheric emissivities in a polluted urban atmosphere at Hamilton, Ontario. *J. Appl. Met.* 12, 798–807.

Sachindra, D. A. et al. 2015. Impact of climate change on urban heat island effect and extreme temperatures: A case study. *Quart. J., Roy. Met Soc.* doi:10.1002/qj.2642.

Seinfeld, J. H. 1989. Air pollution: State of the science. *Science* 243, 745–52.

Shepherd, J. M. 2005. A review of current investigations of urban-induced rainfall and recommendations for the future. *Earth Interactions* 9(12), 1–27.

Stewart, I. D. 2011. A systematic review and scientific critique of methodology in modern urban heat island literature. *Int. J. Climatol.* 31, 200–17.

Wilby, R. L. 2003. Past and projected trends in London's urban heat island. *Weather* 58, 251–60.

Wolf, M. E., and Hidy, G. M. 1997: Aerosols and climate: Anthropogenic emissions and trends for 50 years. *J. Geophys. Res.* 102(D10), 11, 113–21.

Wood, C. R. et al., 2010. Turbulent flow at 190 m height above London during 2006–2008: A climatology and the applicability of similarity theory. *Boundary-Layer Meteorology* 137, 77–96.

Yang J., et al., 2015. Further development of the Regional Boundary Layer Model to study the impacts of greenery on the urban thermal environment. *J. Appl. Met. Clim.* 54, 137–52.

Zhang, X., Xia, X., and Xuan, C. 2015. On the drivers of variability and trend of surface solar radiation in Beijing metropolitan area. *Int. J. Climatol.* 35, 452–61.

11

Topoclimatic Effects on Microclimate

This chapter examines the various effects of topoclimate on microclimates. These include exposition (or aspect), slope angle, and shading, which are all related to incoming solar radiation and thus affect surface and soil temperatures. Exposition and slope angle also modify the wind direction and speed and the precipitation amount. Specific localities such as lakeshores and coasts and spatial subdivisions of urban areas are discussed, as well as mosaic landscapes.

A. Exposition, Slope Angle, and Shade Effects

Slope exposition (or aspect) has significant effects on solar radiation receipts and, therefore, on surface temperature. One of the earliest investigations of the effects of exposition was carried out in Munich, Germany, by Wollny (1878), who showed that south-facing slopes in the Northern Hemisphere received more radiation and had slightly higher soil temperatures at 15 cm than horizontal surfaces and north-facing slopes. At 560 m elevation in the Inn valley and at 780 m elevation in the Gschnitztal in Austria, soil temperatures were measured by Kerner (1891) at a depth of 70 cm over a three-year period with eight different exposures around a small hillock. The slope angles were not reported. Table 11.1 summarizes a few of his results.

The range of soil temperature with exposition at Inntal (Gschnitztal) in February is 2.9 °C (2.8) and in August it is 4.4 °C (2.3). For the yearly average with exposition the range is 3.2 °C (2.7). The annual range on N slopes is 12.5 °C (13.0) and on SE slopes 15.3 °C (7.1). The snow cover melt date averaged February 21 on the south-facing slope and March 12 on the north-facing slope at Inntal and, correspondingly, on March 25 and April 17 at Gschnitztal.

Cantlon (1953) analyzed microclimates during May 1948–January 1950 on 20° N and S slopes of Cushetunk Mountain, New Jersey, at an elevation of 200 m. He recorded temperatures weekly, or daily during the summer, at 4 cm in the duff layer of the soil, at 5 cm in the bryophyte layer, at 20 cm in the herbaceous layer, 1 m in the shrub layer, and 2 m within the trees to represent the air temperature. The stations were substantially screened by trees, and during summer 1949 two additional slope

Table 11.1. Soil Temperatures °C with Exposition at 70 cm Depth (a) in the Inn Valley, and (b) in the Gschnitztal, Austria

Month	N	NE	E	SE	S	SW	W	NW
(a) February	3.5	3.6	3.8	5.2	5.3	6.4	5.5	3.8
(a) August	16.0	17.9	19.5	20.5	20.1	19.6	19.5	17.0
(a) Yearly average	9.5	10.6	11.3	12.6	12.6	12.7	12.2	10.2
(b) February	–0.7	–0.2	–0.3	0.8	1.2	2.1	1.7	0.9
(b) August	12.3	12.6	13.5	14.6	14.5	14.0	13.7	13.0
(b) Yearly average	5.1	5.5	5.9	7.5	7.8	7.8	7.4	6.5

Source: After Kerner, 1891.

Table 11.2. South- Minus North-facing Slope Temperatures (°C) for Stations with Light (*Heavy) Summer Shade at Cushetunk Mountain, New Jersey, in 1949

Month	Soil 4 cm	Soil 4 cm	Air 5 cm	Air 5 cm	Air 20 cm	Air 20 cm	Air 1 m	Air 1 m	Air 2 m	Air 2 m
	Max	Mean	Max	Mean	Max	Mean	Max	Mean	Max	Mean
January	5.8	3.9	5.7	2.8	3.6	1.6	1.8	0.8	1.0	0.6
April	3.9	2.9	6.5	3.1	4.1	2.0	2.2	1.0	1.2	0.1
July	1.0	0.5	4.3	2.0	2.9	1.4	1.7	0.8	1.0	0.5
July*			2.1	1.0			1.3	0.6	1.0	0.4
October	2.9	2.4	9.4	4.9	6.3	3.6	3.8	2.0	2.8	1.6
October*			7.1	3.6			2.7	2.0	1.6	1.1
Average 1949	3.5	2.5	6.8	3.3	4.1	2.0	2.2	2.0	1.3	0.5

Source: From Cantlon, 1953, *Ecological Monographs 23*, p. 249, table 3.

stations received no direct insolation. Table 11.2 summarizes the differences between south- and north-facing slopes and shade effects for selected months in 1949.

The largest slope difference is observed at 5 cm in the air in October, the least in the soil in July. The largest slope difference in the soil is observed in January. In general, the differences diminish with height above the ground surface.

At Armidale, New South Wales, Thompson (1973) observed slope maximum and minimum temperatures at 18 sites during October 1966–September 1967. The north-facing slopes, which receive greater solar radiation, were warmer on 26 days in December by an average of 0.8 °C and on 31 days in July by 1.4 °C. Minimum temperatures averaged 12.3 °C in December and -0.8 °C in July on a north-facing slope, compared with 12.6 °C and –2.7 °C, respectively, on a south-facing slope at the same elevation.

On Mt. Wilhelm, Papua New Guinea, Smith (1977) reported differences between east- and west-facing slopes in temperatures and the altitudinal range of plant species. The flora was richer on east-facing slopes above 4000 m altitude. Ground level

Table 11.3. Values of Calculated Solar Radiation (kW m^{-2}) for 30° Slopes of West and South Direction at 35 °N on Selected Dates

	June 23 30 °W	June 23 30 °S	Sept. 20 30 °W	Sept. 20 30 °S	Dec. 23 30 °W	Dec. 23 30 °S
Horizontal	41.5	41.5	31.3	31.3	275.6	275.6
30 °slope	646.2	585.4	501.0	618.2	278.4	533.1

Source: After Swift, 1976.

Table 11.4. Ratios of Daily Solar Radiation on the 15th of Each Month

Latitude	N Slope						S Slope					
	Apr.	May	June	July	Aug.	Sept.	Apr.	May	June	July	Aug.	Sept.
44°	0.80	0.89	0.91	0.91	0.85	0.66	1.12	1.07	1.02	1.05	1.12	1.24
54°	0.73	0.86	0.89	0.89	0.80	0.59	1.24	1.10	1.05	1.08	1.16	1.39
64°	0.67	0.84	0.86	0.85	0.74	0.52	1.31	1.13	1.06	1.10	1.19	1.44

Source: From Gol'tsberg, 1969, p. 23, table 6.

temperatures were measured with shielded maximum and minimum thermometers. Maximum temperatures were up to 5–8 °C higher on east-facing than west-facing sites at 3720 m in September–October 1975. The difference is due to the diurnal cloud regime with cloud-free mornings and cloudy afternoons. Barry (1978) found east/west differences in 1 cm soil temperatures, measured with thermistors, averaging 6 °C on opposite slopes at about 3600 m in the same area.

The variation of solar radiation receipts with aspect has been widely investigated. Buffo et al. (1972) document direct solar radiation on various slopes from 0 to 60 °N. Swift (1976) provides algorithms to calculate these variations and selected calculations. Table 11.3 illustrates values for west- and south-facing slopes of 30° at latitude 35 °N. Duguay (1993) provides a review of radiation modeling in mountains.

Gol'tsberg (1969) provides ratios of solar radiation on north- and south-facing slopes relative to the horizontal surface values for middle latitudes during April to September. A selection of his results for 20° slopes at latitudes 44°, 54°, and 64° are shown in Table 11.4. The ratios are always greater (less) than 1 for south- (north-) facing slopes. Monthly values depart from the midsummer averages by about 10 percent except in September, when they depart by ±15–20 percent.

Surface temperatures can be greatly in excess of screen air temperatures. For example, on July 22, 1962, at an alpine site at 3050 m on Niwot Ridge, Colorado, with a screen air temperature of 18 °C, Gates and Janke (1966) recorded surface vegetation temperatures of 27–37 °C and bare soil readings of 53–57 °C. Salisbury and Spomer (1964) also reported leaf temperatures in alpine plants 20 °C higher than air temperatures. At Pindaunde (3480 m) on Mt. Wilhelm, Papua New Guinea, one of the authors

Table 11.5. Differences in Mean Minimum Air Temperature (°C) in July with Topographic Setting in Mountain Systems of Russia

Location	Caucasus	Pamir, Alai	Tien Shan	Altai, -Sayan	Transbaikal
Valleys and basins	–2 to –2.5	–3 to –4	–3 to –4	–2.5 to –3.5	–3 to –3.5
Narrow valleys	–1.5 to –2	–2 to –3	–2.5 to –3	–2 to –3.5	–2 to –3.5
Valley bottoms (4 km wide) with cold air lakes	<–1	<–1. 5	<–1.5	<–1.5	—
Summits	1.5 to 2	2 to 3	2 to 3	1.5 to 2.5	2 to 3
Open slopes	1 to 1.5	1.5 to 2	1.5 to 2	1 to 1.5	1.5 to 2

Source: Gol'tsberg, 1969, p. 183, table 49.

(R. G. Barry) recorded a surface temperature of 60 °C with an infrared thermometer in September 1975 compared with an air temperature of 15 °C.

Gol'tsberg (1969) summarizes differences in mean minimum air temperature in July with topographic setting in the major mountain systems of Russia (Table 11.5). The values refer to relative altitudes of 200–400 m between concave and convex morphological settings. The maximum differences between the top and basin floors do not exceed 4–5 °C. The differences tend to be largest in the Pamir and Tien Shan, which are dry continental regions.

New procedures of terrain analysis have become available in recent years making use of high-resolution Digital Elevation Model (DEM) data. For example, MacMillan et al. (2003) used lidar data with 3 × 3 m horizontal resolution and a vertical accuracy of ±0.3 m RMS to map landform elements over an agricultural watershed of 6 × 8 km and a 12 × 14 km area of forested hilly terrain in British Columbia. They mapped a total of 26 million cells at 5 m resolution. While their aim was to obtain meaningful classifications of ecological and landform entities and to extract hydrological spatial entities, the work illustrates an approach that can be applied to identify and classify potential topoclimatic facets. Grosse et al. (2006) further illustrate this application in a study of permafrost karst terrain in eastern Siberia using Landsat-7 and DEM data at 30 m resolution. Thirteen terrain types were extracted and mapped using a maximum likelihood classification over a 2300 km² area. Earlier, MacMillan et al. (2000) reported a procedure for automatic classification of landform facets using DEMs and a fuzzy rule base.

Ten terrain derivatives were calculated from the DEM data. One group of derivatives included slope gradient, aspect, profile and plan curvature, and relative illumination. A second group was used to compute a compound topographic index that provided a relative wetness index. Relative relief was also defined, as well as the absolute difference in elevation and horizontal distance to the nearest divide and channel cells and pit and peak cells to which it is connected by a defined flow path. Fifteen landform elements were defined for two sites in Alberta, covering 68 and 38 ha.

Currently, landform analysis can be performed using DEM data and Geographic Information Systems (GIS) that are readily available. For example, Morgan and Lesh (2005) detail the application of ESRI ModelBuilder, which is based on Hammond's (1964) procedure that uses slope relief and profile type. See proceedings.esri.com/library/userconf/proc05/papers/pap2206.pdf.

Suggitt et al. (2011) examined the effects of differences in topography in the Peak District of northern England and Lake Vyrny, north Wales, in September 2007 and January 2008 and habitat differences at Skipwith Common, north Yorkshire, in September 2008 and January 2009. They found that habitat type (grassland, heathland, deciduous woodland – birch) is a major modifier of the temperature extremes experienced by organisms. They recorded differences among these habitats of more than 5 °C in monthly air temperature maxima and minima, and of 10 °C in thermal range, on a par with the level of warming expected for extreme future climate change scenarios. Ground-level minimum air temperature in heathland and grassland was 4–6 °C lower than in woodland in both winter (January) and late summer (September). The absolute maximum temperature recorded in woodland in September did not exceed 25 °C, whereas temperatures in excess of 35 °C were recorded in grassland. Comparable differences were found in relation to variation in local topography (slope and aspect). In the Peak District, maximum air temperature was estimated to be 7 °C higher on southerly than on northerly hillsides in September, and 5.7 °C higher on southerly than on northerly hillsides in January.

Wind can exert an important cooling effect on alpine plant temperature on sunny days. Gates and Janke (1966) report that fully sunlit needles on the lee side of an Engelman spruce averaged 29.7 °C, while exposed needles on the top averaged 22.5 °C with an air temperature of 20 °C. Fierce winds and freezing temperatures commonly give rise to krummholz growth forms in alpine timberline trees (Holtmeier, 2009; Körner, 2012). Winter desiccation is a major source of stress. The krummholz trees are stunted and deformed and may lack branches on the windward side (see Figure 7.4). Trees can survive in the lee of boulders and where snowdrifts in winter and spring shelter them from the low air temperatures and abrasion by blowing snow particles.

B. Sea and Lake Coastal Influences

Environments of the seacoast and shores of large lakes give rise to horizontal topoclimatic gradients of temperature, moisture, rainfall, snowfall, and wind velocity. These are basically a result of the differential heat capacity of land and water and the contrast in surface roughness. Water has a specific heat capacity of 4186 J $(kg\ K)^{-1}$ compared with only about 850 J$(kg\ K)^{-1}$ for sand and 1400 J$(kg\ K)^{-1}$ for clay. Consequently water heats up and cools down much more slowly than do land surfaces. Winds flowing parallel to shorelines are subject to differential friction over land and water. The roughness length for momentum (z_0) of the ocean is about

5×10^{-3} m compared with about 10^{-1} m over land. In the Northern Hemisphere, if the wind has land on its right (left) side the low-level air tends to converge (diverge) as a result of the frictional contrast, giving rise to ascent and cloud formation (subsidence and cloud dispersal), respectively. The latter situation is common over the northern coast of Venezuela, for example, when the region is affected by easterly trade winds.

Coastlines also generate diurnal air circulations of sea and land breezes in fine weather when the gradient winds are light (Simpson, 2007). By day, the heating of the land sets up low pressure and rising air so that a sea breeze develops and may penetrate up to 50 km or so from the coast as the day progresses. During the day the wind turns gradually clockwise (counterclockwise) in the Northern (southern) Hemisphere as a result of the Coriolis effect on the wind direction. By night the opposite pattern of temperature contrast between land and water gives rise to a weaker land breeze. Their frequency on the Sea of Japan coast can reach 10–15 days per month between April and November, according to Yoshino (1975, p. 159).

The depth of sea breezes is 1300–1400 m in equatorial latitudes and 500–900 m in middle and high latitudes (Yoshino, 1975). Land breezes are much shallower, typically about half the depth of sea breezes although some observations from the southern Baltic coast show depths of 600–800 m. The sea breeze generally starts about 10 a.m. and ends after sunset. It is strongest in midafternoon.

Steele et al. (2015) show that there are three types of sea breeze: (i) pure, where the wind is directly onshore; (ii) corkscrew, where there is an along-shore component to the gradient wind with the land surface to the left in the Northern Hemisphere; the resulting surface divergence created at the coast, due to friction, strengthens the sea breeze and the circulation forms an elongated helix shape; (iii) backdoor, where there is an along-shore component to the gradient wind with the land surface to the right, generating a region of surface convergence at the coast, which weakens the circulation. They show that for coasts in the southern North Sea there is a large variability in sea-breeze frequency between neighboring coastlines of up to a factor of 3. The backdoor type is least common in this area and the corkscrew type is associated with the strongest breezes.

The degree to which coasts affect local climate depends strongly on whether the prevailing winds are onshore or offshore, and the inland extension of coastal influences depends on the nature of the topography. Thus, the west coasts of North and South America are backed by high mountain ranges, whereas much of Western Europe is lowland, allowing the penetration of maritime influences far inland. Five of California's 16 climate zones are coastal with varying characteristics that are mainly a function of latitude and ocean temperatures (Pacific Energy Center, 2006).

The Köppen classification has four oceanic/coastal subtypes reflecting latitude (Peel et al., 2007): Cwb: subtropical oceanic climate; Cwc: oceanic subpolar climate; Cfb: temperate oceanic climate; and Cfc: cool oceanic climate. A small annual temperature range characterizes them. However, these are on a larger scale than local climates.

One important effect of coastal location is found in subtropical deserts of Southwest Africa and South America, in coastal California and Northwest Africa, and in Newfoundland. In these areas, cool offshore currents carry low stratus cloud and fog. Fog affects narrow coastal strips of the Namib and Atacama deserts, for example, transporting vital moisture for plant life and animals in the morning hours via a diurnal sea breeze. Newfoundland is especially affected by coastal fog in summer, when warm waters of the Gulf Stream meet the cold Labrador Current over the Grand Banks. The cool California Current affects the California coast in summer, when cool foggy air flows inland.

The downwind shores of large lakes in northern latitudes experience distinct local climates. In autumn and early winter when the surrounding land is usually snow covered, the lake surface remains open, creating a large temperature difference between the lake surface and cold air streams flowing over it. The instability so generated leads to rapid cloud formation and snowfall. In November and December 2014 recurrent lake effect storms dropped massive amounts of snow on Buffalo, New York, and adjacent areas. Total amounts by November 21 were a record 224 cm in the area compared with an annual average in Buffalo of 240 cm. The effect of the Great Lakes on air temperature is estimated to average about +2.5 °C in January and −1.7 °C in July compared with the surrounding land. Landsberg (1951) cited temperatures for the coldest month of −8.2 °C on the windward side and −5.0 °C on the lee side and for the warmest month of 20.3° and 21.1 °C, respectively.

The effects of the large lakes in East Africa on local climate are analyzed in a modeling study by Thiery et al. (2015). They find that the four major African Great Lakes (Lake Victoria, Lake Tanganyika, Lake Albert, and Lake Kivu) almost double the annual precipitation amounts (+87 percent, or 732 mm) over their surfaces, but hardly exert any influence on precipitation beyond their shores. Lake precipitation occurs mostly at night and in the morning hours, in contrast to the noon maximum over land. The largest lakes, except Lake Kivu, also cool the annual near-surface air by −0.6 to −0.9 °C on average, with pronounced downwind effects as a result of lake breezes that are, however, modified by topography and the local air circulation. The lake-induced cooling occurs during daytime, particularly at the end of the dry season from August to October, when the lakes absorb incoming solar radiation and inhibit upward turbulent heat transport. Evaporative cooling is the dominating factor. At night, when the sensible heat is released, the lakes warm the near-surface air.

C. Urban Local Climates

The broad characteristics of urban climate were discussed in Chapter 10 and urban microclimates in Chapter 8, Section D. Here we address the local climate of urban areas. The definition of an urban versus a rural environment is rarely detailed. Morgan et al. (1977) identified 13 land use categories in Los Angeles and Sacramento,

California, which are here grouped into six categories. The percentage coverage of each is listed.

	Sacramento	Los Angeles
Residential	35.5	38.5
Commercial	7.2	4.1
Industrial	13.5	12.0
Public right-of-way	17.0	21.8
Institutional	3.2	14.5
Open space – recreational	23.6	9.1

Subsequent definitions of such categories were created by Auer (1978) for the city of St. Louis, Missouri, where he recognized 12 "meteorologically significant" land use categories based on the city's vegetation and building characteristics. Ellefsen (1991) derived a system of 17 neighborhood-scale "Urban Terrain Zones" (UTZs) from the geometry, street configuration, and construction materials of 10 U.S. cities. His was the first system to represent city structure and materials. A key feature of Ellefsen's system is the division of building types into "attached" and "detached" forms.

In an attempt to formalize this approach, Stewart and Oke (2012) distinguish 17 "Local Climate Zones" (LCZs) at the local scale (10^2 to 10^4 m) in cities. Figure 11.1 illustrates the classification.

Each LCZ is categorized in terms of sky view factor, aspect ratio (the mean height-to-width ratio of street canyons, building spacing, and tree spacing), building surface fraction, pervious and impervious surface fractions, and roughness element height.

Figure 11.2 shows an example of temperature differences from the mean in eight LCZs in Vancouver, British Columbia, in March 2010. Departures ranged from + 2.5 to – 2.5 °C.

The vertical extent of the temperature effect of cities has been shown to reach 3–5 times the average building height according to Yoshino (1975, p. 86). Bornstein (1968) showed that the vertical extent of the heat island over New York over 42 observations made during 1964–6 was 300 m.

An urban energy budget model was first developed by Myrup (1969). A scheme for use in regional-scale climate models has been developed by Masson (2000). The approach uses a generalization of local canyon geometry. Three surface energy budgets are considered – roofs, roads, and walls. Orientation effects are averaged for roads and walls. Up to two energy budgets are added for snow when this is present on roofs or roads. The turbulent fluxes are computed for each land surface type by the appropriate scheme and then averaged in the atmospheric model grid mesh, with respect to the proportion occupied by each type. For example, partitions can be (1) sea, (2) inland water, (3) natural and cultivated terrestrial surface, (4) towns. The fluxes calculated are latent and sensible heat, upward radiative fluxes, and component momentum fluxes. It is assumed that there are two turbulent heat sources – one from

Built types	Definition	Land cover types	Definition
1. Compact high-rise	Dense mix of tall buildings to tens of stories. Few or no trees. Land cover mostly paved. Concrete, steel, stone, and glass construction materials.	A. Dense trees	Heavily wooded landscape of deciduous and/or evergreen trees. Land cover mostly pervious (low plants). Zone function is natural forest, tree cultivation, or urban park.
2. Compact midrise	Dense mix of midrise buildings (3–9 stories). Few or no trees. Land cover mostly paved. Stone, brick, tile, and concrete construction materials.	B. Scattered trees	Lightly wooded landscape of deciduous and/or evergreen trees. Land cover mostly pervious (low plants). Zone function is natural forest, tree cultivation, or urban park.
3. Compact low-rise	Dense mix of low-rise buildings (1–3 stories). Few or no trees. Land cover mostly paved. Stone, brick, tile, and concrete construction materials.	C. Bush, scrub	Open arrangement of bushes, shrubs, and short, woody trees. Land cover mostly pervious (bare soil or sand). Zone function is natural scrubland or agriculture.
4. Open high-rise	Open arrangement of tall buildings to tens of stories. Abundance of pervious land cover (low plants, scattered trees). Concrete, steel, stone, and glass construction materials.	D. Low plants	Featureless landscape of grass or herbaceous plants/crops. Few or no trees. Zone function is natural grassland, agriculture, or urban park.
5. Open midrise	Open arrangement of midrise buildings (3–9 stories). Abundance of pervious land cover (low plants, scattered trees). Concrete, steel, stone, and glass construction materials.	E. Bare rock or paved	Featureless landscape of rock or paved cover. Few or no trees or plants. Zone function is natural desert (rock) or urban transportation.
6. Open low-rise	Open arrangement of low-rise buildings (1–3 stories). Abundance of pervious land cover (low plants, scattered trees). Wood, brick, stone, tile, and concrete construction materials.	F. Bare soil or sand	Featureless landscape of soil or sand cover. Few or no trees or plants. Zone function is natural desert or agriculture.
7. Lightweight low-rise	Dense mix of single-story buildings. Few or no trees. Land cover mostly hard-packed. Lightweight construction materials (e.g., wood, thatch, corrugated metal).	G. Water	Large, open water bodies such as seas and lakes, or small bodies such as rivers, reservoirs, and lagoons.

VARIABLE LAND COVER PROPERTIES

Variable or ephemeral land cover properties that change significantly with synoptic weather patterns, agricultural practices, and/or seasonal cycles.

Built types	Definition	Land cover types	Definition
8. Large low-rise	Open arrangement of large low-rise buildings (1–3 stories). Few or no trees. Land cover mostly paved. Steel, concrete, metal, and stone construction materials.		
9. Sparsely built	Sparse arrangement of small or medium-sized buildings in a natural setting. Abundance of pervious land cover (low plants, scattered trees).	b. bare trees	Leafless deciduous trees (e.g., winter). Increased sky view factor. Reduced albedo.
		s. snow cover	Snow cover >10 cm in depth. Low admittance. High albedo.
10. Heavy industry	Low-rise and midrise industrial structures (towers, tanks, stacks). Few or no trees. Land cover mostly paved or hard-packed. Metal, steel, and concrete construction materials.	d. dry ground	Parched soil. Low admittance. Large Bowen ratio. Increased albedo.
		w. wet ground	Waterlogged soil. High admittance. Small Bowen ratio. Reduced albedo.

Figure 11.1. Abridged definitions for local climate zones (LCZs) (Stewart and Oke, 2012).
Source: *Bulletin of the American Meteorological Society* 93, table 2, p. 1881. Courtesy of the American Meteorological Society.

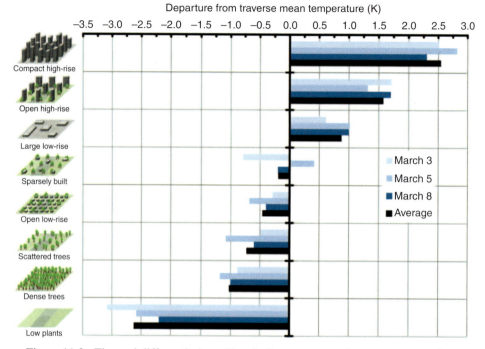

Figure 11.2. Thermal differentiation of local climate zones using temperatures from automobile traverses in Vancouver during calm, clear evenings, March 2010. All measurements were made with an aspirated and insulated copper-constantan thermocouple at 1.5 m above ground. Temperatures are adjusted to a standard base time of 2 h after sunset. Raw temperature data provided by Andreas Christen (University of British Columbia) (Stewart and Oke, 2012).
Source: *Bulletin of the American Meteorological Society* 93, fig. 3, p. 1888. Courtesy of the American Meteorological Society.

the roofs and one from the canyon systems. These are averaged in proportion to their horizontal areas. Momentum flux is computed for the entire town using a roughness length of $h/10$, where h is the average building height. The interception of precipitation by roofs and roads is treated, as well as the runoff. Domestic heating of the atmosphere is based on a fixed internal temperature, while combustion is separated into traffic and industry. Industry effects enter the atmosphere directly, while traffic modifies the canyon air budget. Masson performs sensitivity experiments for annual and diurnal cycles with satisfactory results. Grimmond et al. (2010, 2011) provide a comparative assessment of urban energy balance models.

D. Mosaic Landscapes

In many parts of the world the land surface is not flat or spatially homogeneous, but presents a mosaic of surface slopes and types of land cover (Forman, 1995). There is often a patchwork of fields, some with crops, some fallow, hedgerows, woodland, and settlements, and these are also interrupted by corridors that may be natural (riverine

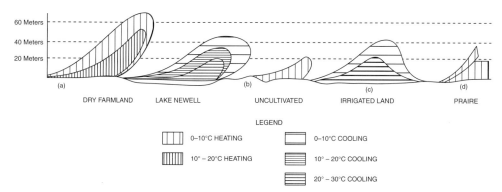

60 Meters
40 Meters
20 Meters

(a) DRY FARMLAND   LAKE NEWELL   (b) UNCULTIVATED   IRRIGATED LAND (c)   (d) PRAIRE

LEGEND

0–10°C HEATING   0–10°C COOLING

10°–20°C HEATING   10°–20°C COOLING

20°–30°C COOLING

Figure 11.3. Profile of boundary layer air temperature over a lake and farmland near Brooks, Alberta, August 6, 1968 (Holmes and Wright, 1978).
Source: *J. Appl. Met.*, 17, p. 1164, fig. 2. Courtesy of the American Meteorological Society.

systems) or artificial (roads, railroads, canals, and power line cuts). Each landscape element has its own characteristic radiative regime, energy balance, and water balance as a result of differences in albedo, emissivity, thermal and aerodynamic properties, soil conductivity, and soil moisture. Especially close to the surface these contrasts give rise to large spatial gradients of temperature, moisture, energy transfer, and wind speed. It is a major problem in climatology to scale from point measurements to local and regional conditions, and vice versa, to downscale from GCMs and regional climate models (RCMs) to localities (see Chapter 6, Section E). Information on interactions of climate between landscape elements is a vital contribution to these scale problems.

The use of aircraft measurements to study the effects of heterogeneous terrain on surface temperatures was pioneered by Lenschow and Dutteon (1964) near Madison, Wisconsin. Holmes and Wright (1978) performed similar work on boundary-layer structure in the late 1960s in the Cypress Hills, Alberta. A light aircraft was flown at heights of 15 and 45 m. Figure 11.3 illustrates plumes of warm and cold air near Brooks, Alberta, on August 6, 1968. The plumes are displaced to the lee shore of the lake and extend upward for about 50 m and occasionally higher.

The related issue of "blending height," where the deviations of concentrations of heat, moisture, and so on, and their fluxes become negligible, was analyzed by Philip (1996a). He showed that flux deviations are much more robust with height than those of concentrations. For a one-dimensional checkerboard pattern, representative heights are ~ 5, 30, and 80 m, respectively, for surface pattern wavelengths of 100, 1000, and 3000 m. These values are independent of wind speed but are somewhat increased by higher values of surface roughness length. Philip (1996b) also showed that when winds are normal or parallel to the boundaries, for square checkerboards, it is only when the pattern wavelength is less than about 100 m that there is a significant reduction from the one-dimensional case. Blending heights for winds parallel to checkerboard boundaries far exceed those for winds normal to them. For oblique wind flow with respect to square checkerboards, Philip (1997) shows that there is a peak in the blending height at a 45° angle (i.e., wind direction parallel to the checkerboard diagonal). For patterns of 100 m (1000 m) the maximum /minimum blending height is 3–5 m (8–14 m). Details for other angles and pattern dimensions are discussed in the paper.

References

Auer, A. H., 1978. Correlation of land use and cover with meteorological anomalies. *J. Appl. Met.* 17, 636–43.

Barry, R. G. 1978. Diurnal effects on topoclimate on an equatorial mountain. *Arbeiten aus der Zentralanstalt für Meteorologie und Geodynamik (Vienna)* Publ. 32(72), 1–8.

Bornstein, R. D. 1968. Observations of the urban heat island effect in New York city. *J. Appl. Met.* 7, 575–82.

Buffo, J., Fritschen, L. J., and Murphy, J. L. 1972. Direct solar radiation on various slopes from 0 to 60 degrees north latitude. Res. Paper PNW-142. Portland, OR: USDA Forest Service, Pacific Northwest Forest and Range Experiment Station.

Cantlon, J.E 1953. Vegetation and microclimates on north and south slopes of Cushetunk Mountain, New Jersey. *Ecol. Monogr.*, 23, 241–70.

Duguay, C. R. 1993. Radiation modeling in mountainous terrain: Review and status. *Mountain Res. Devel.* 13, 339–57.

Ellefsen, R. 1990/91. Mapping and measuring buildings in the urban canopy boundary layer in ten US cities. *Energ. Buildings* 15–16, 1025–49.

Forman, R. T. T. 1995. *Land mosaics: The ecology of landscapes and regions.* Cambridge: Cambridge University Press.

Gates, D. M., and Janke, R. 1966. The energy environment of the alpine tundra. *Oecol. Plantarum* 1, 39–62.

Gol'tsberg, I, A. (ed.) 1969. *Microclimate of the USSR.* Jerusalem: Israel Program for Scientific Translations. (Gidromet Izdatel, Leningrad 1967).

Grimmond, C. S. B. et al. 2010 The International Urban Energy Balance Models Comparison Project: First results from phase 1. *J. Appl. Met. Climatol.* 49, 1268–92.

Grimmond, C. S. B. et al. 2011 Initial results from phase 2 of the International Urban Energy Balance Model comparison. *Int. J. Climatol.* 31, 244–72.

Grosse, G., Schirrmeister, L., and Malthus, T. J. 2006. Application of Landsat-7 satellite data and a DEM for the quantification of thermokarst-affected terrain types in the periglacial Lena–Anabar coastal lowland. *Polar Res.* 25, 51–67.

Hammond, E. H. 1964. Analysis of properties in landform geography: An application to broadscale landform mapping. *Annals Assoc. Amer. Geogr.* 54, 11–19

Holmes, R. M., and Wright, J. L. 1978. Measuring the effects of surface features on the atmospheric boundary layer with instrumented aircraft. *J. Appl. Met.* 17(8), 1163–70.

Holtmeier, F-K. 2009. *Mountain timberlines: Ecology, patchiness, and dynamics,* 2nd ed. New York: Springer

Kerner, A. 1891. Die Änderumg der Bodentemperatur mit der Exposition. *Sitzungsbericht d. Akademie d. Wissensschaft, Wien* 100, 704–29.

Körner, C. 2012. *Alpine treelines: Functional ecology of the global high elevation tree limits.* Basel: Springer.

Landsberg, H. 1951. *Physical climatology.* University Park: Pennsylvania State University.

Lenschow, D. H., and Dutton, J. A. 1964. Surface temperature variations measured from an airplane over several surface types. *J. Appl. Met.* 3, 65–9.

MacMillan, R. A. et al. 2000. A generic procedure for automatically segmenting landforms into landform elements using DEMs, heuristic rules and fuzzy logic. *Fuzzy Sets Systems* 113, 81–109.

MacMillan, R. A. et al. 2003. Automated analysis and classification of landforms using high-resolution digital elevation data: applications and issues. *Can. J. Rem. Sensing* 29, 592–606.

Masson, V. 2000. A physically-based scheme for the urban energy budget in atmospheric models. *Bound.-Layer Met.* 94, 357–97.

Morgan, D. et al. 1977. Microclimates within an urban area. *Annals, Assoc. Amer. Geogr.* 67, 55–65.

Morgan, J. N., and Lesh, A. M. 2005. Developing landform maps using ESRI'S ModelBuilder. http://proceedings.esri.com/library/userconf/proc05/papers/pap2206.pdf.

Myrup, L. O. 1969. A numerical model of the urban heat island. *J. Appl. Met.* 8, 908–18.

Pacific Energy Center. 2006. *Guide to California's climate zones and bioclimatic design.* San Francisco, CA.

Peel, M. C., Finlayson, B. L., and McMahon, T. A. 2007. Updated world map of the Köppen-Geiger climate classification. *Hydrol. Earth System Sci.* 11, 1638–43.

Philip, J. R. 1996a. One-dimensional checkerboards and blending heights. *Bound.-layer Met.* 77, 135–51.

Philip, J. R. 1996b. Two-dimensional checkerboards and blending heights. *Bound.-layer Met.* 80, 1–18.

Philip, J. R. 1997. Blending heights for winds oblique to checkerboards. *Bound.-Layer Met.* 82, 263–81.

Salisbury, F. B., and Spomer, G. G. 1964. Leaf temperatures of alpine plants in the field. *Planta* 60, 497–505.

Simpson, J. E. 2007. *Sea breeze and local winds.* Cambridge: Cambridge University Press. 252 pp.

Smith, J. M. B. 1977. Vegetation and microclimate of east- and west-facing slopes in the grasslands of Mt Wilhelm, Papua New Guinea. *J. Ecol.* 65, 39–53.

Steele, C. J. et al. 2015. Modelling sea-breeze climatologies and interactions on coasts in the southern North Sea: implications for offshore wind energy. *Quart. J. Roy. Met. Soc.* 141, 1821–35.

Stewart, I. D., and Oke, T. R. 2012. Local climate zones for urban temperature studies. *Bull. Amer. Met. Soc.* 93, 1879–1900.

Swift, L. W. 1976. Algorithm for solar radiation on mountain slopes. *Water Resour, Res.,* 12, 108–12.

Suggitt, A. J. et al. 2011. Habitat microclimates drive fine-scale variation in extreme temperatures. *Oikos* 120, 1–8.

Thiery, W. et al. 2015. The impact of the African Great Lakes on the regional climate. *J. Climate* 38, 4061–85.

Thompson, R. D. 1973. The influence of relief on local temperatures: Data from New South Wales, Australia. *Weather* 28, 377–82.

Wollny, E. 1878. Untersuchungen über den Einfluss der Exposition auf die Erwarmung des Bodens. *Forsch. Agr. Phys.* 1, 43–69.

Yoshino, M. M. 1975. *Climate in a small area. An introduction to local meteorology.* Tokyo: University of Tokyo Press.

Part III

Environmental Change

Chapter 12 deals with the effects of climate change on microclimate.

12

Climate Change and Microclimate

Global climate has undergone major changes over the last century. Mountain glaciers have retreated worldwide, exposing new land areas, and biomes have shifted their boundaries poleward and upward. Global temperatures have risen by almost 1 °C since the 1880s and in high northern latitudes the warming has been amplified several-fold. The atmospheric concentration of carbon dioxide increased by almost 40 percent from about 290 ppm in 1900 to 400 ppm in 2015. Atmospheric methane increased since 1750 by 250 percent to 1800 ppb. How these changes have affected microclimates is the topic of this chapter. We begin with an overview of the major changes that have been observed in the climate system.

A. Overview of Global Trends

Global climate is determined by a combination of natural processes and human impacts on the climate system. Natural variability on decadal to century timescales is a result of variations in solar output, volcanic eruptions, air–sea interactions, and atmospheric variability. Solar variability occurs during the solar cycle, which lasts about 11 years on average. Between solar maximum and minimum, total solar output varies by 1.3 W m^{-2}, or 0.1 percent. However, there are larger variations in the UV spectrum. Lean et al. (1997) calculate that the variability is 1.1 percent in the 0.2–0.3 μm range and 0.25 percent in the 0.3–0.4 μm range. Counterintuitively, sunspots (cooler, darker areas on the Sun's photosphere) increase in number toward a solar maximum because the bright faculae that surround the spots are more numerous, increasing the solar output. Explosive volcanic eruptions, especially those occurring in the tropics, force fine ash and sulfur dioxide (SO_2) particles into the stratosphere, where they circle the globe. These exert a cooling effect in the subsequent two to three years, lowering average temperatures significantly, particularly in the boreal summers. Sigi et al. (2015) analyze ice cores from Greenland and Antarctica for ash and SO_2 signatures over the last 2500 years and correlate these with tree ring data. They find that the 5 year average cooling in three regions of the Northern Hemisphere was –0.6 °C for the 19 largest tropical eruptions in the Common Era (CE). Two very large events in the decade 530s

CE led to departures in European temperatures of -1.6 to -2.5 °C and -1.4 to -2.7 °C. Cold conditions from the second event persisted for almost a decade. Air-sea interactions can give rise to climate anomalies on a near-global scale. The best known is the El Niño–Southern Oscillation, or ENSO, which affects ocean temperatures across the equatorial Pacific. In the warm El Niño (cold La Niña) phase, sea surface temperatures are 2–3.5 °C above (1–3 °C below) average from the Dateline to the West Coast of South America. Each phase may last 1–3 years, and they recur about every 5–7 years. Teleconnections to midlatitudes give rise to large-scale anomalies of temperature and precipitation. Natural variability in atmospheric circulation arises principally on weekly to monthly timescales and can be characterized as variations in circulation strength and patterns (blocking, zonal flow, etc.).

Concentrations of greenhouse gases and pollutants have undergone large increases since the Industrial Revolution as a result of fossil fuel burning and deforestation. This time interval has been named the Anthropocene although its time boundary is still being debated. Increases in carbon dioxide concentrations have direct effects on plant growth in addition to the indirect effects on air temperatures as a result of changes in the terrestrial radiative forcing. Carbon dioxide, methane (CH_4), and nitrous oxide (N_2O) are greenhouse gases that trap infrared radiation emitted by the surface (see Chapter 4, Section B). The gain of carbon by natural terrestrial ecosystems is considered to take place mainly through the uptake of CO_2 by enhanced photosynthesis at higher CO_2 levels, with nitrogen deposition and longer growing seasons (Stocker et al., 2013).

In the year 2011 the radiative forcing of the global climate system relative to 1750 due to CO_2 was 1.82 W m^{-2}, to CH_4 it was 0.48 W m^{-2}, and to N_2O it was 0.17 W m^{-2}. Aerosols have radiative interactions that lead to a negative forcing of -0.35 W m^{-2} and cloud interactions that give rise to a forcing of -0.9 W m^{-2}. During 2000–10 the 22 ppm increase in CO_2 concentration was determined to have increased the clear-sky radiative forcing by 0.2 W m^{-2} in the southern Great Plains and the north slope of Alaska (Feldman et al., 2015).

Box 12.1. The Paris Agreement

In December 2015, the Conference of Parties, or COP, met in Paris to discuss an agreement to limit global warming due to anthropogenic CO_2 emissions from the burning of fossil fuels. Following two weeks of meetings, discussions, and negotiations, an historic agreement was reached, referred to as the Paris Agreement. This historic agreement of 195 countries recognizes that a target of keeping the increase in the global mean air temperature to below 2 °C, but preferably 1.5 °C, by the year 2100. All countries were asked to review their greenhouse gas emissions targets every five years starting in 2020, with the goal of achieving carbon neutrality by the year 2050. This legally binding historical agreement essentially aims for the world to abandon the use of fossil fuels by the end of this century replacing them with renewable sources.

Global land-ocean temperatures relative to 1961–90 increased from about –0.3 °C in 1880–1910 to 0.55 °C in the 2000s (representing an increase of 0.85 °C), while land station air temperatures relative to 1961–90 increased from about –0.4 °C in 1890 to 0.7 °C in the 2000s (Hansen et al., 2010; Hartmann et al., 2013). The effect of urban heat islands on global temperature trends is found to be negligible (~ 0.01 °C) as the fraction of land covered by urban areas is small – only about 1 percent.

In association with the temperature increase, the frequency of cold nights has decreased since the 1970s and that of warm days has increased (Alexander et al., 2006; Fischer and Knutti, 2014). For 1951–2003, Alexander et al. showed that 74 per-cent (73 percent) of the near-global land area sampled had a significant decrease (increase) in the annual occurrence of cold nights (warm nights). Globally the annual number of warm nights (cold nights) increased (decreased) by about 25 (20) days since 1951. This increase in the nocturnal air temperatures indicates warming in the longwave portion of the radiation balance, due to the increase in greenhouse gas concentrations.

Urban areas have been especially affected. Mishra et al. (2015) analyzed tempera-ture and wind data for 217 cities worldwide for 1973–2012. They found that almost half of the urban areas experienced significant increases in the number of extreme hot days (>99th percentile), while almost two-thirds showed significant increases in the frequency of extreme hot nights. Extreme windy days (daily mean wind speed that exceeded the 99th percentile threshold of daily mean wind speed) declined substan-tially during the last four decades for three-quarters of the urban areas.

The length of the frost-free season in the contiguous United States has been deter-mined for 1920–2012 by McCabe et al. (2015) on the basis of daily minimum tem-peratures using a 0 °C threshold at 654 sites. The mean date of last spring frost ranges from Day of Year (DOY) 99 for the 20th percentile to DOY 132 for the 80th percen-tile; 88 percent of sites have dates within the months of April–May. The correspond-ing last autumn frost dates are between DOY 272 and 299 with the dates at 85 percent of sites falling in September–October. The trends are positive at 77 percent of sites for frost-free season and at 62 percent of sites for first autumn freeze. They are negative at 81 percent of sites for last spring frost. The overall trends in the twentieth century show earlier dates of last spring frost, beginning about 1983, and later dates of first autumn frost since about 1993.

The longest record of spring Snow Cover Extent (SCE) in the Northern Hemisphere is provided by Brown and Robinson (2011); it spans 1922 to 2010. The data for 1922–91/2 are based on reconstructed daily snow depths at surface climate stations. They find significant reductions with the rate of decrease accelerating over the past 40 years. The rate of decrease in March and April SCE over the 1970–2010 period was 0.8 million km^2 per decade corresponding to a 7 percent and 11 percent decrease in March and April, respectively, from pre-1970 values. These decreases must also have led to changes in spring soil temperatures (cf. Chapter 12, Section B).

Comprehensive and continuous sea ice records for both polar oceans begin in 1978 when multichannel passive microwave were obtained (Barry and Gan, 2011). Data for daily and monthly sea ice concentration from 1978 to present are shown in the Sea Ice Index from the National Snow and Ice Data Center (NSIDC) (nsidc.org/data/seaice_index/). In the Arctic the maximum concentration occurs annually in March and the minimum in September. In the 1980s these concentrations averaged 16 million and 7.5 million km^2, respectively, but they declined, at rates of 2 percent and greater than 10 percent per decade· subsequently. A record minimum occurred in September 2012 when the ice shrank to 3.6 million km^2, less than half of the 1980s average. Ice thickness declined by 1.75 m from 1980, to a mean of only 1.89 m, during winter 2008 of the ICESat record in the data release areas that cover 38 percent of the Arctic Ocean (Kwok and Rothrock, 2009). Over the last three decades there has been a slight increase of approximately 3 percent in Antarctic sea ice area overall, with a maximum in March (the annual minimum), and with large regional variations.

From 1950 to the 1980s, solar radiation receipts globally decreased by between 3 and 9 W m^{-2} as a result of increased sulfate loading in the atmosphere (Wild et al., 2009; Wild 2012). Sulfate emissions in the Northern Hemisphere rose from below 40 Tg in 1950 to 60 Tg in the 1980s. The reduction in solar radiation became known as "global dimming." In many parts of the world, the trend reversed by 1–4 W m^{-2} in the 1990s and 2000s (global brightening), except in China and India, where declines in solar radiation continued as a result of high pollution concentrations.

Black carbon in the surface air at Kevo, Finland, declined by an average of 1.8 percent $year^{-1}$ from 1970 to 2010 (Dutkiewicz et al., 2014). Black carbon decreased from ~300 ng m^{-3} in ~1970 to 82 ng m^{-3} in 2010, similar to trends at other Arctic sites. Mean winter, spring, summer, and autumn black carbon values based on the entire data set were 339, 199, 127, and 213 ng m^{-3}, respectively.

Trends of NOx, VOCs, O_3, and PAN for 1960 to 2010 in Los Angeles demonstrate the reduction in urban pollution in many western cities over the last half-century (Figure 12.1) (see also Pollack et al., 2013).

Growing degree-days for corn (with a base of 10 °C) in the United States have generally shown an increase over the last 100 years. Badh (2011) finds that 154 of 245 stations (63 percent) showed a significant trend and that 81 percent of these trends were positive. The greatest increases were found in North Dakota.

Parmesan and Yohe (2003) find significant range shifts in vegetation species averaging 6.1 km per decade toward the poles, and significant mean advance of spring events by 2.3 days per decade. In central Spain, air temperatures rose by 1.3 °C between 1970 and 2000, equivalent to a 225 m upward shift of the isotherms (Illan et al., 2010). However, microclimatic conditions may moderate plant response to general warming trends. De Frenne et al. (2013) report that in 1400 temperate forest plots in Europe and North America resurveyed over 12–67 years, the increase in species adapted to warmer conditions has lagged in forests where the canopy has become

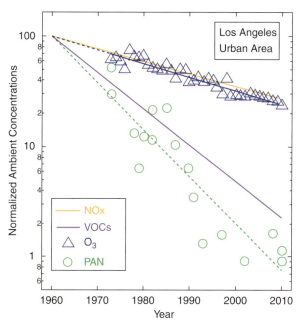

Figure 12.1. Trends in NOx, VOCs, O_3, and PAN for 1960 to 2010 in Los Angeles, normalized concentrations (from Parrish and Stockwell, 2015).
Source: Courtesy of the American Geophysical Union: Eos 96(1), p. 13, fig. 2b.

more dense. This probably reflects cooler growing-season conditions at the ground as a result of increased shading.

In the Arctic tundra, greenness, a measure of vegetation productivity, continues to increase. The trend in peak greenness indicates an average tundra biomass increase of approximately 20 percent during the period (1982–2013) of AVHRR satellite observation. Most of this increase in greenness in arctic, high-latitude, and alpine tundra ecosystems is due to the increase in shrubs, short multistemmed woody plants (Anthelme et al., 2007; Tape et al., 2006). Myers-Smith et al. (2011) state that the key environmental drivers behind this increase in shrub growth and expanded coverage are temperature, soil disturbances, and herbivory. Low temperatures limit the growing season length and reproduction in many shrub species, so any increase in temperature and growing season length would increase shrub growth and extent. Soil disturbances such as those created by fire or permafrost degradation can create microhabitats conducive for the establishment of shrubs. Herbivory could decrease shrub extent or increase shrub extent through increased seed dispersion or through decreased grazing in regions such as Scandinavia where sheep and reindeer use tundra for pasture. Overall, a shift from low-stature perennial grasses to taller, woody shrubs has a pronounced effect on the surface radiation, energy, carbon, and nutrient balance.

In contrast, however, there has also been a recent "browning" of the tundra, suggesting a longer-term decline in growing season length (Epstein et al., 2014). This could be due to hydrological changes associated with permafrost degradation such as a decrease in the near-surface soil moisture (i.e., a deeper active layer) (Jorgenson et al., 2001). Overall, however, many regions are expected to experience an increase in the length of the growing season due to increased temperature. For this century, projections of the annual growing season length in 14 basins of 11 U.S. states show increases by an average of 27–47 days over 2001–99 for three emission scenarios (Chriatianssen et al., 2011).

B. Soil Temperatures

In principle we can expect that the largest temperature changes will be at the ground surface. However, there are few observational data to examine this. A modeling study of soil temperature trends in the coterminous United States has recently been performed by Hao et al. (2014). They use the Soil Thermal Model (STM) – a sub-model of the Terrestrial Ecosystem Model (TEM), which is a process-based bio-geochemistry model that describes carbon, nitrogen, and water dynamics of plants and soils for terrestrial ecosystems. The STM is divided into six layers, including snow cover, moss, upper organic soil, lower organic soil, upper mineral soil, and lower mineral soil layers. To run the STM, simulation depth steps, layer specific information of layer thickness, and thermal properties are prescribed. The model can be operated with either 0.5 h or 0.5 day interval time steps, and driven by either daily or monthly air temperature. They examined spatiotemporal trends in simulated soil temperature at depths of 10, 20, and 50 cm during 1948–2008. They found a warming trend of between 0.2 and 0.4 °C at all depths over the 61 years. Areas that experienced weak cooling in summer soil temperature include Texas, Oklahoma, and Arkansas. Warming was recorded in Arizona, Nevada, and Oregon. In winter, Mississippi, Alabama, and Georgia showed a cooling trend, while Montana, North Dakota, and South Dakota have warmed.

In high northern latitudes, Smith et al. (2004) report soil freeze-thaw cycles using passive microwave data. In Eurasia, the date of thaw occurred first in evergreen coni-fer forests (Day Of Year, DOY 120 ± 25), followed by larch forests (131 ± 15) and tundra (151 ± 21). In North America, the area west of the Hudson Bay in Canada had the latest mean thaw. North American evergreen conifers thawed first (132 ± 20), followed by tundra (157 ± 21). Changes in soil freeze–thaw were investigated for 1988–2002. In Eurasia there was a trend toward earlier thaw dates in tundra (-3.3 ± 1.8 days per decade) and larch biomes (-4.5 ± 1.8 days per decade) over the period. In North America there was a trend toward later freeze dates in evergreen conifer forests by 3.1 ± 1.2 days per decade that led, in part, to a lengthening of the growing season by 5.1 ± 2.9 days per decade. The growing season length in North American tundra increased by 5.4 ± 3.1 days per decade. Despite the trend toward earlier thaw dates in Eurasian larch forests, the growing season length did not increase because of parallel changes in timing of the fall freeze (-5.4 ± 2.1 days per decade), which led to a for-ward shift of the growing season. Thaw timing was negatively correlated with surface air temperatures in the spring, whereas freeze timing was positively correlated with surface air temperatures in the autumn.

C. Permafrost

Data on ground temperatures span only a few decades but there are now networks of monitoring stations, both boreholes and shallow active layer measurements. Permafrost temperatures in Alaska increased dramatically in the last quarter of the twentieth century (Osterkamp, 2008). The permafrost surface warmed with a temperature increase of 3–4 °C on the Arctic Coastal Plain, 1–2 °C in the Brooks Range, and 0.3–1 °C south of the Yukon River. In interior Alaska the increase in the late 1980s–1990s was due to greater snow cover. The increase in near-surface permafrost temperature is most pronounced in winter. At Barrow, Alaska, about half of the recent surface warming was due to an increase in air temperature and half due to snow cover effects. Shiklomanov et al. (2008) report that a borehole at the 14 m depth at Barrow, Alaska, indicates warming of about 1.2 °C between 1950s and contemporary permafrost temperatures.

Smith et al. (2010) show that permafrost has generally been warming in the western Arctic since the 1970s and in parts of eastern Canada since the early 1990s. The increases are generally greater north of the tree line and the magnitude of the change was less in "warm" permafrost (> -2°C) than in colder permafrost.

Despite the warming, there is no apparent long-term trend in active-layer thickness at the Cold Regions Research and Engineering Laboratory plots near Barrow from the 1960s to 2000s, according to Shiklomanov et al. (2010).

In the subarctic, east of Hudson Bay, a peat plateau progressively fragmented during the late nineteenth and twentieth centuries (Payette et al., 2004). About 18 percent of the original surface occupied by permafrost was thawed in 1957, whereas only 13 percent was still surviving in 2003. Rapid permafrost melting over the last 50 years caused the concurrent formation of thermokarst ponds and fen-bog vegetation. The driver for accelerated permafrost thawing was snowfall, which increased from 1957 to the 2000s, while annual and seasonal temperatures remained relatively stable until about the mid-1990s, when annual temperature rose well above the long-term mean.

D. Experimental Studies

Since the 1990s there have been numerous studies of plant responses to elevated temperatures or levels of CO_2 concentration. Polythene tents and infrared heaters have been used to raise canopy temperature, for example (Robinson et al., 1998; Nijs et al. 1996). An alternative approach used in the Swiss Jura was the moving of turf and soil cores from high elevation (1900 m) sites to warmer, drier low elevation (400 m) sites (Gavazov et al., 2014). Many of these studies have been short term. A different approach is adopted by Jia et al. (2009), who monitor tundra greening using AVHRR data collected during 1982 to 2006. Peak annual greenness increased 0.49–0.79 percent yr^{-1} over the High Arctic, where prostrate dwarf shrubs, forbs, mosses, and lichens dominate, and 0.46–0.67 percent yr^{-1} over the Low Arctic, where erect dwarf shrubs and graminoids dominate. Spring temperatures during this period increased by up to 0.5 °C decade^{-1} and about half as much in summer (Serreze et al., 2000).

One of the simplest temperature manipulation experiments in tundra ecosystems was the International Tundra Experiment (ITEX). Initiated in the late 1990s, ITEX had the primary goal of determining the effects of global climate change on major circumpolar plant species in the Northern Hemisphere. Small (50 cm tall, 1.5 m diameter) open-top hexagon fiberglass-sided chambers were placed over vegetation, resulting in an average surface temperature increase of 1–3 °C during the growing season simply due to this decrease in surface wind speed (Henry and Molau, 1997). The increased surface temperatures also changed the growing season length by altering the snowmelt timing. Long-term, continuous monitoring of temperature (air and soil), as well as other standard meteorological variables, together with vegetation phenology measurements (species composition, height, leaf area, etc.) allowed for the simulated effects of temperature increase relative to adjacent control sites to be examined. Presently, roughly 50 ITEX sites exist in 11 countries.

In the roughly 20 years since ITEX began, research synthesizing the many individual warming plots has shown mixed results. Strong regional variations in the response to warming on tundra vegetation have been reported. For example, Elmendorf et al. (2012) found that the response of plant groups to warming differed with ambient summer temperature, soil moisture, and duration of experimental warming. Shrubs expanded most in warm tundra regions, and graminoids and forbs expanded most in colder tundra regions. The taller shrubs found in the milder southern tundra regions generally increased with warming, whereas in the colder northern tundra regions, the shorter shrubs declined with warming. These results illustrate the complexities involved with climate change within one biome. Warming alone cannot be used as a generic predictor of vegetation response since the many other factors such as soil moisture and nutrients, precipitation, winter temperature, and snow cover, and their spatial variation, can also play a significant role.

Wan et al. (2002) observed changes in air temperature, soil temperature, and soil-moisture content under experimental warming and clipping in a tall-grass prairie in the Great Plains, United States. They used a factorial design with warming as the primary factor nested with clipping as the secondary factor. Infrared heaters were used to simulate climatic warming and clipping to mimic mowing or grazing. The warming treatment significantly increased daily mean and minimum air temperatures by 1.1 and 2.3 °C, respectively, but had no effect on daily maximum air temperature, resulting in reduced diurnal air-temperature range. Infrared heaters substantially increased daily maximum (2.5 and 3.5 °C), mean (2.0 and 2.6 °C) and minimum (1.8 and 2.1°C) soil temperatures in both the unclipped and clipped subplots. Clipping also significantly increased daily maximum (3.4 and 4.3 °C) and mean (0.6 and 1.2 °C) soil temperatures, but decreased daily minimum soil temperature (1.0 and 0.6 °C in the control and warmed plots, respectively). Daily maximum, mean, and minimum soil temperatures in the clipped, warmed subplots were 6.8, 3.2, and 1.1 °C higher than those in the unclipped, control subplots. Infrared heaters caused a reduction of 11 percent in soil moisture in the clipped subplots, but not in the unclipped subplots. Clipping reduced soil-moisture content by 17.7 and 22.7 percent in the control and warmed plots, respectively. The warming and clipping interacted to exacerbate soil moisture loss (26.7 percent).

The effects of increased atmospheric CO_2 concentrations on vegetative growth have been evaluated by Poorter and Navas (2003). They analyzed responses at three integration levels: carbon economy parameters, vegetative biomass of isolated plants, and growth in competition. They found an increase in the whole-plant rate of photosynthesis. However, the increased photosynthetic rate per unit leaf area is accompanied by a decrease in specific leaf area (leaf thickness). The net result of these and other changes is that relative growth rate is only marginally stimulated. The biomass enhancement ratio varied substantially across experiments and species. Fast-growing herbaceous C3 species responded more strongly than slow-growing C3 herbs or C4 plants.

This different responses of C3 and C4 species to elevated CO_2 highlights both the fundamental difference in photosynthetic physiological response and possible ecosystem shifts that would favor C3 species over C4 species in an atmosphere with a high CO_2 concentration. The photosynthetic response to increasing CO_2 concentrations for C3 and C4 species (Figure 12.2) shows that net photosynthesis for the C4 plant exceeds that for the C3 plant until concentrations reach ~ 400 ppm$_V$. Figure 12.2 also shows that the net photosynthesis for the C3 plant continues to increase with increasing CO_2 concentrations, yet for the C4 plant, the rate of net photosynthesis decreases as CO_2 concentrations increase. The implications are that above the optimum CO_2 concentration of ~400 ppmv for C4 species (the concentration in the year 2015), further increases will likely favor C3 species over C4 species. Thus, the impacts will be greatest for grass species, of which nearly 50 percent are C4. This includes the important agricultural crop species maize, sugarcane, millet, and sorghum.

To examine further the effects of increasing atmospheric CO_2 concentrations on vegetation growing in natural, open-air conditions, the Free Air Carbon Experiment, or FACE, was initiated in the early 1990s. Enclosure systems, even when placed outside, can produce unwanted effects that compound those due to increasing CO_2 concentrations. Open-top chambers, similar to but larger than those described for ITEX, result in increase in surface temperature (the desired effect for ITEX), but this can also result in an amplification of photosynthesis and productivity (Morgan et al., 2001). FACE experiments overcome this by not using an enclosure at all. CO_2-enriched air (usually twice the ambient concentrations) is injected into the air from emission tubes placed in an octagon patterned around the treatment area (typically 8–30 m in diameter). For agricultural crops of wheat and rice, over a 3-year period, FACE experiments at 200 ppmv above ambient

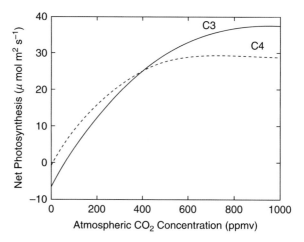

Figure 12.2. Net photosynthesis in relation to atmospheric CO_2 concentration for typical C3 and C4 species. At roughly 400 ppmv, the mean CO_2 concentration in the atmosphere in 2015, the net photosynthetic rates are equal.
Source: P.D. Blanken. Redrawn after M. B. Kirkham, (2011). *Elevated carbon dioxide: Impacts on soil and plant water relations,* p. 256, fig. 12.8. Boca Raton, FL: CRC Press (Taylor & Francis).

revealed only a 5–7 percent and 8 percent yield increase for rice and wheat, respectively. This is much lower than the 19 percent increase revealed by enclosure-based experiments and indicates uncertainty in the expected increase in crop yield with elevated CO_2 concentrations. For other vegetation groups, photosynthesis in trees is generally more responsive than grass, forbs, legumes, or crops. Expectedly, trees responded with the largest increase in the leaf area index. For C4 species, FACE experiments (Ainsworth and Long, 2004) revealed a far smaller response to elevated CO_2 than those shown by chamber studies. In addition to these differences in results when CO_2 enrichment experiments using open-top chambers and free-air techniques are used, FACE experiments have been criticized for their spatial bias (Jones et al., 2014). Most are located in the temperate ecosystems of North America and Europe, and investigation in tropical and boreal forests, and belowground responses, are largely absent.

Modeling Studies of the Future

The Ecosystem–Atmosphere Simulation Scheme (EASS) model results for eastern China show that climate change (1981–2005) contributed more to the changes of land surface energy fluxes than land-cover change, with their contribution ratio reaching 4:1 or even higher (Yan et al., 2014).

In eastern China increased incident solar radiation has the greatest effect on land surface energy exchange. The impacts of land-cover change on the seasonal variations in land surface heat fluxes between 1989, 1995, 2000, and 2005 were large, especially for the sensible heat flux H. The changes in the regional energy fluxes resulting from different land-cover type conversions varied greatly. The conversion from farmland to evergreen coniferous forests had the greatest influence on land surface energy exchange, leading to a decrease in H by 19.4 percent and an increase in λE and R_n by 7.4 and 2.7 percent, respectively. This type of study can be expected to become more common in the future. It is virtually absent from the 2013 IPCC report.

References

Alexander, L. V., et al. 2006. Global observed changes in daily climate extremes of temperature and precipitation. *J. Geophys. Res.* 111, D05109.

Ainsworth, E. A., and Long, S. P. 2004. What have we learned from 15 years of free-air CO_2 enrichment (FACE)? A meta-analytical review of the responses of photosynthesis, canopy properties and plant production to rising CO_2. *Tansley Review, New Phytologist* 165, 351–72.

Anthelme, F., Villaret, J-C, and Brun, J-J. 2007. Shrub encroachment in the Alps gives rise to the convergence of sub-alpine communities on a regional scale. *J. Veg. Sci.* 18, 355–62.

Badh, A. 2011. Understanding the climate-change impact on the growing g degree days for corn in the United States of America. Ph.D. Dissertation. Fargo: North Dakota State University of Agriculture and Applied Science,

Barry, R. G., and Gan, T. Y,. 2011. *The global cryosphere; Past, present and future.* Cambridge: Cambridge University Press, pp. 165–89.

Brown, R. D., and Robinson, D. A. 2011. Northern Hemisphere spring snow cover variability and change over 1922–2010 including an assessment of uncertainty. *The Cryosphere* 5, 219–29.

Christianssen, D. E, Markstrom, S. L., and Hay, L. E. 2011. Impacts of climate change on the growing season in the United States. *Earth Interactions* 15(33), 1–17.

De Frenne, P. 2013. Microclimate moderates plant responses to macroclimate warming. *Proc. Nat. Acad. Sci.* 110(46), 18, 561–6.

Dutkiewicz, V. A. et al., 2014, Forty-seven years of weekly atmospheric black carbon measurements in the Finnish Arctic: Decrease in black carbon with declining emissions. *J. Geophys. Res.*, 119, 7667–683,

Elmendorf, S. C. et al., 2012. Global assessment of experimental climate warming on tundra vegetation: heterogeneity over space and time. *Ecology Letters* 15(2), 164–75.

Epstein, H. E. et al. 2014 Tundra greenness. Arctic Report Card: Update for 2014. www.arctic .noaa.gov/reportcard/tundra_greenness.html.

Feldman, D. R. et al. 2015. Observational determination of surface radiative forcing by CO_2 from 2000 to 2010. *Nature.* doi:10.1038/nature14240

Fischer, E. M., and Knutti, R. 2014. Detection of spatially aggregated changes in temperature and precipitation extremes. *Geophys. Res. Lett.* 41. doi:10.1002/2013GL058499.

Gavazov, K., Spiegelberger, T., and Buttler, A. 2014. Transplantation of subalpine wood-pasture turfs along a natural climatic gradient reveals lower resistance of unwooded pastures to climate change compared to wooded ones. *Oecologia* 174, 1425–35.

Hansen, J. et al. 2010. Global surface temperature change. *Rev. Geophys.* 48(4), RG 4004.

Hao, G-C. et al. 2014. Soil thermal dynamics of terrestrial ecosystems of the conterminous United States from 1948 to 2008: An analysis with a process-based soil physical model and AmeriFlux data. *Clim. Change* 126, 135–50.

Hartmann, D. L., Klein Tank, A. M. G., Rusticucci, M. et al. 2013. Observations: Atmosphere and surface. In T. F. Stocker et al. (eds.), *Climate Change 2013: The Physical Science Basis.* Contribution of Working Group I, Fifth Assessment Report of the Inter- governmental Panel on Climate Change. Cambridge: Cambridge University Press, pp. 159–254.

Henry, G. H. R., and Molau, U. 1997. Tundra plants and climate change: The International Tundra Experiment (ITEX). *Global Change Biology* 3(Suppl. 1), 1–9.

Illán, J. G., Gutiérrez, D., and Wilson, R. 2010. The contributions of topoclimate and land cover to species distributions and abundance: Fine-resolution tests for a mountain butterfly fauna. *Global Ecol., Biogeog.* 19, 159–73.

Jia, G. J., Epstein, H. E., and Walker, D. M. 2009. Vegetation greening in the Canadian Arctic related to decadal warming. *J. Environ. Monit.* 11, 2231–8.

Jones, A. G. et al., 2014. Completing the FACE of elevated CO_2 research. *Environment International* 73, 252–8.

Jorgenson, M. T., Racine, C. H., Walters, J. C., and Osterkamp, T. E. 2001. Permafrost degradation and ecological changes associated with a warming climate in central Alaska. *Climate Change* 48, 551–79.

Kirkham, M. B. 2011. *Elevated carbon dioxide: Impacts on soil and plant water relations.* Boca Raton, FL: CRC Press.

Kwok, R., and Rothrock, D. A. 2009. Decline in Arctic sea ice thickness from submarine and ICESat records: 1958–2008. *Geophysical. Research Letters* 36 (15), L15501.

Lean, J. L. et al. 1997. Detection and parameterization of variations in mid- and near-ultraviolet (200–400 nm). *J. Geophys. Res.* 102 (D25), 29, 939–56.

McCabe, G. J., Betancourt, J. L., and Feng, S. 2015. Variability in the start, end, and length of frost-free periods across the conterminous United States during the past century. *Int. J. Climatol.* doi: 10.1002/joc.4315

Mishra, V. et al. 2015. Changes in observed climate extremes in global urban areas. *Environ. Res. Lett.* 10(2), 02405.

Morgan, J. A. et al. 2001. Elevated CO_2 enhances water relations and productivity and affects gas exchange in C-3 and C-4 grasses of the Colorado shortgrass steppe. *Global Change Biology* 7, 451–66.

Myers-Smith, I. H. et al. 2011. Shrub expansion in tundra ecosystems: Dynamics, impacts, and research priorities. *Environ. Res. Lett.* 6, 045509.

Nijs I. et al. 1996. Free Air Temperature Increase (FATI): A new tool to study global warming effects on plants in the field. *Plant, Cell Environ.* 19, 495–502.

Osterkamp, T. 2008. Thermal state of permafrost in Alaska during the fourth quarter of the twentieth century. In D. L. Kane and K. M. Hinkel (eds.), *Proceedings of the Ninth International Conference on Permafrost.* Fairbanks, Alaska, Vol. 2, 1333–8.

Parmesan, C., and Yohe, G. 2003. A globally coherent fingerprint of climate change impacts across natural systems. *Nature* 421, 37–42.

Parrish, T. D., and Stockwell, W. R. 2015. Urbanization and air pollution then and now. *Eos* 96(1), 12–15.

Payette, S. et al. 2004. Accelerated thawing of subarctic peatland permafrost over the last 50 years. *Geophys. Res. Lett.* 131, L18208.

Pollack, I. B. et al. 2013. Trends in ozone, its precursors, and related secondary oxidation products in Los Angeles, California: A synthesis of measurements from 1960 to 2010. *J. Geophys. Res.* 118(11),

Poorter, H., and Navas, M. L. 2003. Plant growth and competition at elevated CO_2: on winners, losers and functional groups. *New Phytologist* 157, 175–98.

Robinson, C. H. et al. 1998. Plant community responses to simulated environmental change at a high Arctic polar semi-desert. *Ecology* 79, 856–66.

Serreze, M. C. et al. 2000. Observational evidence of recent change in the northern high-latitude environment. *Clim. Change* 46, 159–207

Shiklomanov, N.I. et al, 2008. The Circumpolar Active Layer Monitoring (CALM) program: Data collection, management, and dissemination strategies. In *Proceedings, Ninth International Conference on Permafrost*, 29 June–3 July 2008, Fairbanks, Alaska. University of Alaska, Fairbanks, Alaska. 1 164752.

Shiklomanov, N. I. et al, 2010. Decadal variations of active-layer thickness in moisture-controlled landscapes, Barrow, Alaska. *J. Geophys. Res., Biogeosci.* 115, G4, G00I04. doi:10.1029/2009JG001248.

Sigi, M. et al. 2015. Timing and climate forcing of volcanic eruptions for the past 2,500 years. *Nature* 523, 543–9.

Smith, N. V., Saatchi, S. S., and Rangerson, J. T. 2004. Trends in high northern latitude soil freeze and thaw cycles from 1988 to 2002. *J. Geophys. Res. Atmos.* 109(D12). doi: 10.1029/2003JD004472.

Smith, S. L. 2010. Thermal state of permafrost. In North America – a contribution to the International Polar Year. *Permafrost Periglac. Proc.* 21, 117–35.

Stocker, T. E., Qin, D-H., and Plattner, G-K. (Coordinating lead authors) et al. 2013. Technical Summary. In *Climate Change 2013: The Physical Science Basis*. Contribution of Working Group I to the Fifth Assessment Report of the Intergovernmental Panel on Climate Change. Cambridge, UK: Cambridge University Press.

Tape, K. D., Sturm, M., and Racine, C. H. 2006. The evidence for shrub expansion in Northern Alaska and the Pan-Arctic. *Glob. Change Biol* . 12, 686–702.

Wan, S., Luo, Y., and Wallace, L. L. 2002. Changes in microclimate induced by experimental warming and clipping in tallgrass prairie. *Global Change Biol.* 8, 754–68.

Wild, M. et al. 2009. Global dimming and brightening: A n update beyond 2000. *J. Geophys. Res.*, 114, D 0 0D13.

Wild, M. 2012. Enlightening Global Dimming and Brightening. *Bull. Amer. Meteor. Soc.*, 93, 27–37.

Yan, J. W. et al. 2014. Changes in the land surface energy budget in eastern China over the past three decades: Contributions of land-cover change and climate change. *J. Climate* 27, 9233–52.

Problems

Chapter 1

1. Discuss the possible reasons why the topic of microclimate originated with observations related to agriculture.
2. Microclimate is often taught in Geography Departments. What are the disciplinary connections between microclimate and geography?
3. Describe three methods that could be used to relate the microclimate scale to the macroclimate scale quantitatively.
4. Provide examples of how the air temperature, soil temperature, soil moisture, wind speed, and solar radiation might vary when close to a hummuck in arctic tundra, and a krumholtz in the alpine treeline.

Chapter 2

1. Use the air temperature data in the following table to calculate the statistics shown in Table 2.2.

Example of hourly observations (Hr:Min) observations of air temperature (T_a; °C) observations.

Time	1:00	2:00	3:00	4:00	5:00	6:00
T_a	2.2	2.2	2.8	2.8	3.3	3.3
Time	7:00	8:00	9:00	10:00	11:00	12:00
T_a	3.9	5.0	5.6	5.6	6.1	7.8
Time	13:00	14:00	15:00	16:00	17:00	18:00
T_a	10.0	11.1	11.7	10.6	8.3	6.7
Time	19:00	20:00	21:00	22:00	23:00	24:00
T_a	6.1	5.0	4.4	3.3	3.3	3.3

2. Using the air temperature data in the table, calculate the saturation vapor pressure at 6:00, 12:00, 18:00, and 24:00. If the relative humidity at these times was 93, 76, 80, and 85 percent, respectively, what were the ambient vapor pressures?
3. Discuss how soils with different texture can have the same volumetric soil water content, yet different soil water potentials.

4. Describe how transpiration from a C3, a C4, and a CAM plant would differ during drought conditions. How might each plant cope differently with the water stress conditions?

5. What is the equivalent depth of liquid water contained in new snow 100 cm deep with a density of 50 kg m^{-3} compared to old, settled snow with a density of 200 kg m^{-3}?

6. Many biological chemical reactions increase exponentially with temperature. Why or why not would you expect soil respiration to increase continually with soil temperature?

Chapter 3

1. Discuss accuracy, resolution, and precision in the context of designing a meteorological instrument.

2. If the surface of a lake emits 357 W m^{-2} and has an emissivity of 0.98, what is its surface temperature?

3. What is the area sampled by an infrared thermometer with a 20° degree field of view placed 3 m above the surface at an angle of 40° from the nadir? Estimate the area sampled if the thermometer were tilted so much that the horizon was visible.

4. Using soil thermal properties provided in Table 3.1, what is a mineral soil's volumetric heat capacity having volumetric fractions of air, solids, and water of 10, 70, and 20 percent, respectively?

5. Describe how each component of net radiation as shown in Eq. (3.6) can be measured.

6. Calculate the hourly average wind speed and direction on the basis of two half-hour observations of 2 m s^{-1} at 280° and 6 m s^{-1} at 80°.

7. Use the Bowen ratio energy balance method to calculate the latent and sensible heat fluxes based on these observations: $\gamma = 0.067$ kPa °C^{-1}; $T_{a1} = 20$ °C; $T_{a2} = 16$ °C; $e_{a1} = 1.8$ kPa; $e_{a2} = 1.6$ kPa; $R_n\text{-}G = 600$ W m^{-2}.

Chapter 4

1. Calculate the wavelength of maximum energy emission and the total energy emitted by you with a surface temperature of 35 °C and an emissivity of 0.98. Do the same calculations for the Sun with a surface temperate of 5800 °C and an emissivity of 1. How do these calculations compare?

2. The angle of incidence between the Sun and the line perpendicular to a sloping surface is 20°. What is the solar radiation received on this slope if the incident solar radiation is 800 W m^{-2}?

3. Calculate the amount of reflected shortwave radiation for the surfaces shown in Table 4.4 receiving 1000 W m^{-2} of shortwave radiation.

4. Sketch the profile of PAR through a tropical deciduous forest and a boreal coniferous forest. Discuss how leaf shape, orientation, and location in the canopy can affect the absorption of PAR.

5. Calculate the net radiation of a surface receiving 750 W m^{-2} of incident shortwave radiation with an albedo of 0.30, and a surface temperature, air temperature, and emissivity of 20 °C, 16 °C, and 0.98, respectively.

Chapter 5

1. Explain how the presence of ice in soil would affect its thermal properties.

2. Describe, with the aid of sketches, the wind profile of a corn crop as it progresses from seedlings, to maturity, to harvest.

3. Sketch and explain the contrast between a 24 hour plot of the surface energy balance of a dry prairie grassland and a deciduous forest in the growing season.

4. What are the concerns with the accuracy of pan evaporation and lysimeter measurements?

5. Why did Monteith modify the original Penman evaporation equation, and how was the equation modified?

Chapter 6

1. Sketch a plot of the annual changes in the LAI for a deciduous and a coniferous forest in the Northern Hemisphere. Are there any circumstances where the LAI might not change but the NDVI could?

2. Many remote sensing instruments rely on the visible wavelengths for observations. Over which regions, and during what times, could this pose a problem for such observations?

3. Barriers such as vegetation or buildings can have a large effect on the microclimate due to the alternations they create in the airflow. Currently, large wind turbines, or "wind farms," have also been examined to determine their real or potential impact on microclimate. Discuss the possible alterations in microclimate created by wind farms.

4. The spatial coverage of snow can be fairly easily determined from remote sensing platforms, yet the SWE is much harder to determine from remote sensing. Discuss why this is so.

Chapter 7

1. Describe some of the differences in arctic versus alpine tundra that could translate into differences in microclimate.

2. Many grasslands and agricultural crops have areas of bare soil exposed between the vegetation. As a result, the total evaporation measured above the canopy consists

of both transpiration and the evaporation of water from the bare soil. Discuss some micrometeorological methods that could be used to partition the two evaporation streams.

3. Discuss how a wintertime drought (low SWE) affected the surface energy balance of High Creek Fen, a high elevation wetland in central Colorado, the following summer.
4. The leaf shape and canopy structure for boreal coniferous forests are much different from those for southern deciduous forests. In terms of radiation absorption, are there any benefits for the leaf shape and canopy structure of a coniferous boreal forest at high latitudes?
5. Describe changes in the microclimate that could be expected if a small clearing in a forest suddenly developed as a result of a disturbance such as fire, logging, or disease.

Chapter 8

1. Sketch the annual pattern of the net radiation, latent heat flux (evaporation), sensible heat flux, and heat storage for a shallow and a deep lake located close together. Explain how these differences in lake depth affect the surface energy balance in terms of magnitude and seasonal patterns (timing).
2. Describe the conditions that would result in a large loss of snow through sublimation.
3. Starting at sea level (0 m a.sl.), a parcel of air with a temperature of 24 °C rises over a mountain barrier and descends on the leeward side. The dry adiabatic lapse rate is 10 °C km^{-1}, the dew point temperature is 4 °C, and the moist adiabatic lapse rate is 6 °C km^{-1}. At what height is the dew point temperature reached on the windward side? The air continues to rise following the moist adiabatic lapse rate until the 3 km summit is reached. What is the air temperature at the summit? Since all of the moisture has condensed, the air descends on the leeward side from the summit to sea level following the dry adiabatic lapse rate. What is the air temperature at sea level on the leeward side? Explain why the sea level air temperatures are the same or different on each side of the mountain.
4. Using a commercially available hand-held infrared thermometer, measure the surface temperature of the exterior walls of a building. Explain how the different azimuth (the direction the building is facing) affected the surface temperature you measured. Now, measure the surface temperature for buildings constructed of different materials and/or color but all facing the same direction. Use these data to describe how urban building materials and aspect affect temperature. Describe urban design features that could be used to minimize the urban heat island effect.

Chapter 9

1. Calculate the effective temperature for an organism under the following conditions: $T_a = 30$ °C; $R_{abs} = 600$ W m^{-2}; $\varepsilon = 0.98$; $\rho_a = 1$ kg m^{-3}; and $c_p = 1012$ J kg^{-1}

K^{-1}. Assume the characteristic dimension for heat transfer (the downwind length exposed to heat loss) is equal to 0.17 m (an average person). Do the calculation for a horizontal wind speed of 1 m s^{-1}, then 10 m s^{-1}. Explain how the increase in wind speed affects the effective temperature.

2. Calculate the cutaneous cooling for a person with a total body resistance to vapor of 40 s cm^{-1} under the following conditions: $T_a = 28\ °C$; $T_s = 34\ °C$; $\rho_a = 1\ kg\ m^{-3}$; $e_a = 1.5\ kPa$, and $P_a = 101\ kPa$. What would this cooling be in watts for a person weighing 60 kg and a height of 182 cm, and a person weighing 120 kg and 182 cm tall?

3. For the same conditions provided in question 2, calculate the sensible heat loss in watts per square meter. Then express the sensible heat loss in watts for the 60 and 120 kg person with a height of 182 cm.

4. Use a psychrometer to measure the dry- and wet-bulb temperatures both inside and outside the location where you spend most of your time. Then calculate the discomfort index. Discuss how and why the discomfort index changed from indoor to outdoor environments. If a psychrometer is not available, use data provided by weather stations at different locations.

5. Discuss how the microclimate experienced in the nest of a burrowing owl (*Athene cunicularia*) would compare to that of a northern saw-whet owl (*Aegolius acadicus*).

Chapter 10

1. Nighttime air temperatures tend to be higher in urban compared to rural areas, thus resulting in a larger nighttime urban heat island effect. Discuss why this occurs.

2. Discuss urban planning criteria that could be used to reduce the urban heat island effect.

3. Several studies have shown that urban areas experience more precipitation than the surrounding rural areas. Provide reason(s) that could provide an explanation for this increased urban precipitation.

4. Cities in Europe tend to have a smaller urban heat island effect compared to cities in North America. Present some reasons to explain why.

Chapter 11

1. In the foothills of the Rocky Mountains in Colorado, the valleys are oriented on an east-west aspect. Trees are dominant on the south sides of the valleys compared to the north sides (this is visible on any satellite image). Explain the microclimate conditions that would favor tree growth on the south side (north-facing) slopes. Then describe how the trees, once established, could then further modify the microclimate.

2. Explain how differences in the thermal properties of land compared to water can result in daytime onshore and nighttime offshore breezes in coastal regions.
3. The North American Great Lakes region is known for its lake effect snow. Explain how ice cover on the lakes might influence the occurrence of lake effect snow.

Chapter 12

1. Permafrost degradation in arctic and alpine regions has been observed. Describe the changes in soil thermal properties that would be associated with the loss of permafrost. How might this affect vegetation in these regions?
2. The increasing concentration of atmospheric CO_2 will affect plants with C3 photosynthesis differently than those with C4 photosynthetic pathways. Why and how might the distribution of C3 and C4 species differ in an atmosphere with CO_2 concentrations in excess of 400 ppm?
3. Several long-term studies have been used experimentally to manipulate the microclimate to simulate the effects of warming on ecosystems. Present an argument for or against the validity of these types of experiments as a proxy of future global warming.

Glossary

Absolute humidity	the mixing ratio of water vapor in dry air; grams of water vapor per kilogram of dry air.
Accuracy	how close a measurement is to the actual value.
Advection	horizontal transfer in the atmosphere.
Aerosol optical depth	the total amount of aerosol integrated through an atmospheric column.
Albedo	the reflectivity of a surface to incoming solar radiation (also known as reflection coefficient).
Anabatic wind	an upslope wind due to slope heating.
Anemometer	an instrument used to measure wind speed.
AVHRR	Advanced Very High Resolution Radiometer.
Barometer	an instrument used to measure the barometric pressure, the pressure created by all molecules in the atmosphere above the height of the barometer's location.
BDRF	bi-directional reflectance factor.
BREB: Bowen ratio energy balance	a method used to measure the latent and sensible heat fluxes that is based on the ratio of those fluxes (the Bowen ratio) and the energy balance equation.
Blackbody	a hypothetical body that absorbs all the radiation striking it.
BOREAS	the Boreal Ecosystem-Atmosphere Study.
Buoyancy	an upward force exerted on an object in a fluid or gas that is equal to the weight of the fluid or gas displaced by the object; in air the density of an air parcel is less than that of the surrounding air, causing it to rise.
Celsius	a relative temperature scale of freezing and boiling points of water based on the work of Anders Celsius, 1701–44.
Concentration	amount of a substance in unit volume.
Conductance	the ease with which a fluid or gas enters and flows through a conduit (millimeters per second); the reciprocal of resistance.

Conduction	the transfer of energy by molecular contact.
Convection	predominantly vertical motion in a fluid that gives rise to the transfer and mixing of mass and energy.
Coriolis effect	the apparent effect of the Earth's rotation on wind direction resulting in a turning of wind direction to the right (left) in the Northern (southern) Hemisphere; Coriolis balances the pressure gradient force in balanced geostrophic flow above the friction layer.
COSMOS	COsmic-ray Soil Moisture Observing System, which uses a cosmic ray probe to measure neutrons emitted by soil moisture.
DALR	Dry Adiabatic Lapse Rate, the dry air daytime air temperature decrease of 0.98 $^{\circ}$C per 100 m increae in height.
Dew point	the temperature at which saturation occurs when air is cooled at constant pressure and vapor content.
Diffuse radiation	shortwave radiation that has been scattered as a result of air molecules, aerosols, and cloud droplets from the direct solar beam.
Diffusion	the exchange of fluid particles from regions of high to low concentration due to molecular motion.
Direct beam radiation	the fraction of the shortwave radiation received in a parallel beam directly from the Sun.
Disdrometer	an instrument that measures the droplet size, velocity, and form of precipitation by measuring the attenuation of light as precipitation passes through a laser beam.
Eddy covariance	a method used to measured turbulent fluxes that is based on the covariance of the deviations of the vertical wind speed and a scalar.
Emissivity	the ratio of radiant energy emitted by a surface (at a specified wavelength and temperature) to that emitted by a blackbody under the same conditions.
Evaporation	the process by which liquid water is transformed into water vapor.
Evapotranspiration	the combined transfer of water to the atmosphere by evaporation and transpiration.
Fetch	distance to a source in the upwind direction.
Flux	rate of flow of some quantity.
Flux density	the flux of any quantity through unit surface area.
Föhn wind	a dynamically induced downslope wind that causes air temperatures to rise and humidity to decrease.

Forced convection	vertical motion induced by mechanical forcing due to friction or deflection.
Free convection	vertical motion caused by density differences in a fluid or gas.
FDD - Freezing Degree-Day,	obtained by accumulating the value of mean daily air temperature below the freezing threshold value (0 °C).
GDD	Growing Degree-Day, obtained by accumulating the value of mean daily air temperature above a threshold value for plant growth, typically 10 °C.
Global shortwave radiation	the sum of all shortwave radiation including direct beam, diffuse beam, and downward reflected.
Gravimetric method	a method to measure soil moisture based on the difference between the wet and oven-dried sample weights.
HAPEX	the Hydrological Atmospheric Pilot Experiment, conducted in Southwest France in 1985–6.
Heat capacity	the energy required (release) for a 1 K increase (decrease) in an object's temperature (joules per kelvin). Specific heat capacity (c) is the heat capacity per kilogram (joules per Kelvin per kilogram), and the volumetric heat capacity (C) is the heat capacity per volume (joules per Kelvin per cubic meter).
Infrared gas analyzer (IRGA)	an instrument used to measure the density of a gas in air (typically CO_2 and H_2O) based on measuring the absorption in the infrared wavelengths; typically used in conjunction with a sonic anemometer to measure fluxes using the eddy covariance method.
Infrared thermometer	an instrument used to measure the skin or surface temperature of an object without direct contact; uses the Stefan-Boltmann law and the object's emissivity to determine the skin temperature.
Inversion	a reversal of the usual decrease of temperature (also moisture) with height in the atmosphere.
ISLSCP	International Satellite Land Surface Climatology Project.
Katabatic wind	a downslope wind due to cold air drainage.
Kelvin	the temperature scale based on kinetic energy developed by Lord Kelvin in 1848; all molecular motion ceases at 0 K (absolute zero), and 273.15 K is the triple point of water; there is no "degree" symbol used, and negative values are not possible.

Kriging	An interpolation technique based on the assumption of a Gaussian distribution and the covariance of nearby data.
Krummholz	(German: crooked wood) a stunted and deformed growth form of alpine timberline trees in which branches may be absent on the windward side of the trunk.
Laminar flow	smooth fluid motion in parallel streamlines.
Latent heat	the heat released (or absorbed) per unit mass during a phase change from liquid to solid or gas to liquid (or the reverse).
Latent heat flux	the product of the latent heat of vaporization and the rate of evaporation (or condensation) in watts per square meter.
Leaf area index (LAI)	the one-sided green leaf area per unit ground surface area (m^2 leaf per m^2 ground).
Longwave radiation	electromagnetic radiation with wavelengths of 4.0–100 μm emitted by objects with surface temperature ~ 300 K; almost all terrestrial objects (therefore also known as terrestrial radiation).
Lysimeter	an instrument for measuring evapotranspiration by monitoring the weight changes of a vegetated soil plot.
MAGT	Mean annual ground temperature.
Microclimate	the study of climates near the ground and the factors that affect them.
Mixing ratio	mass of a particular gas divided by the total atmospheric mass (grams per kilogram).
MODIS	Moderate Resolution Imaging Spectroradiometer is operated on Terra launched December 1999 and AQUA launched May 2002 by NASA; it has 36 bands from 0.4 to 14.3 μm with spatial resolution of 250–1000 m.
Monin-Obukhov length (L)	length that defines the effect of buoyancy on turbulent flows, usually negative by day and positive at night; by day it is the height at which the buoyant production of turbulent kinetic energy (TKE) is equal to that produced by the wind shear.
NDVI	Normalized difference vegetation index.
NEXRAD	Next-generation Weather Radar (United States).
Net radiometer	an instrument used to measure the net radiation using an upward- and downward-acing thermopile surface.
Nyquist rate	a sampling rate that is twice the maximum frequency of the measurement's time series.
Obukov stability length (L)	the absolute value is equal to the height where turbulent and shear production of kinetic energy are equal; when

	divided into the height above ground, z/L provides an index of atmospheric stability.
Partial pressure	the pressure of an individual gas; Dalton's law states that the sum of all partial pressures equals the total pressure.
Permafrost	perennially frozen ground (during at least two summers).
Photosynthesis	a process used by plants and other organisms to convert light energy from the Sun into chemical energy.
Photosynthetic active radiation (PAR)	solar radiation in the visible range of the electromagnetic spectrum that is used in photosynthesis; wavelengths are 400–700 nm.
Planetary boundary layer	the atmospheric layer from the surface to a height where frictional effects on airflow are absent.
Potential temperature	the temperature of an unsaturated air parcel changed adiabatically to a standard pressure, usually 1000 hPa.
Precision	the repeatability of a measurement; an instrument that has a high precision is not necessarily an accurate instrument.
Psychrometer	an instrument used to measure humidity on the basis of the difference between a dry-bulb and a wet-bulb thermometer, aspirated by rotating by hand (a sling psychrometer) or by a mechanical fan (an aspirated psychrometer).
Pyranometer	an instrument used to measure shortwave radiation on the basis of a thermopile situated beneath a glass dome, or a silicon photoelectric diode.
Pyrgeometer	an instrument used to measure longwave radiation on the basis of a thermopile.
Relative humidity	ratio of the ambient vapor pressure to the saturation vapor pressure; often multiplied by 100 to give a percentage.
Resistance (of plant leaves)	the resistance of leaves to the in- or out-flow of moisture or gas via stomata (seconds per meter); the reciprocal of conductance.
Resolution	the number of significant figures that an instrument is capable of measuring.
Respiration	a biological process in plants and animals whereby sugars and carbohydrates are oxidized to release energy, heat, and CO_2.
Richardson number	a dimensionless number used to classify atmospheric stability that relates the ratio of buoyancy to inertial forces.

Roughness length (z_o)	in the log wind profile it is equivalent to the height at which the wind speed theoretically becomes zero; it is approximately one-tenth the height of the surface roughness elements.
Scalar	a quantity that has magnitude only, such as temperature (a vector has magnitude and direction).
SDMs	Statistical Downscaling Models
SeaWiFS	Sea-viewing Wide Field-of-view Sensor; an Earth-orbiting ocean color sensor launched in August 1997 and operated by NASA.
Seebeck effect	the relationship between the temperature difference between two junctions of two different metals and the resultant electromagnetic field that is created; discovered by Thomas Johann Seebeck in 1821, this is the basis for thermocouple temperature measurements.
Sensible heat flux	energy exchange in watts per square meter resulting from convection of air due to a vertical temperature gradient.
Shortwave radiation	electromagnetic radiation with wavelengths of 0.1–4.0 μm emitted by objects with surface temperature ~ 6000 K; the Sun (therefore also known as solar radiation).
Sky view factor	a measure of the degree to which the sky is obscured by the surroundings of a given point.
SMMR	Scanning Multichannel Microwave Radiometer flown on NASA's Nimbus 7 satellite, October 25, 1978, through August 20, 1987; the five dual-polarized (horizontal, vertical) frequencies ranged from 6.6 gigahertz (GHz) to 37.0 GHz.
Soil heat flux	the energy flux density (watts per square meter) in a soil (or any substrate) created by the temperature gradient and thermal conductivity.
Soil heat flux plate	a thermopile with a high thermal conductivity that is used to measure the soil heat flux.
Solar zenith angle	the angle of the Sun's beam measured from the zenith.
Sonic anemometer	an instrument that measures wind speed by measuring changes in the transit time of sound (sonic) waves between opposing pairs of transducers' typically used to measure the vertical wind speed at a high sampling rate for eddy covariance flux measurements.
Specific leaf area	the ratio of the leaf area to its dry mass weight; a measure of leaf thickness.

Specific humidity	the mixing ratio for water vapor; grams of water vapor per kilogram of air including water vapor.
SSM/I	Special Sensor Microwave Imager flown on Defense Meteorological Satellite Program (DMSP) satellites (United States) from late 1987 to present; it has four frequencies from 19.35 to 85.5 GHz that operate in horizontal and vertical polarizations, except at 22.35 GHz, which is only vertical.
Stomata	pores typically located in the epidermis of vascular plant leaves where CO_2 and water vapor are exchanged between the plant and atmosphere.
SWE	Snow Water Equivalent; the amount of liquid water contained in a volume of snow.
Time constant	the response time of an instrument to changes in the variable it is measuring.
Time domain reflectrometry (TDR)	a method of measuring soil moisture that is based on the dielectric constant of soils versus liquid water.
Thermal admittance	a measure a material's ability to transfer heat in the presence of a temperature difference on opposite sides of the material.
Thermocouple	a device used to measure temperature; thin wires consisting of two dissimilar metals when connected produce a small voltage proportional to the temperature difference between two junctions.
Thermistor	a device used to measure temperature; Temperature is proportional to the resistance produced when a small electrical current is passed through the thermistor.
Thermopile	several thermocouples wired in series to increase the voltage output of an individual thermocouple; commonly used in soil heat flux plates and radiometers.
Thiessen polygons	a technique developed by the American meteorologist Alfred H. Thiessen to interpolate spatially between sparse precipitation observations.
Topoclimates	local climates primarily controlled by elements of topography.
Transpiration	the evaporation of water from within the leaves (the stomata) of vascular plants.
Turbidity	a measure of atmospheric clarity: how much the material suspended in the atmosphere decreases the passage of light through the air; it represents the combined effect of scattering and absorption of solar radiation by aerosols, air molecules, and water vapor.

Turbulence   or turbulent flow: a flow regime characterized by chaotic property changes.

Vector   a quantity that includes both magnitude and direction, such as wind (a scalar has magnitude only).

Virtual temperature   the temperature, typically measured by a sonic anemometer, that includes and then should be corrected to account for the effects of humidity.

Wind vane   an instrument used to measure wind direction.

Zenith   the highest point reached by the Sun above an observation point.

Zero plane displacement   the height above the ground at which zero wind speed is achieved as a result of obstacles such as trees or buildings; it is approximately two-thirds of the average height of the obstacles.

Symbols

Italic Uppercase

Symbol	Quantity	Units
A	Area	m^2
C_P	Air volumetric heat capacity	$J\ m^{-3}\ K^{-1}$
C_s	Soil volumetric heat capacity	$J\ m^{-3}\ K^{-1}$
DI	Discomfort index	$^\circ C$
E	Mass flux of H_2O per unit area; evaporation rate	$g\ m^{-2}\ s^{-1}$; $m\ s^{-1}$
E_i	Activation energy	eV
F	Force	$N\ (kg\ m\ s^{-2})$
F	Fog deposition velocity	$kg\ m^{-2}\ s^{-1}$
F_c	Mass flux of CO_2 per unit area	$g\ m^{-2}\ s^{-1}$
G	Soil (ground) heat flux	$W\ m^{-2}$
G_o	Soil surface (ground) heat flux	$W\ m^{-2}$
G_z	Soil (ground) heat flux at depth z	$W\ m^{-2}$
H	Sensible heat flux	$W\ m^{-2}$
H_R	Hydraulic resistance	$Pa\ s\ m^{-2}$
I	Electrical current	amperes
J_s	Heat storage	$W\ m^{-2}$
K	Eddy diffusivity	$m\ s^{-1}$
K	Thermal conductivity	$W\ m^{-1}\ K$
K	Hydraulic conductivity	$m\ s^{-1}$
K_C	Eddy diffusivity for CO_2	$m\ s^{-1}$
K_H	Eddy diffusivity for heat	$m\ s^{-1}$
K_M	Eddy diffusivity for momentum	$m\ s^{-1}$
K_V	Eddy diffusivity for water vapor	$m\ s^{-1}$
K_o	Cooling rate	$kg\ cal\ hr^{-1}\ m^{-2}$
L	Obukov stability length	m
L_{WC}	Liquid water content	$kg\ m^{-3}$
L_*	Net longwave (terrestrial) radiation	$W\ m^{-2}$
$L\downarrow$	Down-welling longwave (terrestrial) radiation	$W\ m^{-2}$
$L\uparrow$	Up-welling longwave (terrestrial) radiation	$W\ m^{-2}$
M	Metabolic heat flux or rate	W; $W\ m^{-2}$

(continued)

Symbol	Quantity	Units
M	Molecular weight of dry air (28.9)	g mol^{-1}
M_b	Basal metabolic rate	W
M_w	Work rate	W
O_e	Oxygen partial pressure of exhaled air	Pa (N m^{-2} = kg s^{-2} m^{-1})
O_i	Oxygen partial pressure of inhaled air	Pa (N m^{-2} = kg s^{-2} m^{-1})
P	Pressure	Pa (N m^{-2} = kg s^{-2} m^{-1})
P_a	Air pressure	Pa (N m^{-2} = kg s^{-2} m^{-1})
P_h	Air pressure at height h above sea level	Pa (N m^{-2} = kg s^{-2} m^{-1})
P_o	Air pressure at a reference height (sea level)	Pa (N m^{-2} = kg s^{-2} m^{-1})
Q	Discharge	m^3 s^{-1}
R	Rainfall rate	m s^{-1}
R	Universal gas constant (8.31)	J mol^{-1} K^{-1}
R	Electrical resistance	ohms
R_{abs}	Absorbed net radiation	W m^{-2}
Ri	Richardson number	–
R_n	Net (all-wave) radiation	W m^{-2}
R_T	Respiration flux at temperature T	g m^{-2} s^{-1}
S	Shortwave (solar) radiation	W m^{-2}
$S_{b,o}$	Extraterrestrial shortwave (solar) radiation incident on a horizontal surface	W m^{-2}
S_d	Diffuse shortwave (solar) radiation	W m^{-2}
S_o	Solar constant (1372)	W m^{-2}
S_t	Shortwave (solar) radiation at the earth's surface	W m^{-2}
S_*	Net shortwave (solar) radiation	W m^{-2}
$S\downarrow$	Down-welling shortwave (solar) radiation	W m^{-2}
$S\uparrow$	Up-welling shortwave (solar) radiation	W m^{-2}
T	Temperature	K; ºC
T_a	Temperature of air	K; ºC
T_d	Diffuse atmospheric transmission coefficient	–
T_e	Effective temperature	K; ºC
$Tmax$	Minimum air temperature over a 24-hr period	K; ºC
$Tmin$	Maximum air temperature over a 24-hr period	K; ºC
T_n	Nasal temperature	K; ºC
T_o	Temperature of air at sea level	K; ºC
T_H	Humidex temperature	–
T_s	Temperature at surface (skin temperature)	K; ºC
T_s	Temperature of soil	K; ºC
T_T	Total atmospheric transmission coefficient	–
T_v	Virtual temperature	K; ºC
T_w	Temperature of a wet-bulb thermometer	K; ºC
T_{WC}	Wind chill temperature	K; ºC
V	Electrical voltage	volts
V_d	Deposition velocity	m s^{-1}
W	Mass	kg
Z	Radar reflectivity	dBZ

Italic Lowercase

Symbol	Quantity	Units
a	Acceleration	m s^{-2}
b_o	Species-specific normalization constant	W kg$^{-3/4}$
c	Specific heat capacity	J kg^{-1} K^{-1}
c_a	CO_2 concentration in ambient air	g m^{-3}
c_P	Specific heat capacity of air	J kg^{-1} K^{-1}
c_s	CO_2 concentration in the leaf stomata	g m^{-3}
c_s	Specific heat capacity of soil	J kg^{-1} K^{-1}
c_w	Specific heat capacity of water	J kg^{-1} K^{-1}
d	Zero plane displacement	m
d	Characteristic dimension for heat transfer	m
d	Distance through which heat is transferred	m
e_a	Vapor pressure	Pa (N m^{-2} = kg s^{-2} m^{-1})
e_s	Saturation vapor pressure ($e_s(T)$ notes the e_s calculated at temperature T)	Pa (N m^{-2} = kg s^{-2} m^{-1})
g	Acceleration due to gravity (9.81)	m s^{-2}
h	Height above sea level	m
h	Canopy height	m
h	Human height	m
h	Solar elevation angle	Degrees or radians
k	Boltzmann constant (8.62×10^{-5})	eV K^{-1}
k	von Karman's Constant (0.41)	–
k	Extinction coefficient	m^{-1}
q	Specific humidity	g kg^{-1}
q	Water flux	m s^{-1}
r_a	Aerodynamic resistance	s m^{-1}; s cm^{-1}
r_c	Canopy resistance	s m^{-1}; s cm^{-1}
r_s	Stomatal resistance	s m^{-1}; s cm^{-1}
r_H	Convective boundary layer resistance	s m^{-1}; s cm^{-1}
r_R	Radiative boundary layer resistance	s m^{-1}; s cm^{-1}
r_{SV}	Skin's resistance to vapor transfer	s m^{-1}; s cm^{-1}
r_V	Water vapor mixing ratio	g kg^{-1}
t	Time	s
u	Horizontal component wind speed	m s^{-1}
u_z	Horizontal wind speed at height z above ground	m s^{-1}
u_*	Friction velocity	m s^{-1}
v	Cross-wind component wind speed	m s^{-1}
w	Vertical component wind speed	m s^{-1}
z	Height above ground or depth below surface	m
z_o	Roughness length	m

Greek Uppercase

Symbol	Quantity	Units
Δ	Finite difference	
Φ_M	Dimensionless stability function for momentum	–
Φ_X	Dimensionless stability function for scalar X	–
Γ	Lapse rate	$^{\circ}C \ m^{-1}$
Ψ	Soil water potential	$J \ kg^{-1}$; Pa

Greek Lowercase

Symbol	Quantity	Units
α	Reflection coefficient (albedo)	–
β	Bowen ratio ($H / \lambda E$)	
ε	Emissivity	–
ε_w	Ratio of the molecular weight of water vapor/dry air (0.622)	–
γ	Psychrometric "constant" ($\gamma = PC_p/\lambda\varepsilon_w$)	
κ	Hydraulic permeability	m^2
λ	Latent heat of vaporization	$J \ kg^{-1}$
λ	Wavelength	m
λ_{max}	Wavelength of maximum energy emission	m
λ_O	Energy produced per kilogram of oxygen consumed (3.00×10^4)	$kJ \ kg^{-1}$
λE	Latent heat flux; product of λ and E	$W \ m^{-2}$
λE_R	Latent heat flux from the evaporation of water from the respiratory tract	$W \ m^{-2}$
λE_S	Latent heat flux from sweating	$W \ m^{-2}$
μ	Fluid dynamic viscosity	Pa s
θ	Potential temperature	K; $^{\circ}C$
θ	Volumetric soil moisture; angle	$m^3 \ m^{-3}$
θ	Angle	Degrees; radians
θ_v	Virtual potential temperature	K; $^{\circ}C$
ρ	Density	$kg \ m^{-3}$
ρ_a	Density of air	$kg \ m^{-3}$
ρ_c	Density of CO_2 in air	$kg \ m^{-3}$
ρ_s	Density of soil	$kg \ m^{-3}$
ρ_v	Density of water vapor	$kg \ m^{-3}$
ρ_w	Density of water	$kg \ m^{-3}$
σ	Stefan-Boltzmann constant (5.67×10^{-8})	$W \ m^{-2} \ K^{-4}$
σ	Population standard deviation	–
τ	Momentum flux	$kg \ m^{-1} \ s^{-2}$
τ_λ	Atmospheric optical depth at a given wavelength λ	m
ζ	Monin-Obukov stability parameter	–

System International (SI) Units and Conversions

Length m (1 m = 3.2808 ft)
Mass kg (1 kg = 2.2046 lb)
Density (1 kg m^{-3} = 0.0624 lb ft^{-3})
Pressure Pa (pascal) = N m^{-2}
1 bar = 10^5 Pa
1 hPa = 1 millibar (mb)
Mean sea level pressure is 1013.25 hPa or 1013.25 mb
Speed m s^{-1} (1 m s^{-1} = 3.50 km hr^{-1}; 1 m s^{-1} = 2.24 mph; 1 m s^{-1} = 1.94 knots)

Energy

watt = J s^{-1}
joule (J) = kg m^2 s^{-2}
1 cal cm^{-2} day^{-1} = 4.1868 10^{-2} MJ m^{-2} day^{-1}

Carbon

1 Gt of carbon = 3.665 Gt of CO_2
1 mg m^{-2} s^{-1} of CO_2 = 22.73 μmol m^{-2} s^{-1} of CO_2

Conductance (at standard temperature and pressure)

H_2O; 1 mm s^{-1} = 55.6 mol m^{-2} s^{-1}
CO_2; 1 mm s^{-1} = 0.0449 mol m^{-2} s^{-1}

Index

Italics indicate reference to an illustration.